Tasty Bits of Several Complex Variables

A Whirlwind Tour of the Subject

Jiří Lebl

January 1, 2026
(version 4.3)

2

Typeset in LATEX.

Acknowledgments:
I would like to thank Debraj Chakrabarti, Anirban Dawn, Alekzander Malcom, John Treuer, Jianou Zhang, Liz Vivas, Trevor Fancher, Nicholas Lawson McLean, Alan Sola, Achinta Nandi, Sivaguru Ravisankar, Tomas Rodriguez, Mina Farag, Frank Wikström, Isak Ellmer, George Roman, and students in my classes for pointing out typos/errors and helpful suggestions. Some of the new material in version 4.0 was inspired by the comments and lecture notes from Richard Lärkäng and Elizabeth Wulcan.

During some of the writing of this book, the author was in part supported by NSF grant DMS-1362337 and Simons Foundation collaboration grant 710294.

More information:
See https://www.jirka.org/scv/ or https://jirilebl.github.io/scv/ for more information (including contacts).

Contents

Introduction

This book is a polished version of my course notes for Math 6283, Several Complex Variables, given in Spring 2014, Spring 2016, Spring 2019, and Fall 2023 semesters at the Oklahoma State University. There is more material than can fit in a one-semester class allowing for several versions of the course. In fact, I did a different selection each semester I taught it. Quite a few exercises of various difficulty are sprinkled throughout the text, and I hope a reader is at least attempting or thinking about most of them. Many are required later in the text. The reader should attempt exercises in sequence; earlier exercises can help or even be required to solve later ones.

The prerequisites are a decent knowledge of vector calculus, basic real analysis, and a working knowledge of complex analysis in one variable. Measure theory (Lebesgue integral and its convergence theorems) is useful, but it is not essential except in a couple of places later in the book. The first two chapters and most of the third are accessible to beginning graduate students after one semester of a standard single-variable complex analysis graduate course. From time to time (e.g. proof of Baouendi–Trèves in chapter 3, and most of chapter 4, and chapter 5), basic knowledge of differential forms is useful, and in chapter 6 we use some basic ring theory from algebra. By design, it can replace the second semester of complex analysis, with the first semester perhaps taught with my one-variable book [L].

This book is not intended as an exhaustive reference. It is simply a whirlwind tour of several complex variables. See the end of the book for a list of books for reference and further reading. There are also appendices for a list of one-variable results, an overview of differential forms, some basic algebra, measure theory, and other bits and pieces of analysis. See appendix B, appendix C, appendix D, and appendix E.

Changes in edition 4: The major addition of this edition is the greatly expanded chapter on the $\bar{\partial}$-problem, chapter 4. Many minor changes and additions throughout, especially in chapters 1, 2, and 6, resulted in some renumberings, including some renumbering of exercises. Finally, I've added a short appendix listing some useful results from analysis, including the very basics of measure theory. See the detailed listing of changes on the book website: https://www.jirka.org/scv/ or https://jirilebl.github.io/scv/.

0.1 \ Motivation, single variable, and Cauchy's formula

We start with some standard notation. We use $\mathbb{C}$ for complex numbers, $\mathbb{R}$ for real numbers, $\mathbb{Z}$ for integers, $\mathbb{N} = \{1, 2, 3, \ldots\}$ for natural numbers, $i = \sqrt{-1}$. Throughout this book, the standard terminology of *domain* means a connected open set. We try to avoid using it if connectedness is not needed, but sometimes we use it just for simplicity.

As complex analysis deals with complex numbers, perhaps we should begin with $\sqrt{-1}$. Start with the real numbers, $\mathbb{R}$, and add $\sqrt{-1}$ into our field. Call this square root i, and write the complex numbers, $\mathbb{C}$, by identifying $\mathbb{C}$ with $\mathbb{R}^2$ using

$$z = x + iy,$$

where $z \in \mathbb{C}$ and $(x, y) \in \mathbb{R}^2$. A subtle philosophical issue is that there are two square roots of -1. Two chickens are running around in our yard, and because we like to know which is which, we catch one and write "i" on it. If we happened to have caught the other chicken, we would have got an exactly equivalent theory, which we could not tell apart from the original.

Given a complex number z, its "opposite" is the *complex conjugate* of z and is defined as

$$\bar{z} \overset{\text{def}}{=} x - iy.$$

The size of z is measured by the so-called *modulus*, which is just the *Euclidean distance*:

$$|z| \overset{\text{def}}{=} \sqrt{z\bar{z}} = \sqrt{x^2 + y^2}.$$

If $z = x + iy \in \mathbb{C}$ for $x, y \in \mathbb{R}$, then x is called the *real part* and y is called the *imaginary part*. We write

$$\operatorname{Re} z = \operatorname{Re}(x + iy) = \frac{z + \bar{z}}{2} = x, \qquad \operatorname{Im} z = \operatorname{Im}(x + iy) = \frac{z - \bar{z}}{2i} = y.$$

A function $f \colon U \subset \mathbb{R}^n \to \mathbb{C}$ for an open set U is said to be continuously differentiable, or C^1 if the first (real) partial derivatives exist and are continuous. Similarly, it is C^k or C^k-*smooth* if the first k partial derivatives all exist and are continuous. Finally, a function is said to be C^∞ or simply *smooth** if it is *infinitely differentiable*, or in other words, if it is C^k for all $k \in \mathbb{N}$.

Complex analysis is the study of holomorphic (or complex-analytic) functions. Holomorphic functions are a generalization of polynomials, and to get there one leaves the land of algebra to arrive in the realm of analysis. One can do an awful lot with polynomials, but sometimes they are just not enough. For example, there is no nonzero polynomial function that solves the simplest of differential equations, $f' = f$. We need the exponential function, which is holomorphic.

---

*While C^∞ is a common definition of *smooth*, not everyone always means the same thing by the word *smooth*. I have seen it mean differentiable, C^1, piecewise-C^1, C^∞, holomorphic, ...

We start with polynomials. A nonzero polynomial in z is an expression

$$P(z) = \sum_{k=0}^{d} c_k z^k,$$

where $c_k \in \mathbb{C}$ and $c_d \neq 0$. The number d is called the *degree* of the polynomial P. We can plug in some number z and compute $P(z)$, to obtain a function $P \colon \mathbb{C} \to \mathbb{C}$.

We try to write

$$f(z) = \sum_{k=0}^{\infty} c_k z^k,$$

and all is very fine until we wish to know what $f(z)$ is for some number $z \in \mathbb{C}$. We usually mean

$$\sum_{k=0}^{\infty} c_k z^k = \lim_{d \to \infty} \sum_{k=0}^{d} c_k z^k.$$

As long as the limit exists, we have a function. You know all this; it is your one-variable complex analysis. We typically start with the functions and prove that we can expand them as series.

Let $U \subset \mathbb{C}$ be open. A function $f \colon U \to \mathbb{C}$ is *holomorphic* (or *complex-analytic*) if it is *complex-differentiable* at every point, that is, if

$$f'(z) = \lim_{\xi \in \mathbb{C} \to 0} \frac{f(z + \xi) - f(z)}{\xi} \qquad \text{exists for all } z \in U.$$

Importantly, the limit is taken with respect to complex ξ. Another vantage point is to start with a continuously differentiable* f, and say $f = u + iv$ is holomorphic if it satisfies the *Cauchy–Riemann equations*:

$$\frac{\partial u}{\partial x} = \frac{\partial v}{\partial y}, \qquad \frac{\partial u}{\partial y} = -\frac{\partial v}{\partial x}.$$

The so-called *Wirtinger operators*,

$$\frac{\partial}{\partial z} \overset{\text{def}}{=} \frac{1}{2} \left(\frac{\partial}{\partial x} - i \frac{\partial}{\partial y} \right), \qquad \frac{\partial}{\partial \bar{z}} \overset{\text{def}}{=} \frac{1}{2} \left(\frac{\partial}{\partial x} + i \frac{\partial}{\partial y} \right),$$

provide an easier way to understand the Cauchy–Riemann equations. These operators are determined by insisting on

$$\frac{\partial}{\partial z} z = 1, \qquad \frac{\partial}{\partial z} \bar{z} = 0, \qquad \frac{\partial}{\partial \bar{z}} z = 0, \qquad \frac{\partial}{\partial \bar{z}} \bar{z} = 1.$$

---

*Holomorphic functions end up being infinitely differentiable anyway, so this hypothesis is not overly restrictive.

The function f is holomorphic if and only if

$$\frac{\partial f}{\partial \bar{z}} = 0.$$

That seems a far nicer statement of the Cauchy–Riemann equations; it is just one complex equation. It says a function is holomorphic if and only if it depends on z but not on $\bar{z}$ (perhaps that does not make a whole lot of sense at first glance). We check:

$$\frac{\partial f}{\partial \bar{z}} = \frac{1}{2}\left(\frac{\partial f}{\partial x} + i\frac{\partial f}{\partial y}\right) = \frac{1}{2}\left(\frac{\partial u}{\partial x} + i\frac{\partial v}{\partial x} + i\frac{\partial u}{\partial y} - \frac{\partial v}{\partial y}\right) = \frac{1}{2}\left(\frac{\partial u}{\partial x} - \frac{\partial v}{\partial y}\right) + \frac{i}{2}\left(\frac{\partial v}{\partial x} + \frac{\partial u}{\partial y}\right).$$

This expression is zero if and only if the real parts and the imaginary parts are zero. In other words,

$$\frac{\partial u}{\partial x} - \frac{\partial v}{\partial y} = 0 \qquad \text{and} \qquad \frac{\partial v}{\partial x} + \frac{\partial u}{\partial y} = 0.$$

That is, the Cauchy–Riemann equations are satisfied.

If f is holomorphic, the derivative in z is the standard complex derivative you know and love:

$$\frac{\partial f}{\partial z}(z_0) = f'(z_0) = \lim_{\xi \to 0} \frac{f(z_0 + \xi) - f(z_0)}{\xi}.$$

That is because

$$\frac{\partial f}{\partial z} = \frac{1}{2}\left(\frac{\partial u}{\partial x} + \frac{\partial v}{\partial y}\right) + \frac{i}{2}\left(\frac{\partial v}{\partial x} - \frac{\partial u}{\partial y}\right) = \frac{\partial u}{\partial x} + i\frac{\partial v}{\partial x} = \frac{\partial f}{\partial x}$$

$$= \frac{1}{i}\left(\frac{\partial u}{\partial y} + i\frac{\partial v}{\partial y}\right) = \frac{\partial f}{\partial (iy)}.$$

A function on $\mathbb{C}$ is a function defined on $\mathbb{R}^2$ as identified above, and so it is a function of x and y. Writing $x = \frac{z+\bar{z}}{2}$ and $y = \frac{z-\bar{z}}{2i}$, think of it as a function of two complex variables, z and $\bar{z}$. Pretend for a moment as if $\bar{z}$ did not depend on z. The Wirtinger operators work as if z and $\bar{z}$ really were independent variables. For instance:

$$\frac{\partial}{\partial z}\left[z^2\bar{z}^3 + z^{10}\right] = 2z\bar{z}^3 + 10z^9 \qquad \text{and} \qquad \frac{\partial}{\partial \bar{z}}\left[z^2\bar{z}^3 + z^{10}\right] = z^2(3\bar{z}^2) + 0.$$

A holomorphic function is a function "not depending on $\bar{z}$."

The most important theorem in one variable is the *Cauchy integral formula*.

Theorem 0.1.1 (Cauchy integral formula). *Let $U \subset \mathbb{C}$ be a bounded domain where the boundary ∂U is a piecewise smooth simple closed path (a Jordan curve). Let $f: \bar{U} \to \mathbb{C}$ be a continuous function, holomorphic in U. Orient ∂U positively (going around counterclockwise). Then*

$$f(z) = \frac{1}{2\pi i}\int_{\partial U}\frac{f(\zeta)}{\zeta - z}\,d\zeta \qquad \text{for all } z \in U.$$

The Cauchy formula is the essential ingredient we need from one complex variable. It follows from Green's theorem* (Stokes' theorem in two dimensions). You can look forward to Theorem 4.1.1 for a proof of a more general formula, the Cauchy–Pompeiu integral formula.

As a differential form, $dz = dx + i\, dy$. If you are uneasy about differential forms, you possibly defined the path integral above directly using the Riemann–Stieltjes integral in your one-complex-variable class. Let us write down the formula in terms of the standard Riemann integral in a special case. Take the *unit disc*

$$\mathbb{D} \stackrel{\text{def}}{=} \{z \in \mathbb{C} : |z| < 1\}.$$

The boundary is the unit circle $\partial \mathbb{D} = \{z \in \mathbb{C} : |z| = 1\}$ oriented positively, that is, counter-clockwise. Parametrize $\partial \mathbb{D}$ by e^{it}, where t goes from 0 to 2π. If $\zeta = e^{it}$, then $d\zeta = ie^{it}dt$, and

$$f(z) = \frac{1}{2\pi i} \int_{\partial \mathbb{D}} \frac{f(\zeta)}{\zeta - z}\, d\zeta = \frac{1}{2\pi} \int_0^{2\pi} \frac{f(e^{it})e^{it}}{e^{it} - z}\, dt.$$

If you are not completely comfortable with path integrals, try to think about how you would parametrize the path, and write the integral as an integral any calculus student would recognize.

I venture a guess that 90% of what you learned in a one-variable complex analysis course (depending on who taught it) is more or less a straightforward consequence of the Cauchy integral formula. An important theorem from one variable that follows from the Cauchy formula is the *maximum modulus principle* (or just the *maximum principle*). Let us give its simplest version.

Theorem 0.1.2 (Maximum modulus principle). *Suppose $U \subset \mathbb{C}$ is a domain and $f : U \to \mathbb{C}$ is holomorphic. If for some $z_0 \in U$*

$$\sup_{z \in U} |f(z)| = |f(z_0)|,$$

then f is constant, that is, $f \equiv f(z_0)$.

That is, if the supremum is attained in the interior of the domain, then the function must be constant. Another way to state the maximum principle is to say: If f extends continuously to the boundary of a bounded domain, then the supremum of $|f(z)|$ is attained on the boundary. In one variable you learned that the maximum principle is really a property of harmonic functions.

---

*If you wish to feel inadequate, note that this theorem, on which all of complex analysis (and all of physics) rests, was proved by George Green, who was the son of a miller and had one year of formal schooling.

Theorem 0.1.3 (Maximum principle). *Let $U \subset \mathbb{C}$ be a domain and $h\colon U \to \mathbb{R}$ harmonic, that is,*

$$\nabla^2 h = \frac{\partial^2 h}{\partial x^2} + \frac{\partial^2 h}{\partial y^2} = 0.$$

If for some $z_0 \in U$

$$\sup_{z \in U} h(z) = h(z_0) \qquad or \qquad \inf_{z \in U} h(z) = h(z_0),$$

then h is constant, that is, $h \equiv h(z_0)$.

In one variable, if $f = u + iv$ is holomorphic for real-valued u and v, then u and v are harmonic. Similarly, outside the zero set of f, $\log|f|$ is harmonic. Locally, a harmonic function is the real (or imaginary) part of a holomorphic function, so in one complex variable, studying harmonic functions is almost equivalent to studying holomorphic functions. Things are decidedly different in two or more variables.

Holomorphic functions admit a power series representation in z at each point a:

$$f(z) = \sum_{k=0}^{\infty} c_k (z - a)^k.$$

No $\bar{z}$ is necessary, since $\frac{\partial f}{\partial \bar{z}} = 0$.

Let us see the proof using the Cauchy integral formula, as we will require this computation in several variables as well. Given $a \in \mathbb{C}$ and $\rho > 0$, define the disc of radius ρ around a

$$\Delta_\rho(a) \overset{\text{def}}{=} \left\{ z \in \mathbb{C} : |z - a| < \rho \right\}.$$

Suppose $U \subset \mathbb{C}$ is open, $f\colon U \to \mathbb{C}$ is holomorphic, $a \in U$, and $\overline{\Delta_\rho(a)} \subset U$ (that is, the closure of the disc is in U, and so its boundary $\partial\Delta_\rho(a)$ is also in U).

For $z \in \Delta_\rho(a)$ and $\zeta \in \partial\Delta_\rho(a)$,

$$\left| \frac{z - a}{\zeta - a} \right| = \frac{|z - a|}{\rho} < 1.$$

In fact, if $|z - a| \le \rho' < \rho$, then $\left| \frac{z-a}{\zeta-a} \right| \le \frac{\rho'}{\rho} < 1$. Therefore, the geometric series

$$\sum_{k=0}^{\infty} \left(\frac{z - a}{\zeta - a} \right)^k = \frac{1}{1 - \frac{z-a}{\zeta-a}} = \frac{\zeta - a}{\zeta - z}$$

converges uniformly absolutely for $(z, \zeta) \in \overline{\Delta_{\rho'}(a)} \times \partial\Delta_\rho(a)$ (that is, $\sum_k \left| \frac{z-a}{\zeta-a} \right|^k$ converges uniformly).

Let γ be the path going around $\partial \Delta_\rho(a)$ once in the positive direction. Compute

$$f(z) = \frac{1}{2\pi i} \int_\gamma \frac{f(\zeta)}{\zeta - z} \, d\zeta$$

$$= \frac{1}{2\pi i} \int_\gamma \frac{f(\zeta)}{\zeta - a} \frac{\zeta - a}{\zeta - z} \, d\zeta$$

$$= \frac{1}{2\pi i} \int_\gamma \frac{f(\zeta)}{\zeta - a} \sum_{k=0}^\infty \left(\frac{z - a}{\zeta - a} \right)^k d\zeta$$

$$= \sum_{k=0}^\infty \left(\frac{1}{2\pi i} \int_\gamma \frac{f(\zeta)}{(\zeta - a)^{k+1}} \, d\zeta \right) (z - a)^k.$$

In the last equality, we may interchange the limit on the sum with the integral either via Fubini's theorem or via uniform convergence: z is fixed and if M is the supremum of $\left| \frac{f(\zeta)}{\zeta - a} \right| = \frac{|f(\zeta)|}{\rho}$ on $\partial \Delta_\rho(a)$, then

$$\left| \frac{f(\zeta)}{\zeta - a} \left(\frac{z - a}{\zeta - a} \right)^k \right| \le M \left(\frac{|z - a|}{\rho} \right)^k \quad \text{and} \quad \frac{|z - a|}{\rho} < 1.$$

The key point is writing the *Cauchy kernel* $\frac{1}{\zeta - z}$ as

$$\frac{1}{\zeta - z} = \frac{1}{\zeta - a} \frac{\zeta - a}{\zeta - z},$$

and then using the geometric series.

Not only have we proved that f has a power series, but we computed that the radius of convergence is at least R, where R is the maximum R such that $\Delta_R(a) \subset U$. We also obtained a formula for the coefficients

$$c_k = \frac{1}{2\pi i} \int_\gamma \frac{f(\zeta)}{(\zeta - a)^{k+1}} \, d\zeta.$$

For a set K, denote the *supremum norm*:

$$\|f\|_K \overset{\text{def}}{=} \sup_{z \in K} |f(z)|.$$

By a brute force estimation, we obtain the very useful *Cauchy estimates*:

$$|c_k| = \left| \frac{1}{2\pi i} \int_\gamma \frac{f(\zeta)}{(\zeta - a)^{k+1}} \, d\zeta \right| \le \frac{1}{2\pi} \int_\gamma \frac{\|f\|_\gamma}{\rho^{k+1}} |d\zeta| = \frac{\|f\|_\gamma}{\rho^k}.$$

We differentiate Cauchy's formula k times (using the Wirtinger $\frac{\partial}{\partial z}$ operator),

$$f^{(k)}(z) = \frac{\partial^k f}{\partial z^k}(z) = \frac{1}{2\pi i} \int_\gamma \frac{k! f(\zeta)}{(\zeta - z)^{k+1}} \, d\zeta,$$

and therefore

$$k!\, c_k = f^{(k)}(a) = \frac{\partial^k f}{\partial z^k}(a).$$

Hence, we can control derivatives of f by the size of the function:

$$\left| f^{(k)}(a) \right| = \left| \frac{\partial^k f}{\partial z^k}(a) \right| \leq \frac{k!\, \|f\|_\gamma}{\rho^k}.$$

This estimate is one of the key properties of holomorphic functions, and the reason why the correct topology for the set of holomorphic functions is the same as the topology for continuous functions. Consequently, obstructions to solving problems in complex analysis are often topological in character.

For a further review of one-variable results, see appendix B.

1 Holomorphic Functions in Several Variables

1.1 Onto several variables

Let $\mathbb{C}^n = \overbrace{\mathbb{C} \times \mathbb{C} \times \cdots \times \mathbb{C}}^{n \text{ times}}$ denote the n-dimensional *complex Euclidean space*. Denote by $z = (z_1, z_2, \ldots, z_n)$ the coordinates of $\mathbb{C}^n$. Let $x = (x_1, x_2, \ldots, x_n)$ and $y = (y_1, y_2, \ldots, y_n)$ denote the coordinates in $\mathbb{R}^n$. Identify $\mathbb{C}^n$ with $\mathbb{R}^{2n}$ by letting $z = x + iy$, that is, $z_k = x_k + iy_k$ for every k. As in one complex variable, write $\bar{z} = x - iy$. We call z the *holomorphic coordinates* and $\bar{z}$ the *antiholomorphic coordinates*.

Definition 1.1.1. For $\rho = (\rho_1, \rho_2, \ldots, \rho_n)$ where $\rho_k > 0$ and $a \in \mathbb{C}^n$, define a *polydisc*

$$\Delta_\rho(a) \overset{\text{def}}{=} \left\{ z \in \mathbb{C}^n : |z_k - a_k| < \rho_k \text{ for } k = 1, 2, \ldots, n \right\}.$$

Call a the *center* and ρ the *polyradius* or simply the *radius* of the polydisc $\Delta_\rho(a)$. If $\rho > 0$ is a number, then

$$\Delta_\rho(a) \overset{\text{def}}{=} \left\{ z \in \mathbb{C}^n : |z_k - a_k| < \rho \text{ for } k = 1, 2, \ldots, n \right\}.$$

In two variables, a polydisc is sometimes called a *bidisc*. As there is the unit disc $\mathbb{D}$ in one variable, so is there the *unit polydisc* in several variables:

$$\mathbb{D}^n = \mathbb{D} \times \mathbb{D} \times \cdots \times \mathbb{D} = \Delta_1(0) = \left\{ z \in \mathbb{C}^n : |z_k| < 1 \text{ for } k = 1, 2, \ldots, n \right\}.$$

In more than one complex dimension, it is difficult to draw exact pictures for lack of real dimensions on our paper. We visualize the unit polydisc in two variables (bidisc) as in Figure 1.1 by plotting against the modulus of the variables.

Recall the *Euclidean inner product* on $\mathbb{C}^n$:

$$\langle z, w \rangle \overset{\text{def}}{=} z_1 \bar{w}_1 + z_2 \bar{w}_2 + \cdots + z_n \bar{w}_n.$$

The inner product gives us the standard *Euclidean norm* on $\mathbb{C}^n$:

$$\|z\| \overset{\text{def}}{=} \sqrt{\langle z, z \rangle} = \sqrt{|z_1|^2 + |z_2|^2 + \cdots + |z_n|^2}.$$

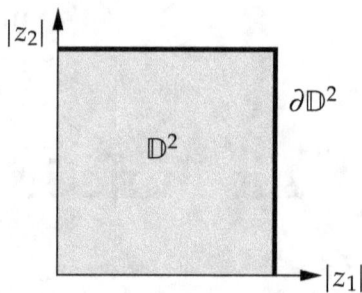

Figure 1.1: The bidisc.

This norm agrees with the standard Euclidean norm on $\mathbb{R}^{2n}$. Define *balls* as in $\mathbb{R}^{2n}$:

$$B_\rho(a) \overset{\text{def}}{=} \{z \in \mathbb{C}^n : \|z - a\| < \rho\},$$

And the *unit ball*,

$$\mathbb{B}_n \overset{\text{def}}{=} B_1(0) = \{z \in \mathbb{C}^n : \|z\| < 1\}.$$

A ball centered at the origin can also be pictured by plotting against the modulus of the variables, since the inequality defining the ball only depends on the moduli of the variables. Not every domain can be drawn like this, but if it can, it is called a *Reinhardt domain*, more on this later. A picture of $\mathbb{B}_2$ is in Figure 1.2.

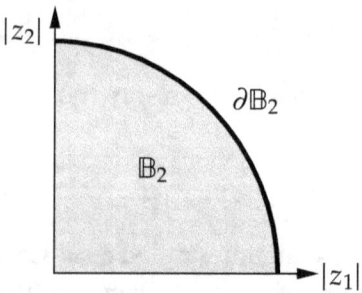

Figure 1.2: The ball $\mathbb{B}_2$ as a Reinhardt domain.

Definition 1.1.2. Let $U \subset \mathbb{C}^n$ be open. A function $f \colon U \to \mathbb{C}$ is *holomorphic* if it is *locally bounded** and holomorphic in each variable separately. That is, f is holomorphic if it is locally bounded and complex-differentiable in each variable separately:

$$\lim_{\xi \in \mathbb{C} \to 0} \frac{f(z_1, \ldots, z_k + \xi, \ldots, z_n) - f(z)}{\xi} \qquad \text{exists for all } z \in U \text{ and all } k = 1, 2, \ldots, n.$$

*For every $p \in U$, there is a neighborhood N of p such that $f|_N$ is bounded. Equivalently, f is bounded on compact subsets of U. It is a deep result of Hartogs that we might in fact just drop "locally bounded" from the definition and obtain the same set of functions.

In this book, the words "differentiable" and "derivative" (without the "complex-") refer to plain-vanilla real differentiability.

As in one variable, we define the *Wirtinger operators*

$$\frac{\partial}{\partial z_k} \overset{\text{def}}{=} \frac{1}{2}\left(\frac{\partial}{\partial x_k} - i\frac{\partial}{\partial y_k}\right), \qquad \frac{\partial}{\partial \bar{z}_k} \overset{\text{def}}{=} \frac{1}{2}\left(\frac{\partial}{\partial x_k} + i\frac{\partial}{\partial y_k}\right).$$

An alternative definition is to say that a continuously differentiable function $f \colon U \to \mathbb{C}$ is *holomorphic* if it satisfies the *Cauchy–Riemann equations*

$$\frac{\partial f}{\partial \bar{z}_k} = 0 \qquad \text{for } k = 1, 2, \ldots, n.$$

For holomorphic functions, using the natural definition for partial derivatives yields the Wirtinger $\frac{\partial}{\partial z_k}$. Namely, if f is holomorphic, then

$$\frac{\partial f}{\partial z_k}(z) = \lim_{\xi \in \mathbb{C} \to 0} \frac{f(z_1, \ldots, z_k + \xi, \ldots, z_n) - f(z)}{\xi}.$$

Due to the following proposition, the alternative definition using the Cauchy–Riemann equations is just as good as the definition we gave.

Proposition 1.1.3. *Let $U \subset \mathbb{C}^n$ be an open set and suppose $f \colon U \to \mathbb{C}$ is holomorphic. Then f is infinitely differentiable.*

Proof. Suppose $\Delta = \Delta_\rho(a) = \Delta_1 \times \cdots \times \Delta_n$ is a polydisc centered at a, where each Δ_k is a disc, and suppose $\overline{\Delta} \subset U$, that is, f is holomorphic on a neighborhood of the closure of Δ. Let z be in Δ. Orient $\partial\Delta_1$ positively and apply the Cauchy formula (after all f is holomorphic in z_1):

$$f(z) = \frac{1}{2\pi i} \int_{\partial\Delta_1} \frac{f(\zeta_1, z_2, \ldots, z_n)}{\zeta_1 - z_1} \, d\zeta_1.$$

Apply it again on the second variable, again orienting $\partial\Delta_2$ positively:

$$f(z) = \frac{1}{(2\pi i)^2} \int_{\partial\Delta_1} \int_{\partial\Delta_2} \frac{f(\zeta_1, \zeta_2, z_3, \ldots, z_n)}{(\zeta_1 - z_1)(\zeta_2 - z_2)} \, d\zeta_2 \, d\zeta_1.$$

Applying the formula n times, we obtain

$$f(z) = \frac{1}{(2\pi i)^n} \int_{\partial\Delta_1} \int_{\partial\Delta_2} \cdots \int_{\partial\Delta_n} \frac{f(\zeta_1, \zeta_2, \ldots, \zeta_n)}{(\zeta_1 - z_1)(\zeta_2 - z_2)\cdots(\zeta_n - z_n)} \, d\zeta_n \cdots d\zeta_2 \, d\zeta_1. \quad (1.1)$$

As f is bounded on the compact set $\partial\Delta_1 \times \cdots \times \partial\Delta_n$, we find that f is continuous in Δ, and hence on U. We may differentiate underneath the integral via the standard Leibniz rule, because the integrand and its partial derivatives with respect to x_k and y_k, where $z_k = x_k + iy_k$, are all continuous, as long as z is a positive distance away from $\partial\Delta_1 \times \cdots \times \partial\Delta_n$. We may differentiate as many times as we wish. $\qquad\square$

In (1.1) above, we derived the Cauchy integral formula in several variables. To write the formula more concisely, we apply Fubini's theorem to write it as a single integral. We will write it down using differential forms. If you are unfamiliar with differential forms, think of the integral as the iterated integral above, and you can read the next few paragraphs a little lightly. It is enough to understand real differential forms; we simply allow complex coefficients here. See appendix C for an overview of differential forms, or Rudin [R1] for an introduction with all the details.

Given real coordinates $x = (x_1, \ldots, x_n)$, a one-form dx_k is a linear functional on tangent vectors such that $\langle dx_k, \frac{\partial}{\partial x_k} \rangle = 1$ and $\langle dx_k, \frac{\partial}{\partial x_\ell} \rangle = 0$ if $k \neq \ell$, where we use the pairing notation $\langle \omega, v \rangle$ instead of the functional notation $\omega(v)$ as is traditional to indicate multilinearity. As $z_k = x_k + iy_k$ and $\bar{z}_k = x_k - iy_k$,

$$dz_k = dx_k + i\, dy_k, \qquad d\bar{z}_k = dx_k - i\, dy_k.$$

Let δ_k^ℓ be the Kronecker delta, that is, $\delta_k^k = 1$, and $\delta_k^\ell = 0$ if $k \neq \ell$. Then, as expected,

$$\left\langle dz_k, \frac{\partial}{\partial z_\ell} \right\rangle = \delta_k^\ell, \qquad \left\langle dz_k, \frac{\partial}{\partial \bar{z}_\ell} \right\rangle = 0, \qquad \left\langle d\bar{z}_k, \frac{\partial}{\partial z_\ell} \right\rangle = 0, \qquad \left\langle d\bar{z}_k, \frac{\partial}{\partial \bar{z}_\ell} \right\rangle = \delta_k^\ell.$$

One-forms are the objects

$$\sum_{k=1}^{n} \alpha_k\, dz_k + \beta_k\, d\bar{z}_k,$$

where α_k and β_k are functions (of z). Two-forms are combinations of wedge products, $\omega \wedge \eta$, of one-forms. A wedge of a two-form and a one-form is a three-form, etc. An m-form is an object that can be integrated on a so-called m-chain, for example, a m-dimensional surface. The wedge product takes care of the orientation as it is anticommutative on one-forms: For one-forms ω and η, we have $\omega \wedge \eta = -\eta \wedge \omega$.

At this point, we need to talk about orientation in $\mathbb{C}^n$, that is, the ordering of the real coordinates. There are two natural real-linear isomorphisms of $\mathbb{C}^n$ and $\mathbb{R}^{2n}$. We identify $z = x + iy$ as either

$$(x, y) = (x_1, \ldots, x_n, y_1, \ldots, y_n) \qquad \text{or} \qquad (x_1, y_1, x_2, y_2, \ldots, x_n, y_n).$$

If we take the natural orientation of $\mathbb{R}^{2n}$, it is possible that we obtain two opposite orientations on $\mathbb{C}^n$ (the real linear map that takes one ordering to the other has determinant $(-1)^{n(n-1)/2}$). The orientation we take as the natural orientation of $\mathbb{C}^n$ (in this book) corresponds to the second ordering above, that is, $(x_1, y_1, \ldots, x_n, y_n)$. Either isomorphism may be used in computation as long as it is used consistently, and the underlying orientation is kept in mind.

Theorem 1.1.4 (Cauchy integral formula). *Let $\Delta \subset \mathbb{C}^n$ be a polydisc. Suppose $f \colon \overline{\Delta} \to \mathbb{C}$ is a continuous function holomorphic in Δ. Write $\Gamma = \partial \Delta_1 \times \cdots \times \partial \Delta_n$ oriented appropriately (each $\partial \Delta_k$ oriented positively). Then for $z \in \Delta$*

$$f(z) = \frac{1}{(2\pi i)^n} \int_\Gamma \frac{f(\zeta_1, \zeta_2, \ldots, \zeta_n)}{(\zeta_1 - z_1)(\zeta_2 - z_2) \cdots (\zeta_n - z_n)}\, d\zeta_1 \wedge d\zeta_2 \wedge \cdots \wedge d\zeta_n.$$

We stated a more general result where f is only continuous on $\overline{\Delta}$ and holomorphic in Δ. The proof of this slight generalization is contained within the next two exercises.

Exercise 1.1.1: *Suppose $f \colon \overline{\mathbb{D}^2} \to \mathbb{C}$ is continuous and holomorphic on $\mathbb{D}^2$. For every $\theta \in \mathbb{R}$, prove*

$$g_1(\xi) = f(\xi, e^{i\theta}) \qquad and \qquad g_2(\xi) = f(e^{i\theta}, \xi)$$

are holomorphic in $\mathbb{D}$.

Exercise 1.1.2: *Prove the theorem above, that is, the slightly more general Cauchy integral formula where f is only continuous on $\overline{\Delta}$ and holomorphic in Δ.*

The Cauchy integral formula shows an important and subtle point about holomorphic functions in several variables: The value of the function f on Δ is completely determined by the values of f on the set Γ, which is much smaller than the boundary of the polydisc $\partial\Delta$. In fact, Γ is of real dimension n, while the boundary of the polydisc is of real dimension $2n - 1$. The set $\Gamma = \partial\Delta_1 \times \cdots \times \partial\Delta_n$ is called the *distinguished boundary*. See Figure 1.3 for the distinguished boundary of the bidisc.

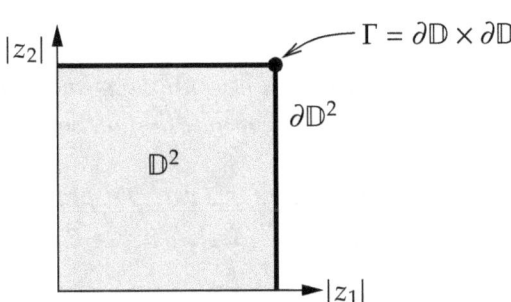

Figure 1.3: The distinguished boundary of $\mathbb{D}^2$.

The set Γ is a 2-dimensional torus, like the surface of a donut. Whereas the set $\partial\mathbb{D}^2 = (\partial\mathbb{D} \times \overline{\mathbb{D}}) \cup (\overline{\mathbb{D}} \times \partial\mathbb{D})$ is the union of two filled donuts, or more precisely, it is both the inside and the outside of the donut put together, and these two things meet on the surface of the donut. So the set Γ is quite small in comparison to the entire boundary $\partial\mathbb{D}^2$.

Exercise 1.1.3: *Suppose Δ is a polydisc, Γ its distinguished boundary, and $f \colon \overline{\Delta} \to \mathbb{C}$ is continuous on $\overline{\Delta}$ and holomorphic on Δ. Prove $|f(z)|$ achieves its maximum on Γ.*

Exercise 1.1.4: *A ball is different from a polydisc. Prove that for every $p \in \partial\mathbb{B}_n$ there exists a continuous $f \colon \overline{\mathbb{B}_n} \to \mathbb{C}$, holomorphic on $\mathbb{B}_n$, such that $|f(z)|$ achieves a strict maximum at p.*

Exercise 1.1.5: *Show that in the real setting, differentiable in each variable separately does not imply differentiable even if the function is locally bounded. Let* $f(x,y) = \frac{xy}{x^2+y^2}$ *outside the origin and* $f(0,0) = 0$. *Prove that* f *is a locally bounded function in* $\mathbb{R}^2$, *which is differentiable in each variable separately (all partial derivatives exist at every point), but* f *is not even continuous. There is something very special about the holomorphic category.*

Exercise 1.1.6: *Suppose* $U \subset \mathbb{C}^n$ *is open. Prove that* $f \colon U \to \mathbb{C}$ *is holomorphic if and only if* f *is locally bounded and for every* $a, b \in \mathbb{C}^n$, *the function* $\zeta \mapsto f(\zeta a + b)$ *is holomorphic on the open set* $\{\zeta \in \mathbb{C} : \zeta a + b \in U\}$.

Exercise 1.1.7: *Prove a several complex variables version of Morera's theorem (see Theorem B.4). A triangle* $T \subset \mathbb{C}^n$ *is the closed convex hull of three points, so including the inside. Orient* T *in some way and orient* ∂T *accordingly. A triangle* T *lies in a complex line if its vertices* a, b, c *satisfy* $\zeta(b-a) = c - a$ *for some* $\zeta \in \mathbb{C}$. *Suppose* $U \subset \mathbb{C}^n$ *is open and* $f \colon U \to \mathbb{C}$ *is continuous. Prove that* f *is holomorphic if and only if*

$$\int_{\partial T} f(z)\, dz_k = 0$$

for every triangle $T \subset U$ *that lies in a complex line, and every* $k = 1, 2, \ldots, n$. *Hint: The previous exercise may be useful.*

Exercise 1.1.8: *Let* $f \colon \overline{\mathbb{D}^2} \setminus \{0\} \to \mathbb{C}$ *be continuous and holomorphic on* $\mathbb{D}^2 \setminus \{0\}$.
   a) *Prove that* f *is bounded. Hint: Consider the functions* $\xi \mapsto f(\xi, a)$ *and* $\xi \mapsto f(a, \xi)$ *for different* a.
   b) *Using the Riemann extension in one variable, prove that there exists a continuous* $F \colon \overline{\mathbb{D}^2} \to \mathbb{C}$, *holomorphic on* $\mathbb{D}^2$, *such that* $f = F$ *on* $\overline{\mathbb{D}^2} \setminus \{0\}$.

1.2   Power series representation

As you noticed, writing out all the components can be a pain. Just as we write vectors as z instead of $(z_1, z_2, \ldots, z_n)$, we similarly define the so-called *multi-index notation* to deal with more complicated formulas such as the ones above.

Let $\alpha \in \mathbb{N}_0^n$ be a vector of nonnegative integers (where $\mathbb{N}_0 = \mathbb{N} \cup \{0\}$). We write

$$z^\alpha \overset{\text{def}}{=} z_1^{\alpha_1} z_2^{\alpha_2} \cdots z_n^{\alpha_n}, \qquad\qquad |z|^\alpha \overset{\text{def}}{=} |z_1|^{\alpha_1} |z_2|^{\alpha_2} \cdots |z_n|^{\alpha_n},$$

$$\frac{1}{z} \overset{\text{def}}{=} \frac{1}{z_1 z_2 \cdots z_n}, \qquad\qquad \frac{z}{w} \overset{\text{def}}{=} \left(\frac{z_1}{w_1}, \frac{z_2}{w_2}, \ldots, \frac{z_n}{w_n} \right),$$

$$\frac{\partial^{|\alpha|}}{\partial z^\alpha} \overset{\text{def}}{=} \frac{\partial^{\alpha_1}}{\partial z_1^{\alpha_1}} \frac{\partial^{\alpha_2}}{\partial z_2^{\alpha_2}} \cdots \frac{\partial^{\alpha_n}}{\partial z_n^{\alpha_n}}, \qquad\qquad dz \overset{\text{def}}{=} dz_1 \wedge dz_2 \wedge \cdots \wedge dz_n,$$

$$|\alpha| \overset{\text{def}}{=} \alpha_1 + \alpha_2 + \cdots + \alpha_n, \qquad\qquad \alpha! \overset{\text{def}}{=} \alpha_1! \alpha_2! \cdots \alpha_n!.$$

We can also make sense of this notation, especially the notation z^α, if $\alpha \in \mathbb{Z}^n$, that is, if it includes negative integers. Although α is usually assumed to be in $\mathbb{N}_0^n$. Furthermore, when we use 1 as a vector, it means $(1, 1, \ldots, 1)$. If $z \in \mathbb{C}^n$, then

$$1 - z = (1 - z_1, 1 - z_2, \ldots, 1 - z_n), \quad \text{or} \quad z^{\alpha+1} = z_1^{\alpha_1+1} z_2^{\alpha_2+1} \cdots z_n^{\alpha_n+1}.$$

It goes without saying that when using this notation it is important to be careful to always realize which symbol lives where, and most of all, to not get carried away. For instance, we can interpret $\frac{1}{z}$ in different ways depending on whether we interpret 1 as a vector or not, and whether we expect a vector or a number. Best to just keep to the limited set of cases as given above, and only use it when it is clear what is meant. In this notation, the Cauchy formula becomes the perhaps deceptively simple

$$f(z) = \frac{1}{(2\pi i)^n} \int_\Gamma \frac{f(\zeta)}{\zeta - z} \, d\zeta.$$

Let us move on to power series. For simplicity, we start with power series at the origin. Using the multi-index notation, we write such a series as

$$\sum_{\alpha \in \mathbb{N}_0^n} c_\alpha z^\alpha.$$

You must admit that the above is far nicer to write than writing, for example, in $\mathbb{C}^3$,

$$\sum_{k=0}^{\infty} \sum_{\ell=0}^{\infty} \sum_{m=0}^{\infty} c_{k\ell m} z_1^k z_2^\ell z_3^m, \tag{1.2}$$

which is not even exactly the definition of the series sum (see below). When it is clear from context that we are talking about a power series and all the powers are nonnegative, we write simply

$$\sum_\alpha c_\alpha z^\alpha.$$

It is important to note what this means. The sum does not have a natural ordering. We are summing over $\alpha \in \mathbb{N}_0^n$, and there is no natural ordering of $\mathbb{N}_0^n$. It makes no sense to talk about conditional convergence. When we say the series *converges*, we mean absolutely. Fortunately, power series do converge absolutely, so the ordering does not matter. If you want to write the limit in terms of partial sums, you pick some ordering of the multi-indices, $\alpha(1), \alpha(2), \ldots$, and then

$$\sum_\alpha c_\alpha z^\alpha = \lim_{m \to \infty} \sum_{k=1}^{m} c_{\alpha(k)} z^{\alpha(k)}.$$

By the Fubini theorem (for sums), this limit is equal to the iterated sum such as (1.2).

A power series $\sum_\alpha c_\alpha z^\alpha$ converges *uniformly absolutely* for $z \in X$ when $\sum_\alpha |c_\alpha z^\alpha|$ converges uniformly for $z \in X$. The *geometric series in several variables* is the series $\sum_\alpha z^\alpha$. For $z \in \mathbb{D}^n$ (unit polydisc),

$$\frac{1}{1-z} = \frac{1}{(1-z_1)(1-z_2)\cdots(1-z_n)} = \left(\sum_{k=0}^{\infty} z_1{}^k\right)\left(\sum_{k=0}^{\infty} z_2{}^k\right)\cdots\left(\sum_{k=0}^{\infty} z_n{}^k\right)$$

$$= \sum_{k_1=0}^{\infty}\sum_{k_2=0}^{\infty}\cdots\sum_{k_n=0}^{\infty}\left(z_1{}^{k_1} z_2{}^{k_2}\cdots z_n{}^{k_n}\right) = \sum_\alpha z^\alpha.$$

The series converges uniformly absolutely on all compact subsets of the unit polydisc: Any compact set in the unit polydisc is contained in a closed polydisc $\overline{\Delta}$ centered at 0 of radius $1 - \epsilon$ for some $\epsilon > 0$. The convergence is uniformly absolute on $\overline{\Delta}$. This claim follows by noting that the same fact holds for each factor in one dimension.

Holomorphic functions are precisely those that allow a power series expansion:

Theorem 1.2.1. *Let $\Delta = \Delta_\rho(a) \subset \mathbb{C}^n$ be a polydisc. Suppose $f : \overline{\Delta} \to \mathbb{C}$ is a continuous function holomorphic in Δ. Then on Δ, f is equal to a power series converging uniformly absolutely on compact subsets of Δ:*

$$f(z) = \sum_\alpha c_\alpha (z-a)^\alpha. \tag{1.3}$$

Conversely, if $f : \Delta \to \mathbb{C}$ is defined by (1.3) converging uniformly absolutely on compact subsets of Δ, then f is holomorphic on Δ.

The hypothesis that f is continuous on $\overline{\Delta}$ is not necessary. We will prove in a moment that the power series is unique and hence we could have used an arbitrary smaller polydisc centered at a for the development.

Proof. Suppose a continuous $f : \overline{\Delta} \to \mathbb{C}$ is holomorphic on Δ. Let $\Gamma = \partial\Delta_1 \times \cdots \times \partial\Delta_n$ be oriented positively. Take $z \in \Delta$ and $\zeta \in \Gamma$. As in one variable, write the Cauchy kernel as

$$\frac{1}{\zeta - z} = \frac{1}{\zeta - a}\left(\frac{1}{1 - \frac{z-a}{\zeta-a}}\right) = \frac{1}{\zeta - a}\sum_\alpha \left(\frac{z-a}{\zeta-a}\right)^\alpha.$$

Interpret the formulas as $\frac{1}{\zeta - z} = \frac{1}{(\zeta_1 - z_1)\cdots(\zeta_n - z_n)}$, $\frac{1}{\zeta - a} = \frac{1}{(\zeta_1 - a_1)\cdots(\zeta_n - a_n)}$ and $\frac{z-a}{\zeta - a} = \left(\frac{z_1 - a_1}{\zeta_1 - a_1}, \ldots, \frac{z_n - a_n}{\zeta_n - a_n}\right)$. The multivariable geometric series is a product of the geometric series in one variable, and the geometric series in one variable is uniformly absolutely convergent on compact subsets of the unit disc. So the series above converges uniformly absolutely for $(z, \zeta) \in K \times \Gamma$ for every compact subset K of Δ.

For $z \in \Delta$,

$$f(z) = \frac{1}{(2\pi i)^n} \int_\Gamma \frac{f(\zeta)}{\zeta - z} d\zeta$$

$$= \frac{1}{(2\pi i)^n} \int_\Gamma \frac{f(\zeta)}{\zeta - a} \sum_\alpha \left(\frac{z - a}{\zeta - a}\right)^\alpha d\zeta$$

$$= \sum_\alpha \left(\frac{1}{(2\pi i)^n} \int_\Gamma \frac{f(\zeta)}{(\zeta - a)^{\alpha+1}} d\zeta\right) (z - a)^\alpha.$$

The last equality follows by Fubini or uniform convergence just as it does in one variable. Uniform absolute convergence (as z moves) on compact subsets of the final series follows from the uniform absolute convergence of the geometric series. It is also a direct consequence of the Cauchy estimates below. We have shown that

$$f(z) = \sum_\alpha c_\alpha (z - a)^\alpha, \quad \text{where} \quad c_\alpha = \frac{1}{(2\pi i)^n} \int_\Gamma \frac{f(\zeta)}{(\zeta - a)^{\alpha+1}} d\zeta.$$

Notice how strikingly similar the computation is to one variable.

Let us prove the converse statement. The limit of the series is continuous, as it is a uniform-on-compact-sets limit of continuous functions, and hence it is locally bounded in Δ. Next, we restrict to each variable in turn (fixing the others),

$$z_k \mapsto \sum_\alpha c_\alpha (z - a)^\alpha.$$

This one-variable function is holomorphic as it is a uniform limit on compact subsets of holomorphic functions. Thus f is holomorphic by definition. □

The converse statement also follows by applying the Cauchy–Riemann equations to the series termwise. We leave that as an exercise. First, one must show that the term-by-term derivative series also converges uniformly absolutely on compact subsets. Then one applies the theorem from real analysis about derivatives of limits: If a sequence of functions and the sequences of its derivatives converge uniformly, then the derivatives converge to the derivative of the limit.

> *Exercise* 1.2.1: *Prove the claim above that if a power series converges uniformly absolutely on compact subsets of a polydisc Δ, then the term-by-term derivative converges. Do the proof without using the analogous result for single-variable series.*

A third way to prove the converse statement of the theorem is to note that partial sums are holomorphic and write them using the Cauchy formula. Uniform convergence shows that the limit also satisfies the Cauchy formula, and differentiating under the integral obtains the result.

Exercise 1.2.2: *Follow the logic above to prove the converse of the theorem without using the analogous result for single-variable series. Hint: Let $\Delta'' \subset \Delta' \subset \Delta$ be polydiscs with the same center a such that $\overline{\Delta''} \subset \Delta'$ and $\overline{\Delta'} \subset \Delta$. Apply Cauchy formula on Δ' for $z \in \overline{\Delta''}$.*

Exercise 1.2.3: *Suppose that $\Delta \subset \mathbb{C}^n$ is a possibly unbounded polydisc centered at $a \in \mathbb{C}^n$, where by possibly unbounded we mean that some of the factors can be all of $\mathbb{C}$ (that is, some components of the polyradius are allowed to be ∞). Prove that if $f \colon \Delta \to \mathbb{C}$ is holomorphic, then there is a power series representation $\sum_\alpha c_\alpha (z - a)^\alpha$ converging uniformly on compact subsets to f on Δ.*

Exercise 1.2.4: *One can also do a **Laurent** series expansion. Suppose $a \in \mathbb{C}^n$ and $U = \Delta_1 \times \cdots \times \Delta_k \times \Delta_{k+1}^* \times \cdots \times \Delta_n^* \subset \mathbb{C}^n$, where each Δ_ℓ is a disc centered at a_ℓ or $\mathbb{C}$, and $\Delta_\ell^* = \Delta_\ell \setminus \{a_\ell\}$. Prove that if $f \colon U \to \mathbb{C}$ is holomorphic, then there is a series representation $\sum_\alpha c_\alpha (z - a)^\alpha$, where $\alpha_{k+1}, \ldots, \alpha_n$ now range over all integers, converging uniformly on compact subsets to f on U.*

Proposition 1.2.2. *Let $\Delta = \Delta_\rho(a) \subset \mathbb{C}^n$ be a polydisc, and Γ its distinguished boundary. Suppose $f \colon \overline{\Delta} \to \mathbb{C}$ is a continuous function holomorphic in Δ. Then, for $z \in \Delta$,*

$$\frac{\partial^{|\alpha|} f}{\partial z^\alpha}(z) = \frac{1}{(2\pi i)^n} \int_\Gamma \frac{\alpha! f(\zeta)}{(\zeta - z)^{\alpha+1}} \, d\zeta.$$

In particular, if f is given by (1.3), then

$$c_\alpha = \frac{1}{\alpha!} \frac{\partial^{|\alpha|} f}{\partial z^\alpha}(a),$$

and we have the Cauchy estimates:

$$|c_\alpha| \leq \frac{\|f\|_\Gamma}{\rho^\alpha}.$$

Consequently, the coefficients of the power series depend only on the derivatives of f at a (and so on the values of f in an arbitrarily small neighborhood of a) and not the specific polydisc used in the theorem.

Proof. By the Leibniz rule, if $z \in \Delta$ (not on the boundary), we can differentiate under the integral in the Cauchy formula. We are talking regular real partial differentiation, and we use it to apply the Wirtinger operator. The point is that

$$\frac{\partial}{\partial z_\ell} \left[\frac{1}{(\zeta_\ell - z_\ell)^k} \right] = \frac{k}{(\zeta_\ell - z_\ell)^{k+1}}.$$

Let us do a single derivative to get the idea:

$$\frac{\partial f}{\partial z_1}(z) = \frac{\partial}{\partial z_1}\left[\frac{1}{(2\pi i)^n}\int_\Gamma \frac{f(\zeta_1, \zeta_2, \ldots, \zeta_n)}{(\zeta_1 - z_1)(\zeta_2 - z_2)\cdots(\zeta_n - z_n)}\, d\zeta_1 \wedge d\zeta_2 \wedge \cdots \wedge d\zeta_n\right]$$

$$= \frac{1}{(2\pi i)^n}\int_\Gamma \frac{f(\zeta_1, \zeta_2, \ldots, \zeta_n)}{(\zeta_1 - z_1)^2(\zeta_2 - z_2)\cdots(\zeta_n - z_n)}\, d\zeta_1 \wedge d\zeta_2 \wedge \cdots \wedge d\zeta_n.$$

How about we do it a second time:

$$\frac{\partial^2 f}{\partial z_1^2}(z) = \frac{1}{(2\pi i)^n}\int_\Gamma \frac{2f(\zeta_1, \zeta_2, \ldots, \zeta_n)}{(\zeta_1 - z_1)^3(\zeta_2 - z_2)\cdots(\zeta_n - z_n)}\, d\zeta_1 \wedge d\zeta_2 \wedge \cdots \wedge d\zeta_n.$$

Notice the 2 before the f. Next derivative, a 3 is coming out. After m derivatives in z_1, you get the constant $m!$. It is exactly the same thing that happens in one variable. A moment's thought will convince you that the following formula is correct for $\alpha \in \mathbb{N}_0^n$:

$$\frac{\partial^{|\alpha|} f}{\partial z^\alpha}(z) = \frac{1}{(2\pi i)^n}\int_\Gamma \frac{\alpha!\, f(\zeta)}{(\zeta - z)^{\alpha+1}}\, d\zeta.$$

Therefore,

$$\alpha!\, c_\alpha = \frac{\partial^{|\alpha|} f}{\partial z^\alpha}(a).$$

We obtain the Cauchy estimates as before:

$$\left|\frac{\partial^{|\alpha|} f}{\partial z^\alpha}(a)\right| = \left|\frac{1}{(2\pi i)^n}\int_\Gamma \frac{\alpha!\, f(\zeta)}{(\zeta - a)^{\alpha+1}}\, d\zeta\right| \le \frac{1}{(2\pi)^n}\int_\Gamma \frac{\alpha!\,|f(\zeta)|}{\rho^{\alpha+1}}\,|d\zeta| \le \frac{\alpha!}{\rho^\alpha}\|f\|_\Gamma. \qquad \square$$

As in one-variable theory, the Cauchy estimates prove the following proposition.

Proposition 1.2.3. *Let $U \subset \mathbb{C}^n$ be an open set. Suppose the sequence $f_\ell \colon U \to \mathbb{C}$ converges uniformly on compact subsets to $f \colon U \to \mathbb{C}$. If every f_ℓ is holomorphic, then f is holomorphic, and for every α, the sequence $\left\{\frac{\partial^{|\alpha|} f_\ell}{\partial z^\alpha}\right\}$ converges to $\frac{\partial^{|\alpha|} f}{\partial z^\alpha}$ uniformly on compact subsets.*

Exercise 1.2.5: Prove the proposition above.

Given a power series, let $W \subset \mathbb{C}^n$ be the set of all points where the series converges absolutely. The interior of W is called the *domain of convergence* of the series. In one variable, every domain of convergence is a disc and hence is described with a single number (the radius). In several variables, the domain of convergence is not as easy to describe. For the multivariable geometric series, the domain of convergence is the unit polydisc, but in general, the domain of convergence is more complicated.

Example 1.2.4: In $\mathbb{C}^2$, the series

$$\sum_{k=0}^{\infty} z_1 z_2^k$$

converges absolutely exactly on the set

$$\{z \in \mathbb{C}^2 : |z_2| < 1\} \cup \{z \in \mathbb{C}^2 : z_1 = 0\}.$$

This set is not quite a polydisc. It is neither an open set nor a closed set, and its closure is not the closure of the domain of convergence, which is the set $\{z \in \mathbb{C}^2 : |z_2| < 1\}$.

Example 1.2.5: The series

$$\sum_{k=0}^{\infty} z_1^k z_2^k$$

converges absolutely exactly on the set

$$\{z \in \mathbb{C}^2 : |z_1 z_2| < 1\}.$$

The picture is definitely more complicated than a polydisc. See Figure 1.4.

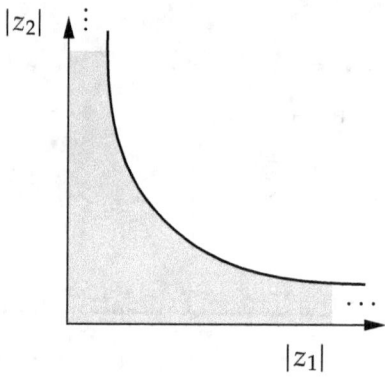

Figure 1.4: Domain of convergence of $\sum_k z_1^k z_2^k$.

Exercise 1.2.6: *Find the domain of convergence of $\sum_{k,\ell} \frac{1}{\ell!} z_1^k z_2^\ell$ and draw the corresponding picture.*

Exercise 1.2.7: *Find the domain of convergence of $\sum_{k,\ell} c_{k\ell} z_1^k z_2^\ell$ and draw the corresponding picture if $c_{\ell\ell} = 2^\ell$, $c_{0\ell} = c_{k0} = 1$ and $c_{k\ell} = 0$ otherwise.*

Exercise 1.2.8: *Suppose a power series in two variables can be written as a sum of a power series in z_1 and a power series in z_2. Show that the domain of convergence is a polydisc.*

A domain $U \subset \mathbb{C}^n$ is a *Reinhardt domain* if whenever $z \in U$ and $|z_k| = |w_k|$ for all k, then $w \in U$. The domains we were drawing so far are Reinhardt domains. They are exactly the domains that you can draw by plotting what happens for the moduli of the variables. A domain is a *complete Reinhardt domain* if $z \in U$, then $\overline{\Delta_r(0)} \subset U$ where $r = (r_1, \ldots, r_n)$ and $r_k = |z_k|$ for all k. So a complete Reinhardt domain is a union (possibly infinite) of polydiscs centered at the origin.

Proposition 1.2.6. *Let $\sum_\alpha c_\alpha z^\alpha$ be a convergent power series. Prove that its domain of convergence is a complete Reinhardt domain.*

> *Exercise* **1.2.9:** *Prove Proposition 1.2.6.*

Theorem 1.2.7 (Identity theorem). *Let $U \subset \mathbb{C}^n$ be a domain (connected open set) and let $f : U \to \mathbb{C}$ be holomorphic. If $f|_N \equiv 0$ for a nonempty open subset $N \subset U$, then $f \equiv 0$.*

Proof. Let Z be the set where all derivatives of all orders of f are zero; then $N \subset Z$, so Z is nonempty. The set Z is closed in U as all derivatives are continuous. Take an arbitrary $a \in Z$. Expand f in a power series around a converging to f in a polydisc $\Delta_\rho(a) \subset U$. As the coefficients are given by derivatives of f, the power series is the zero series. Hence, f is identically zero in $\Delta_\rho(a)$. Therefore, Z is open. As Z is also closed and nonempty, and U is connected, we have $Z = U$. $\qquad\square$

The theorem is often used to show that if two holomorphic functions f and g are equal on a small open set, then $f \equiv g$. In one variable (see Theorem B.7), the hypothesis that N has a limit point in U (rather than being open) is sufficient. In several variables, things are not so simple: $f(z_1, z_2) = z_1$ is zero on the set $\{z \in \mathbb{C}^2 : z_1 = 0\}$, all of whose points are its limit points. When $n \geq 2$, zeros are never isolated, see Exercise 1.2.21. For now, let us move on.

Theorem 1.2.8 (Maximum principle). *Let $U \subset \mathbb{C}^n$ be a domain. Let $f : U \to \mathbb{C}$ be holomorphic and suppose $|f(z)|$ attains a local maximum at some $a \in U$. Then $f \equiv f(a)$.*

Proof. Suppose $|f(z)|$ attains a local maximum at $a \in U$. Consider a polydisc $\Delta = \Delta_1 \times \cdots \times \Delta_n \subset U$ centered at a. The function

$$z_1 \mapsto f(z_1, a_2, \ldots, a_n)$$

is holomorphic on the disc Δ_1 and its modulus attains the maximum at the center. Therefore, it is constant by the maximum principle in one variable, that is, $f(z_1, a_2, \ldots, a_n) = f(a)$ for all $z_1 \in \Delta_1$. For any fixed $z_1 \in \Delta_1$, consider the function

$$z_2 \mapsto f(z_1, z_2, a_3, \ldots, a_n).$$

This function, holomorphic on the disc Δ_2, again attains its maximum modulus at the center of Δ_2 and hence is constant on Δ_2. Iterating this procedure, we obtain that $f(z) = f(a)$ for all $z \in \Delta$. The identity theorem says that $f(z) = f(a)$ for all $z \in U$. $\qquad\square$

Exercise 1.2.10: Let V be the volume measure on $\mathbb{R}^{2n}$ and hence on $\mathbb{C}^n$. Suppose Δ centered at $a \in \mathbb{C}^n$, and f is a function holomorphic on a neighborhood of $\overline{\Delta}$. Prove

$$ f(a) = \frac{1}{V(\Delta)} \int_\Delta f(\zeta)\, dV(\zeta), $$

where $V(\Delta)$ is the volume of Δ and dV is the volume measure. That is, $f(a)$ is an average of the values on a polydisc centered at a.

Exercise 1.2.11: Prove the maximum principle by using the Cauchy formula instead. Hint: Use the previous exercise.

Exercise 1.2.12: Prove a several variables analogue of the Schwarz's lemma: Suppose f is holomorphic in a neighborhood of $\overline{\mathbb{B}_n}$, $f(0) = 0$, and for some $k \in \mathbb{N}$ we have $\frac{\partial^{|\alpha|} f}{\partial z^\alpha}(0) = 0$ whenever $|\alpha| < k$. Further suppose for all $z \in \mathbb{B}_n$, $|f(z)| \le M$ for some M. Show that

$$ |f(z)| \le M \|z\|^k \qquad \text{for all } z \in \overline{\mathbb{B}_n}. $$

Exercise 1.2.13: Apply the one-variable Liouville's theorem to prove it for several variables. That is, suppose $f \colon \mathbb{C}^n \to \mathbb{C}$ is holomorphic and bounded. Prove f is constant.

Exercise 1.2.14: Improve Liouville's theorem slightly in $\mathbb{C}^2$. A complex line through the origin is the image of a nonzero linear map $L \colon \mathbb{C} \to \mathbb{C}^n$.
  a) Prove that for every collection of finitely many complex lines through the origin, there exists an entire nonconstant holomorphic function ($n \ge 2$) bounded (hence constant) on these complex lines.
  b) Prove that if an entire holomorphic function in $\mathbb{C}^2$ is bounded on infinitely many distinct complex lines through the origin, then it is constant.
  c) Find a nonconstant entire holomorphic function in $\mathbb{C}^3$ that is bounded on infinitely many distinct complex lines through the origin.

Exercise 1.2.15: Prove the several variables version of Montel's theorem: Suppose $\{f_k\}$ is a uniformly bounded sequence of holomorphic functions on an open set $U \subset \mathbb{C}^n$. Show that there exists a subsequence $\{f_{k_j}\}$ that converges uniformly on compact subsets to some holomorphic function f. Hint: Mimic the one-variable proof.

Exercise 1.2.16: Prove a several variables version of Hurwitz's theorem: Suppose $\{f_k\}$ is a sequence of nowhere zero holomorphic functions on a domain $U \subset \mathbb{C}^n$ converging uniformly on compact subsets to a function f. Show that either f is identically zero or that f is nowhere zero. Hint: Feel free to use the one-variable result.

Exercise 1.2.17: Suppose $p \in \mathbb{C}^n$ is a point and $D \subset \mathbb{C}^n$ is a ball centered at $p \in D$. A holomorphic function $f : D \to \mathbb{C}$ can be analytically continued along a path $\gamma : [0, 1] \to \mathbb{C}^n$, $\gamma(0) = p$, if for every $t \in [0, 1]$ there exists a ball D_t centered at $\gamma(t)$, where $D_0 = D$, and a holomorphic function $f_t : D_t \to \mathbb{C}$, where $f_0 = f$, and for each $t_0 \in [0, 1]$ there is an $\epsilon > 0$ such that if $|t - t_0| < \epsilon$, then $f_t = f_{t_0}$ in $D_t \cap D_{t_0}$. Prove a several variables version of the Monodromy theorem: If $U \subset \mathbb{C}^n$ is a simply connected domain, $D \subset U$ a ball, and $f : D \to \mathbb{C}$ a holomorphic function that can be analytically continued from $p \in D$ to every $q \in U$, then there exists a unique holomorphic function $F : U \to \mathbb{C}$ such that $F|_D = f$.

Definition 1.2.9. Let $U \subset \mathbb{C}^n$ be an open set. Define $\mathcal{O}(U)$ to be the *ring of holomorphic functions* $f : U \to \mathbb{C}$. The letter $\mathcal{O}$ is used to recognize the fundamental contribution to several complex variables by Kiyoshi Oka*.

The set $\mathcal{O}(U)$ really is a commutative ring under pointwise addition and multiplication (exercise below). For us, $\mathcal{O}(U)$ will always mean the set of $\mathbb{C}$-valued functions, however, in the literature the notation is sometimes used to simply denote holomorphicity no matter the codomain.

Exercise 1.2.18: Prove that $\mathcal{O}(U)$ is actually a commutative ring with the operations

$$(f + g)(z) = f(z) + g(z), \qquad (fg)(z) = f(z)g(z).$$

Exercise 1.2.19: Show that $\mathcal{O}(U)$ is an integral domain (no zero divisors) if and only if U is connected. That is, show that U being connected is equivalent to the following: If $h(z) = f(z)g(z)$ is identically zero for $f, g \in \mathcal{O}(U)$, then either f or g is identically zero.

A function F defined on a dense open subset of U is *meromorphic* if locally near every $p \in U$, $F = f/g$ for f and g holomorphic in some neighborhood of p. It is a deep result of Oka that, for domains $U \subset \mathbb{C}^n$, every meromorphic function can be represented as f/g globally. That is, the ring of meromorphic functions is the field of fractions of $\mathcal{O}(U)$. This problem is the so-called *Poincaré problem*, and its solution is no longer positive once one generalizes U to complex manifolds. The points of U through which F does not extend holomorphically are called the *poles* of F. Namely, poles are the points where $g = 0$ for every possible representation f/g. Unlike in one variable, in several variables, poles are never isolated points. There is also a new type of singular point for meromorphic functions in more than one variable:

Exercise 1.2.20: In two variables, one can no longer think of a meromorphic function F having the value ∞ when the denominator vanishes. Show that $F(z, w) = z/w$ achieves all values of $\mathbb{C}$ in every neighborhood of the origin. We call the origin a point of indeterminacy.

---

*See https://en.wikipedia.org/wiki/Kiyoshi_Oka.

Exercise 1.2.21: *Prove that zeros are never isolated in $\mathbb{C}^n$ for $n \geq 2$. Hint: Consider $z_1 \mapsto f(z_1, z_2, \ldots, z_n)$ as you move $z_2, \ldots, z_n$ around, and use, perhaps, Hurwitz.*

1.3 \ Derivatives

Given a function $f = u + iv$, the complex conjugate is $\bar{f} = u - iv$, defined simply by $z \mapsto \overline{f(z)}$. Suppose f is holomorphic. Then $\bar{f}$ is called an *antiholomorphic function*. An antiholomorphic function is a function that depends on $\bar{z}$ but not on z. So if we write the variable, we write $\bar{f}$ as $\bar{f}(\bar{z})$. Let us see why this makes sense. Using the definitions of the Wirtinger operators,

$$\frac{\partial \bar{f}}{\partial z_\ell} = \overline{\frac{\partial f}{\partial \bar{z}_\ell}} = 0, \qquad \frac{\partial \bar{f}}{\partial \bar{z}_\ell} = \overline{\left(\frac{\partial f}{\partial z_\ell} \right)}, \qquad \text{for all } \ell = 1, \ldots, n.$$

For functions that are neither holomorphic nor antiholomorphic, we pretend they depend on both z and $\bar{z}$. Since we want to write functions in terms of z and $\bar{z}$, let us figure out how the chain rule works for Wirtinger derivatives, rather than writing derivatives in terms of x and y.

Proposition 1.3.1 (Complex chain rule). *Suppose $U \subset \mathbb{C}^n$ and $V \subset \mathbb{C}^m$ are open, and suppose $f \colon U \to V$ and $g \colon V \to \mathbb{C}$ are (real) differentiable mappings. Write the variables as $z = (z_1, \ldots, z_n) \in U \subset \mathbb{C}^n$ and $w = (w_1, \ldots, w_m) \in V \subset \mathbb{C}^m$. Then for $\ell = 1, \ldots, n$,*

$$\frac{\partial}{\partial z_\ell} [g \circ f] = \sum_{k=1}^m \left(\frac{\partial g}{\partial w_k} \frac{\partial f_k}{\partial z_\ell} + \frac{\partial g}{\partial \bar{w}_k} \frac{\partial \bar{f}_k}{\partial z_\ell} \right),$$

$$\frac{\partial}{\partial \bar{z}_\ell} [g \circ f] = \sum_{k=1}^m \left(\frac{\partial g}{\partial w_k} \frac{\partial f_k}{\partial \bar{z}_\ell} + \frac{\partial g}{\partial \bar{w}_k} \frac{\partial \bar{f}_k}{\partial \bar{z}_\ell} \right).$$

$$(1.4)$$

Proof. Write $f = u + iv$, $z = x + iy$, $w = s + it$. Let f be a function of z and g be a function of w. The composition plugs in f for w, and so it plugs in u for s, and v for t. Using the standard chain rule,

$$\frac{\partial}{\partial z_\ell} [g \circ f] = \frac{1}{2} \left(\frac{\partial}{\partial x_\ell} - i \frac{\partial}{\partial y_\ell} \right) [g \circ f]$$

$$= \frac{1}{2} \sum_{k=1}^m \left(\frac{\partial g}{\partial s_k} \frac{\partial u_k}{\partial x_\ell} + \frac{\partial g}{\partial t_k} \frac{\partial v_k}{\partial x_\ell} - i \left(\frac{\partial g}{\partial s_k} \frac{\partial u_k}{\partial y_\ell} + \frac{\partial g}{\partial t_k} \frac{\partial v_k}{\partial y_\ell} \right) \right)$$

$$= \sum_{k=1}^m \left(\frac{\partial g}{\partial s_k} \frac{1}{2} \left(\frac{\partial u_k}{\partial x_\ell} - i \frac{\partial u_k}{\partial y_\ell} \right) + \frac{\partial g}{\partial t_k} \frac{1}{2} \left(\frac{\partial v_k}{\partial x_\ell} - i \frac{\partial v_k}{\partial y_\ell} \right) \right)$$

$$= \sum_{k=1}^m \left(\frac{\partial g}{\partial s_k} \frac{\partial u_k}{\partial z_\ell} + \frac{\partial g}{\partial t_k} \frac{\partial v_k}{\partial z_\ell} \right).$$

For $k = 1, \ldots, m$,

$$\frac{\partial}{\partial s_k} = \frac{\partial}{\partial w_k} + \frac{\partial}{\partial \bar{w}_k}, \qquad \frac{\partial}{\partial t_k} = i\left(\frac{\partial}{\partial w_k} - \frac{\partial}{\partial \bar{w}_k}\right).$$

Continuing:

$$\frac{\partial}{\partial z_\ell}[g \circ f] = \sum_{k=1}^{m}\left(\frac{\partial g}{\partial s_k}\frac{\partial u_k}{\partial z_\ell} + \frac{\partial g}{\partial t_k}\frac{\partial v_k}{\partial z_\ell}\right)$$

$$= \sum_{k=1}^{m}\left(\left(\frac{\partial g}{\partial w_k}\frac{\partial u_k}{\partial z_\ell} + \frac{\partial g}{\partial \bar{w}_k}\frac{\partial u_k}{\partial z_\ell}\right) + i\left(\frac{\partial g}{\partial w_k}\frac{\partial v_k}{\partial z_\ell} - \frac{\partial g}{\partial \bar{w}_k}\frac{\partial v_k}{\partial z_\ell}\right)\right)$$

$$= \sum_{k=1}^{m}\left(\frac{\partial g}{\partial w_k}\left(\frac{\partial u_k}{\partial z_\ell} + i\frac{\partial v_k}{\partial z_\ell}\right) + \frac{\partial g}{\partial \bar{w}_k}\left(\frac{\partial u_k}{\partial z_\ell} - i\frac{\partial v_k}{\partial z_\ell}\right)\right)$$

$$= \sum_{k=1}^{m}\left(\frac{\partial g}{\partial w_k}\frac{\partial f_k}{\partial z_\ell} + \frac{\partial g}{\partial \bar{w}_k}\frac{\partial \bar{f}_k}{\partial z_\ell}\right).$$

The $\bar{z}$ derivative works similarly. $\qquad\qquad\square$

Because of the proposition, when we deal with a possibly nonholomorphic function f, we often write $f(z, \bar{z})$ and treat f as a function of z and $\bar{z}$.

Remark 1.3.2. It is good to notice the subtlety of what we just said. Formally it seems as if z and $\bar{z}$ are independent variables when taking derivatives, but in reality, they are not independent if we actually wish to evaluate the function. Under the hood, a smooth function that is not necessarily holomorphic is really a function of the real variables x and y, where $z = x + iy$.

Remark 1.3.3. We could have swapped z and $\bar{z}$, by flipping the bars everywhere. There is no difference between the two, they are twins in effect. We just need to know which one is which. After all, it all starts with taking the two square roots of -1 and deciding which one is i (remember the chickens?). There is no "natural choice" for that, but once we make that choice we must be consistent. And once we picked which root is i, we also picked what is holomorphic and what is antiholomorphic. This is a subtle philosophical as much as a mathematical point.

Definition 1.3.4. Let $U \subset \mathbb{C}^n$ be open. A mapping $f\colon U \to \mathbb{C}^m$ is said to be *holomorphic* if each component is holomorphic. That is, if $f = (f_1, \ldots, f_m)$, then each f_k is a holomorphic function.

As in one variable, the composition of holomorphic functions (mappings) is holomorphic.

Theorem 1.3.5. *Let $U \subset \mathbb{C}^n$ and $V \subset \mathbb{C}^m$ be open sets, and suppose $f\colon U \to V$ and $g\colon V \to \mathbb{C}^q$ are both holomorphic. Then the composition $g \circ f$ is holomorphic.*

Proof. The proof is almost trivial by chain rule. Again let g be a function of $w \in V$ and f be a function of $z \in U$. For $\ell = 1, \dots, n$ and $v = 1, \dots, q$, compute

$$\frac{\partial}{\partial \bar{z}_\ell} \left[g_v \circ f \right] = \sum_{k=1}^{m} \left(\frac{\partial g_v}{\partial w_k} \frac{\partial f_k^{\,0}}{\partial \bar{z}_\ell} + \frac{\partial g_v}{\partial \bar{w}_k} \frac{\partial \bar{f}_k^{\,0}}{\partial \bar{z}_\ell} \right) = 0. \qquad \square$$

For holomorphic mappings the chain rule simplifies, and it formally looks like the familiar vector calculus rule. Suppose again $U \subset \mathbb{C}^n$ and $V \subset \mathbb{C}^m$ are open, and $f \colon U \to V$ and $g \colon V \to \mathbb{C}$ are holomorphic. Name the variables $z = (z_1, \dots, z_n) \in U \subset \mathbb{C}^n$ and $w = (w_1, \dots, w_m) \in V \subset \mathbb{C}^m$. In formula (1.4) for the z_ℓ derivative, the $\bar{w}_\ell$ derivative of g is zero and the z_ℓ derivative of $\bar{f}_k$ is also zero because f and g are holomorphic. Therefore, for $\ell = 1, \dots, n$,

$$\frac{\partial}{\partial z_\ell} \left[g \circ f \right] = \sum_{k=1}^{m} \frac{\partial g}{\partial w_k} \frac{\partial f_k}{\partial z_\ell}.$$

Exercise **1.3.1:** *Using only the Wirtinger derivatives, prove that a holomorphic function that is real-valued must be constant.*

Exercise **1.3.2:** *Let f be a holomorphic function on $\mathbb{C}^n$. When we write $\bar{f}$ we mean the function $z \mapsto \overline{f(z)}$, and we usually write $\bar{f}(\bar{z})$ as the function is antiholomorphic. However, if we write $\bar{f}(z)$ we really mean $z \mapsto \overline{f(\bar{z})}$, that is, composing both the function and the argument with conjugation. Prove $z \mapsto \bar{f}(z)$ is holomorphic, and prove f is real-valued on $\mathbb{R}^n$ (when $y = 0$) if and only if $f(z) = \bar{f}(z)$ for all $z \in \mathbb{C}^n$.*

For a $U \subset \mathbb{C}^n$, a holomorphic mapping $f \colon U \to \mathbb{C}^m$, and a point $p \in U$, define the holomorphic derivative, sometimes called the *(holomorphic) Jacobian matrix*,

$$Df(p) \overset{\text{def}}{=} \left[\frac{\partial f_k}{\partial z_\ell}(p) \right]_{k\ell}.$$

The notation $f'(p) = Df(p)$ is also used. Unless otherwise stated, if the mapping is holomorphic, *Jacobian* will refer to the holomorphic Jacobian.

Exercise **1.3.3:** *Suppose $U \subset \mathbb{C}^n$ is open, $\mathbb{R}^n$ is naturally embedded in $\mathbb{C}^n$. Consider a holomorphic mapping $f \colon U \to \mathbb{C}^m$ and suppose that $f|_{U \cap \mathbb{R}^n}$ maps into $\mathbb{R}^m \subset \mathbb{C}^m$. Prove that given $p \in U \cap \mathbb{R}^n$, the real Jacobian matrix at p of the map $f|_{U \cap \mathbb{R}^n} \colon U \cap \mathbb{R}^n \to \mathbb{R}^m$ is equal to the holomorphic Jacobian matrix of the map f at p. In particular, $Df(p)$ is a matrix with real entries.*

By the holomorphic chain rule above, as in the theory of real functions, the derivative of the composition is the composition of derivatives (multiplied as matrices).

Proposition 1.3.6 (Chain rule for holomorphic mappings). *Let $U \subset \mathbb{C}^n$ and $V \subset \mathbb{C}^m$ be open sets. Suppose $f\colon U \to V$ and $g\colon V \to \mathbb{C}^k$ are both holomorphic, and $p \in U$. Then*

$$D(g \circ f)(p) = Dg(f(p)) \, Df(p).$$

In shorthand, we often simply write $D(g \circ f) = DgDf$.

Exercise **1.3.4**: *Prove the proposition.*

Suppose $U \subset \mathbb{C}^n$, $p \in U$, and $f\colon U \to \mathbb{C}^m$ is differentiable at p. Since $\mathbb{C}^n$ is identified with $\mathbb{R}^{2n}$, the mapping f takes $U \subset \mathbb{R}^{2n}$ to $\mathbb{R}^{2m}$. The normal vector-calculus Jacobian at p of this mapping (a $2m \times 2n$ real matrix) is called the *real Jacobian*, and we write it as $D_{\mathbb{R}}f(p)$.

Proposition 1.3.7. *Let $U \subset \mathbb{C}^n$ be open, $p \in U$, and $f\colon U \to \mathbb{C}^n$ be holomorphic. Then*

$$\left| \det Df(p) \right|^2 = \det D_{\mathbb{R}}f(p).$$

The expression $\det Df(p)$ is called the *(holomorphic) Jacobian determinant* and clearly it is important to know if we are talking about the holomorphic Jacobian determinant or the standard real Jacobian determinant $\det D_{\mathbb{R}}f(p)$. Recall from vector calculus that if the real Jacobian determinant $\det D_{\mathbb{R}}f(p)$ of a smooth mapping is positive, then the mapping preserves orientation. In particular, the proposition says that holomorphic mappings preserve orientation.

Proof. Write f as $(\operatorname{Re} f_1, \operatorname{Im} f_1, \ldots, \operatorname{Re} f_n, \operatorname{Im} f_n)$ as a function of $(x_1, y_1, \ldots, x_n, y_n)$, using our identification of $\mathbb{C}^n$ and $\mathbb{R}^{2n}$. The statement is about the two Jacobians at p, that is, the derivatives at p. Hence, we can assume that $p = 0$ and f is complex linear, $f(z) = Az$ for some $n \times n$ matrix A. It is just a statement about matrices. The matrix A is the (holomorphic) Jacobian matrix of f. Let B be the real Jacobian matrix of f.

We change the basis of B to be $(z, \bar{z})$ using $z = x + iy$ and $\bar{z} = x - iy$ on both the target and the source. The change of basis is some invertible complex matrix M such that $M^{-1}BM$ (the real Jacobian matrix B in this new basis) is a matrix of the derivatives of $(f_1, \ldots, f_n, \bar{f}_1, \ldots, \bar{f}_n)$ in terms of $(z_1, \ldots, z_n, \bar{z}_1, \ldots, \bar{z}_n)$. That is,

$$M^{-1}BM = \begin{bmatrix} A & 0 \\ 0 & \bar{A} \end{bmatrix}.$$

Thus

$$\det(B) = \det(M^{-1}MB) = \det(M^{-1}BM)$$

$$= \det(A)\det(\bar{A}) = \det(A)\,\overline{\det(A)} = |\det(A)|^2. \quad \square$$

The regular (real) implicit function theorem and the chain rule give that the implicit function theorem holds in the holomorphic setting. The main thing to check is to verify that the solution given by the standard implicit function theorem is holomorphic, which follows by the chain rule.

Theorem 1.3.8 (Implicit function theorem). *Let $U \subset \mathbb{C}^n \times \mathbb{C}^m$ be an open set, let $(z, w) \in \mathbb{C}^n \times \mathbb{C}^m$ be our coordinates, and let $f: U \to \mathbb{C}^m$ be a holomorphic mapping. Let $(z^0, w^0) \in U$ be a point such that $f(z^0, w^0) = 0$ and such that the $m \times m$ matrix*

$$\left[\frac{\partial f_k}{\partial w_\ell}(z^0, w^0) \right]_{k\ell}$$

is invertible. Then there exists an open set $V \subset \mathbb{C}^n$ with $z^0 \in V$, open set $W \subset \mathbb{C}^m$ with $w^0 \in W$, $V \times W \subset U$, and a holomorphic mapping $g: V \to W$, with $g(z^0) = w^0$ such that for every $z \in V$, the point $g(z)$ is the unique point in W such that

$$f\big(z, g(z)\big) = 0.$$

Exercise **1.3.5**: *Prove the holomorphic implicit function theorem above. Hint: Check that the normal implicit function theorem for C^1 functions applies, and then show that the g you obtain is holomorphic.*

Exercise **1.3.6**: *State and prove a holomorphic version of the inverse function theorem.*

Exercise **1.3.7**: *Suppose $U \subset \mathbb{C}^n$ is a domain and $f: U \to \mathbb{C}^m$ a holomorphic mapping.*
  a) *Prove the vector-valued version of the maximum principle: If $\| f(z)\|$ achieves a (local) maximum at $p \in U$, then f is constant.*
  b) *Find a counterexample to a vector-valued minimum principle: Find an f such that $\| f(z)\|$ achieves a nonzero minimum, but where f is not constant.*

1.4   Inequivalence of ball and polydisc

Definition 1.4.1. Two domains $U \subset \mathbb{C}^n$ and $V \subset \mathbb{C}^n$ are said to be *biholomorphic* or *biholomorphically equivalent* if there exists a one-to-one and onto holomorphic map $f: U \to V$ such that the inverse $f^{-1}: V \to U$ is holomorphic. The mapping f is said to be a *biholomorphic map* or a *biholomorphism*.

As function theory on two biholomorphic domains is the same, one of the main questions in complex analysis is to classify domains up to biholomorphic transformations. In one variable, there is the rather striking theorem due to Riemann:

Theorem 1.4.2 (Riemann mapping theorem). *If $U \subset \mathbb{C}$ is a nonempty simply connected domain such that $U \neq \mathbb{C}$, then U is biholomorphic to $\mathbb{D}$.*

In one variable, a topological property on U is enough to classify a whole class of domains. It is one of the reasons why studying the disc is so important in one variable, and why many theorems are stated for the disc only. There is no such theorem in several variables. We will show momentarily that the unit ball and the polydisc,

$$\mathbb{B}_n = \big\{ z \in \mathbb{C}^n : \|z\| < 1 \big\} \quad \text{and} \quad \mathbb{D}^n = \big\{ z \in \mathbb{C}^n : |z_k| < 1 \text{ for } k = 1, \ldots, n \big\},$$

are *not* biholomorphically equivalent. Both are simply connected (have no holes), and they are the two most obvious generalizations of the disc to several variables. They are homeomorphic, that is, topology does not see any difference.

Exercise 1.4.1: *Prove that there exists a* homeomorphism $f \colon \mathbb{B}_n \to \mathbb{D}^n$, *that is, f is a bijection, and both f and f^{-1} are continuous.*

Let us stick with $n = 2$. Instead of proving that $\mathbb{B}_2$ and $\mathbb{D}^2$ are biholomorphically inequivalent we will prove a stronger theorem. First a definition.

Definition 1.4.3. Suppose $f \colon X \to Y$ is a continuous map between two topological spaces. Then f is a *proper map* if for every compact $K \subset\subset Y$, the set $f^{-1}(K)$ is compact.

The notation "$\subset\subset$" is a common notation for a relatively compact subset, that is, the closure is compact in the relative (subspace) topology. Often the distinction between compact and relatively compact is not important. For instance, in the definition above we can replace compact with relatively compact; $f^{-1}(K)$ is relatively compact in X for every K relatively compact in Y. Sometimes, simply stating "$K \subset\subset Y$" means that K is compact if it is clear from context.

Vaguely, "proper" means that "boundary goes to the boundary." As a continuous map, f pushes compacts to compacts; a proper map is one where the inverse does so too. If the inverse is a continuous function, then clearly f is proper, but not every proper map is invertible. For example, the map $f \colon \mathbb{D} \to \mathbb{D}$ given by $f(z) = z^2$ is proper, but not invertible. The codomain of f is important. If we replace f by $g \colon \mathbb{D} \to \mathbb{C}$, still given by $g(z) = z^2$, then the map is no longer proper. Let us state the main result of this section.

Theorem 1.4.4 (Rothstein 1935). *There exists no proper holomorphic mapping of the unit bidisc $\mathbb{D}^2 = \mathbb{D} \times \mathbb{D} \subset \mathbb{C}^2$ to the unit ball $\mathbb{B}_2 \subset \mathbb{C}^2$.*

As a biholomorphic mapping is proper, the unit bidisc is not biholomorphically equivalent to the unit ball in $\mathbb{C}^2$. The inequivalence of the ball and the polydisc was first proved by Poincaré by computing the automorphism groups of $\mathbb{D}^2$ and $\mathbb{B}_2$, although his proof assumed the maps extended past the boundary. The first complete proof was by Henri Cartan in 1931, though the theorem is popularly attributed to Poincaré. It seems standard practice that any general audience talk about several complex variables contains a mention of Poincaré, and often the reference is to this exact theorem.

We need some lemmas before we get to the proof of the result. First, a certain one-dimensional object plays an important role in the geometry of several complex variables. It allows us to apply one-variable results in several variables. It is especially important in understanding the boundary behavior of holomorphic functions. It also prominently appears in complex geometry.

Definition 1.4.5. A nonconstant holomorphic mapping $\varphi \colon \mathbb{D} \to \mathbb{C}^n$ is called an *analytic disc*. If the mapping φ extends continuously to the closed unit disc $\overline{\mathbb{D}}$, then the mapping $\varphi \colon \overline{\mathbb{D}} \to \mathbb{C}^n$ is called a *closed analytic disc*.

Often we call the image $\Delta = \varphi(\mathbb{D})$ the analytic disc rather than the mapping. For a closed analytic disc we write $\partial\Delta = \varphi(\partial\mathbb{D})$ and call it the boundary of the analytic disc.

In some sense, analytic discs play the role of line segments in $\mathbb{C}^n$. It is important to always keep in mind that there is a mapping defining the disc, even if we are more interested in the set. Obviously for a given image, the mapping φ is not unique.

Consider the boundaries of the unit bidisc $\mathbb{D} \times \mathbb{D} \subset \mathbb{C}^2$ and the unit ball $\mathbb{B}_2 \subset \mathbb{C}^2$. Notice the boundary of the unit bidisc contains analytic discs $\{p\} \times \mathbb{D}$ and $\mathbb{D} \times \{p\}$ for $p \in \partial\mathbb{D}$. That is, through every point in the boundary, except for the distinguished boundary $\partial\mathbb{D} \times \partial\mathbb{D}$, there exists an analytic disc lying entirely inside the boundary. On the other hand, the ball contains no analytic discs in its boundary.

Proposition 1.4.6. *The unit sphere $S^{2n-1} = \partial\mathbb{B}_n \subset \mathbb{C}^n$ contains no analytic discs.*

Proof. Suppose there is a holomorphic function $g \colon \mathbb{D} \to \mathbb{C}^n$ such that the image $g(\mathbb{D})$ is contained in the unit sphere. In other words, for all $z \in \mathbb{D}$,

$$\|g(z)\|^2 = |g_1(z)|^2 + |g_2(z)|^2 + \cdots + |g_n(z)|^2 = 1.$$

Without loss of generality (after composing with a unitary matrix), assume that $g(0) = (1, 0, 0, \ldots, 0)$. Consider the first component and notice that $g_1(0) = 1$. If a sum of nonnegative numbers is less than or equal to 1, then they all are. Hence, $|g_1(z)| \leq 1$. The maximum principle says that $g_1(z) = 1$ for all $z \in \mathbb{D}$. But then $g_k(z) = 0$ for all $k = 2, \ldots, n$ and all $z \in \mathbb{D}$. Therefore, g is constant and thus not an analytic disc. $\square$

The fact that the sphere contains no analytic discs is the most important geometric distinction between the boundary of the polydisc and the sphere.

Exercise 1.4.2: *Modify the proof to show some stronger results.*
  a) *Let Δ be an analytic disc and $\Delta \cap \partial\mathbb{B}_n \neq \emptyset$. Prove Δ contains points not in $\overline{\mathbb{B}_n}$.*
  b) *Let Δ be an analytic disc. Prove that $\Delta \cap \partial\mathbb{B}_n$ is nowhere dense in Δ.*
  c) *Find an analytic disc in $\mathbb{C}^2$, such that $(1, 0) \in \Delta$, $\Delta \cap \mathbb{B}_2 = \emptyset$, and locally near $(1, 0) \in \partial\mathbb{B}_2$, the set $\Delta \cap \partial\mathbb{B}_2$ is the curve defined by $\operatorname{Im} z_1 = 0$, $\operatorname{Im} z_2 = 0$, $(\operatorname{Re} z_1)^2 + (\operatorname{Re} z_2)^2 = 1$.*

Before we prove the theorem, let us make the statement about proper maps taking boundary to boundary precise.

Lemma 1.4.7. *Let $U \subset \mathbb{R}^n$ and $V \subset \mathbb{R}^m$ be bounded domains and let $f \colon U \to V$ be continuous. Then f is proper if and only if for every sequence $\{p_k\}$ in U such that $p_k \to p \in \partial U$, the set of limit points of $\{f(p_k)\}$ lies in ∂V.*

Proof. Suppose f is proper. Let $\{p_k\}$ be a sequence in U such that $p_k \to p \in \partial U$. Take any convergent subsequence $\{f(p_{k_\ell})\}$ of $\{f(p_k)\}$ converging to some $q \in \overline{V}$. Consider $E = \{f(p_{k_\ell})\}$ as a set. Let $\overline{E}$ be the closure of E in V (subspace topology). If $q \in V$, then $\overline{E} = E \cup \{q\}$ and $\overline{E}$ is compact. Otherwise, if $q \notin V$, then $\overline{E} = E$ and $\overline{E}$ is not compact. The inverse image $f^{-1}(\overline{E})$ is not compact (it contains a sequence going to $p \in \partial U$) and hence $\overline{E}$ is not compact either as f is proper. Thus $q \notin V$, and hence $q \in \partial V$. As we took an arbitrary convergent subsequence of $\{f(p_k)\}$, q was an arbitrary limit point. Therefore, all limit points are in ∂V.

Let us prove the converse. Suppose that for every sequence $\{p_k\}$ in U such that $p_k \to p \in \partial U$, the set of limit points of $\{f(p_k)\}$ lies in ∂V. Take a closed set $E \subset V$ (subspace topology) and suppose $f^{-1}(E)$ is not compact. Then there exists a sequence $\{p_k\}$ in $f^{-1}(E)$ such that $p_k \to p \in \partial U$, because $f^{-1}(E)$ is closed (in U), bounded, but not compact. The hypothesis then says that the limit points of $\{f(p_k)\}$ are in ∂V. Hence E has limit points in ∂V and is not compact. $\qquad\square$

Exercise 1.4.3: *Let $U \subset \mathbb{R}^n$ and $V \subset \mathbb{R}^m$ be bounded domains and let $f \colon \overline{U} \to \overline{V}$ be continuous. Suppose $f(U) \subset V$, and $g \colon U \to V$ is defined by $g(x) = f(x)$ for all $x \in U$. Prove that g is proper if and only if $f(\partial U) \subset \partial V$.*

Exercise 1.4.4: *Let $f \colon X \to Y$ be a continuous function of locally compact Hausdorff topological spaces. Let X_∞ and Y_∞ be the one-point compactifications of X and Y. Then f is a proper map if and only if it extends as a continuous map $f_\infty \colon X_\infty \to Y_\infty$ by letting $f_\infty|_X = f$ and $f_\infty(\infty) = \infty$.*

We now have all the lemmas needed to prove the theorem of Rothstein.

Proof of Theorem 1.4.4. Suppose there is a proper holomorphic map $f \colon \mathbb{D}^2 \to \mathbb{B}_2$. Fix some $e^{i\theta}$ in the boundary of the disc $\mathbb{D}$. Take a sequence $w_k \in \mathbb{D}$ such that $w_k \to e^{i\theta}$. The functions $g_k(\zeta) = f(\zeta, w_k)$ map the unit disc into $\mathbb{B}_2$. By Montel's theorem and by passing to a subsequence, assume that the sequence of functions converges (uniformly on compact subsets) to a limit $g \colon \mathbb{D} \to \overline{\mathbb{B}}_2$. As $(\zeta, w_k) \to (\zeta, e^{i\theta}) \in \partial \mathbb{D}^2$, then by Lemma 1.4.7, $g(\mathbb{D}) \subset \partial \mathbb{B}_2$, and hence g is constant by Proposition 1.4.6.

Let g'_k denote the derivative (we differentiate each component). The functions g'_k converge to $g' = 0$. So for an arbitrary fixed $\zeta \in \mathbb{D}$, $\frac{\partial f}{\partial z_1}(\zeta, w_k) \to 0$. This limit holds for all $e^{i\theta}$ and some subsequence of an arbitrary sequence $\{w_k\}$ where $w_k \to e^{i\theta}$. The holomorphic mapping $w \mapsto \frac{\partial f}{\partial z_1}(\zeta, w)$, therefore, extends continuously to the closure $\overline{\mathbb{D}}$ and is zero on $\partial \mathbb{D}$. We apply the maximum principle or the Cauchy formula and the fact that ζ was arbitrary to find $\frac{\partial f}{\partial z_1} \equiv 0$. By symmetry $\frac{\partial f}{\partial z_2} \equiv 0$. Therefore, f is constant, which is a contradiction as f was proper.

The proof is illustrated in Figure 1.5. In the picture, on the left-hand side is the bidisc, and we restrict f to the horizontal gray lines (where the second component is fixed to be w_k) and take a limit to produce an analytic disc in the boundary of $\mathbb{B}_2$. We then show that $\frac{\partial f}{\partial z_1} = 0$ on the vertical gray line (where the first component is fixed to be ζ). The right-hand side shows the disc where $z_1 = \zeta$ is fixed, which corresponds to the vertical gray line on the left.   □

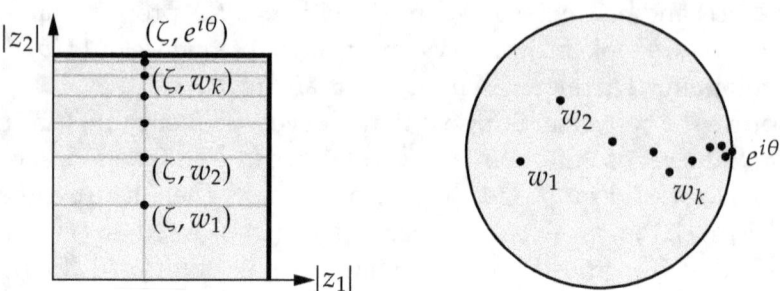

Figure 1.5: The proof of Rothstein's theorem.

The proof says that the reason why there is not even a proper mapping is the fact that the boundary of the polydisc contains analytic discs, while the sphere does not. The proof extends easily to higher dimensions as well, and the proof of the generalization is left as an exercise.

Theorem 1.4.8. *Let $U = U' \times U'' \subset \mathbb{C}^n \times \mathbb{C}^k$, $n, k \geq 1$, and $V \subset \mathbb{C}^m$, $m \geq 1$, be bounded domains such that ∂V contains no analytic discs. Then there exists no proper holomorphic mapping $f : U \to V$.*

Exercise **1.4.5:** *Prove Theorem 1.4.8.*

The key takeaway from this section is that in several variables, to see if two domains are equivalent, the geometry of the boundaries makes a difference, not just the topology of the domains.

The following is a fun exercise in one dimension about proper maps of discs:

Exercise **1.4.6:** *Let $f : \mathbb{D} \to \mathbb{D}$ be a proper holomorphic map. Then*

$$f(z) = e^{i\theta} \prod_{k=1}^{m} \frac{z - a_k}{1 - \bar{a}_k z},$$

where $\theta \in \mathbb{R}$ and $a_k \in \mathbb{D}$ (that is, f is a finite Blaschke product). Hint: Consider $f^{-1}(0)$.

In several variables, when $\mathbb{D}$ is replaced by a ball, this question (what are the proper maps) becomes far more involved, and if the dimensions of the balls are different, it is not solved in general.

> *Exercise 1.4.7:* Suppose $f \colon U \to \mathbb{D}$ is a proper holomorphic map where $U \subset \mathbb{C}^n$ is a nonempty domain. Prove that $n = 1$. Hint: Consider the same idea as in Exercise 1.2.21.

> *Exercise 1.4.8:* Suppose $f \colon \overline{\mathbb{B}_n} \to \mathbb{C}^m$ is a nonconstant continuous map such that $f|_{\mathbb{B}_n}$ is holomorphic and $\|f(z)\| = 1$ whenever $\|z\| = 1$. Prove that $f|_{\mathbb{B}_n}$ maps into $\mathbb{B}_m$ and furthermore that this map is proper.

1.5 Cartan's uniqueness theorem

The following theorem is another analogue of Schwarz's lemma in several variables. It says that for a bounded domain, a self mapping that is the identity up to first order at a single point must, in fact, be the identity. As there are quite a few theorems named for Cartan, this one is referred to as the *Cartan's uniqueness theorem*. It is useful in computing the automorphism groups of certain domains. An *automorphism* of U is a biholomorphic map from U onto U. Automorphisms form a group under composition, called the *automorphism group*. As exercises, you will use the theorem to compute the automorphism groups of $\mathbb{B}_n$ and $\mathbb{D}^n$.

Theorem 1.5.1 (Cartan). *Suppose $U \subset \mathbb{C}^n$ is a bounded domain, $a \in U$, $f \colon U \to U$ is a holomorphic mapping, $f(a) = a$, and $Df(a)$ is the identity. Then $f(z) = z$ for all $z \in U$.*

> *Exercise 1.5.1:* Find a counterexample to the theorem if U is unbounded. Hint: For simplicity take $a = 0$ and $U = \mathbb{C}^n$.

Before we get into the proof, we write the Taylor series of a function in a nicer way, splitting it up into parts of different degree. A polynomial $P \colon \mathbb{C}^n \to \mathbb{C}$ is *homogeneous of degree d* if

$$P(sz) = s^d P(z)$$

for all $s \in \mathbb{C}$ and $z \in \mathbb{C}^n$. A homogeneous polynomial of degree d is a polynomial whose every monomial is of total degree d. For instance, $z^2 w - iz^3 + 9zw^2$ is homogeneous of degree 3 in the variables $(z, w) \in \mathbb{C}^2$. A polynomial vector-valued mapping is homogeneous of degree d if each component is. If f is holomorphic near $a \in \mathbb{C}^n$, then write the power series of f at a as

$$\sum_{m=0}^{\infty} f_m(z - a),$$

where f_m is a homogeneous polynomial of degree m. The f_m is called the *degree m homogeneous part* of f at a. The f_m would be vector-valued if f is vector-valued, such as in the statement of the theorem. In the proof, we will require the vector-valued Cauchy estimates (exercise below)*.

Exercise **1.5.2:** *Prove a vector-valued version of the Cauchy estimates. Suppose $f \colon \overline{\Delta_r(a)} \to \mathbb{C}^m$ is a continuous function holomorphic on a polydisc $\Delta_r(a) \subset \mathbb{C}^n$. Let Γ denote the distinguished boundary of Δ. Show that for every multi-index α,*

$$\left\| \frac{\partial^{|\alpha|} f}{\partial z^{\alpha}}(a) \right\| \leq \frac{\alpha!}{r^{\alpha}} \sup_{z \in \Gamma} \|f(z)\| .$$

Proof of Cartan's uniqueness theorem. Without loss of generality, assume $a = 0$. Write f as a power series at the origin, written in homogeneous parts:

$$f(z) = z + f_k(z) + \sum_{m=k+1}^{\infty} f_m(z) = z + f_k(z) + \text{higher order terms,}$$

where $k \geq 2$ is an integer such that $f_2(z) = f_3(z) = \cdots = f_{k-1}(z) = 0$. The degree-one homogeneous part is simply the vector z, because the derivative of f at the origin is the identity. Compose f with itself ℓ times:

$$f^{\ell}(z) = \underbrace{f \circ f \circ \cdots \circ f(z).}_{\ell \text{ times}}$$

As $f(U) \subset U$, we have that f^{ℓ} is a holomorphic map of U to U. As U is bounded, there is an M such that $\|z\| \leq M$ for all $z \in U$. Therefore, $\|f(z)\| \leq M$ for all $z \in U$, and $\|f^{\ell}(z)\| \leq M$ for all $z \in U$.

Note that

$$f_k(f(z)) = f_k(z + \text{higher order terms}) = f_k(z) + \text{higher order terms.}$$

Therefore,

$$f^2(z) = f(f(z)) = f(z) + f_k(f(z)) + \text{higher order terms}$$
$$= z + 2f_k(z) + \text{higher order terms.}$$

Continuing this procedure,

$$f^{\ell}(z) = z + \ell f_k(z) + \text{higher order terms.}$$

---

*The normal Cauchy estimates could also be used in the proof of Cartan by applying them componentwise.

Suppose $\Delta_r(0)$ is a polydisc whose closure is in U. Via Cauchy estimates, for every multi-index α with $|\alpha| = k$,

$$\frac{\alpha!}{r^\alpha} M \geq \left\| \frac{\partial^{|\alpha|} f^\ell}{\partial z^\alpha}(0) \right\| = \ell \left\| \frac{\partial^{|\alpha|} f}{\partial z^\alpha}(0) \right\|.$$

The inequality holds for all $\ell \in \mathbb{N}$, and so $\frac{\partial^{|\alpha|} f}{\partial z^\alpha}(0) = 0$. Therefore, $f_k \equiv 0$. On the domain of convergence of the expansion, we get $f(z) = z$, as there is no other nonzero homogeneous part in the expansion of f. As U is connected, the identity theorem says $f(z) = z$ for all $z \in U$. $\qquad \square$

As an application, let us classify all biholomorphisms of all bounded circular domains that fix a point. A *circular domain* is a domain $U \subset \mathbb{C}^n$ such that if $z \in U$, then $e^{i\theta} z \in U$ for all $\theta \in \mathbb{R}$.

Corollary 1.5.2. *Suppose $U, V \subset \mathbb{C}^n$ are bounded circular domains with $0 \in U$, $0 \in V$, and $f : U \to V$ is a biholomorphic map such that $f(0) = 0$. Then f is linear.*

For example, $\mathbb{B}_n$ is circular and bounded. So a biholomorphism of $\mathbb{B}_n$ (an automorphism) that fixes the origin is linear. Similarly, a polydisc centered at zero is also circular and bounded. In fact, every Reinhardt domain is circular.

Proof. The map $g(z) = f^{-1}\left(e^{-i\theta} f(e^{i\theta} z)\right)$ is an automorphism of U and via the chain rule, $g'(0) = I$. Therefore, Cartan says that $f^{-1}\left(e^{-i\theta} f(e^{i\theta} z)\right) = z$, or in other words,

$$f(e^{i\theta} z) = e^{i\theta} f(z).$$

Write f near zero as $f(z) = \sum_{m=1}^{\infty} f_m(z)$ where f_m are homogeneous polynomials of degree m (notice $f_0 = 0$). Then

$$\sum_{m=1}^{\infty} e^{i\theta} f_m(z) = e^{i\theta} \sum_{m=1}^{\infty} f_m(z) = \sum_{m=1}^{\infty} f_m(e^{i\theta} z) = \sum_{m=1}^{\infty} e^{im\theta} f_m(z).$$

By the uniqueness of the Taylor expansion, $e^{i\theta} f_m(z) = e^{im\theta} f_m(z)$, or $f_m(z) = e^{i(m-1)\theta} f_m(z)$, for all m, all z, and all θ. If $m \neq 1$, we obtain that $f_m \equiv 0$, which proves the claim. $\qquad \square$

Exercise 1.5.3: *Show that every automorphism f of $\mathbb{D}^n$ (that is, a biholomorphism $f : \mathbb{D}^n \to \mathbb{D}^n$) is given as*

$$f(z) = P\left(e^{i\theta_1} \frac{z_1 - a_1}{1 - \bar{a}_1 z_1}, e^{i\theta_2} \frac{z_2 - a_2}{1 - \bar{a}_2 z_2}, \dots, e^{i\theta_n} \frac{z_n - a_n}{1 - \bar{a}_n z_n}\right)$$

for $\theta \in \mathbb{R}^n$, $a \in \mathbb{D}^n$, and a permutation matrix P.

Exercise 1.5.4: Given $a \in \mathbb{B}_n$, define the linear map $P_a z = \frac{\langle z, a \rangle}{\langle a, a \rangle} a$ if $a \neq 0$ and $P_0 z = 0$. Let $s_a = \sqrt{1 - \|a\|^2}$. Show that every automorphism f of $\mathbb{B}_n$ (that is, a biholomorphism $f : \mathbb{B}_n \to \mathbb{B}_n$) can be written as

$$f(z) = U \frac{a - P_a z - s_a(I - P_a)z}{1 - \langle z, a \rangle}$$

for a unitary matrix U and some $a \in \mathbb{B}_n$.

Exercise 1.5.5: Using the previous two exercises, show that $\mathbb{D}^n$ and $\mathbb{B}_n$, $n \geq 2$, are not biholomorphic via a method more in the spirit of what Poincaré used: Show that the groups of automorphisms of the two domains are different groups when $n \geq 2$.

Exercise 1.5.6: Suppose $U \subset \mathbb{C}^n$ is a bounded open set, $a \in U$, and $f : U \to U$ is a holomorphic mapping such that $f(a) = a$. Show that every eigenvalue λ of the matrix $Df(a)$ satisfies $|\lambda| \leq 1$.

Exercise 1.5.7 (Tricky): For any n, find a domain $U \subset \mathbb{C}^n$ such that the only biholomorphism $f : U \to U$ is the identity $f(z) = z$. Hint: Take the polydisc (or the ball) and remove some number of points (be careful in how you choose them). Then show that f extends to a biholomorphism of the polydisc. Then see what happens to those points you took out.

Exercise 1.5.8:
   a) Show that Cartan's uniqueness theorem is not true in the real case, even for rational functions. That is, find a rational function $R(t)$ of a real variable t, such that R takes $(-1, 1)$ to $(-1, 1)$, $R'(0) = 1$, and $R(t)$ is not the identity. You can even make R bijective.
   b) Show that Exercise 1.5.6 is not true in the real case. For every $\alpha \in \mathbb{R}$, find a rational function $R(t)$ of a real variable t, such that R takes $(-1, 1)$ to $(-1, 1)$ and $R'(0) = \alpha$.

Exercise 1.5.9: Suppose $U \subset \mathbb{C}^n$ is an open set, $a \in U$, $f : U \to U$ is a holomorphic mapping, $f(a) = a$, and suppose that $|\lambda| < 1$ for every eigenvalue λ of $Df(a)$. Prove that there exists a neighborhood W of a, such that $\lim_{\ell \to \infty} f^{\ell}(z) = a$ for all $z \in W$.

Exercise 1.5.10: Let $U \subset \mathbb{C}^n$ be a bounded open set and $a \in U$. Show that the mapping $\varphi \mapsto (\varphi(a), D\varphi(a))$ from the set $\mathrm{Aut}(U)$ of automorphisms of U to $\mathbb{C}^n \times \mathbb{C}^{n^2}$ is injective.

1.6 \ Riemann extension, zero sets, and injective maps

In one dimension, if a function is holomorphic in $U \setminus \{p\}$ and locally bounded* in U, in particular bounded near p, then the function extends holomorphically to U (see

---

*Recall that $f : U \setminus X \to \mathbb{C}$ is *locally bounded in U* if for every $q \in U$, there is a neighborhood W of q such that f is bounded on $W \cap (U \setminus X)$.

Proposition B.22 (i)). In several variables, the same theorem holds, and the analogue of a single point is the zero set of a holomorphic function.

Theorem 1.6.1 (Riemann extension theorem). *Let $U \subset \mathbb{C}^n$ be a domain, $g \in \mathcal{O}(U)$, and suppose g is not identically zero. Let $N = g^{-1}(0)$ be the zero set of g. If $f \in \mathcal{O}(U \setminus N)$ is locally bounded in U, then there exists a unique $F \in \mathcal{O}(U)$ such that $F|_{U \setminus N} = f$.*

The proof is an application of the Riemann extension theorem from one dimension. And just as in one dimension, the boundedness condition is necessary to expect an extension. For instance, $\frac{1}{g(z)}$ is not bounded near N and indeed does not extend through N.

Proof. Take any $p \in N$, and let L be a complex line through p. That is, L is an image of an affine mapping $\varphi \colon \mathbb{C} \to \mathbb{C}^n$ defined by $\varphi(\xi) = a\xi + p$, for a vector $a \in \mathbb{C}^n$. The composition $g \circ \varphi$ is a holomorphic function of one variable, and it is either identically zero, or the zero at $\xi = 0$ is isolated. The function g is not identically zero in any neighborhood of p by the identity theorem. So there is some line L such that $g \circ \varphi$ is not identically zero, or in other words, p is an isolated point of $L \cap N$.

Write $z' = (z_1, \ldots, z_{n-1})$ and $z = (z', z_n)$. Without loss of generality, $p = 0$ and L is the line obtained by $z' = 0$. So $g \circ \varphi$ is $\xi \mapsto g(0, \xi)$. There is a small $r > 0$ such that g is nonzero on the set given by $|z_n| = r$ and $z' = 0$. By continuity, g is nonzero on the set given by $|z_n| = r$ and $\|z'\| < \epsilon$ for some $\epsilon > 0$. In particular, for any fixed $s \in \mathbb{C}^{n-1}$, with $\|s\| < \epsilon$, setting $z' = s$, the zeros of $\xi \mapsto g(s, \xi)$ are isolated. See Figure 1.6.

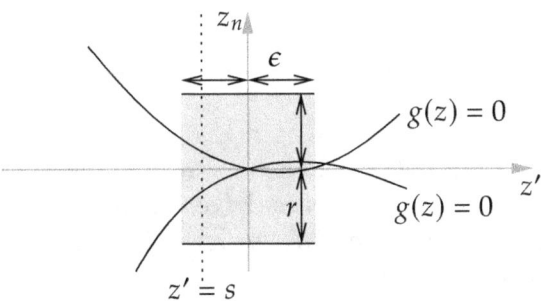

Figure 1.6: Good neighborhood of the origin with respect to the zero set of g.

For $\|z'\| < \epsilon$ and $|z_n| < r$, write

$$F(z', z_n) = \frac{1}{2\pi i} \int_{|\xi|=r} \frac{f(z', \xi)}{\xi - z_n} \, d\xi.$$

The function $\xi \to f(z', \xi)$ is bounded and thus extends holomorphically to the entire disc of radius r by the Riemann extension from one dimension. By the Cauchy integral formula, F is equal to f at the points where they are both defined. By differentiating under the integral, the function F is holomorphic in all variables.

In a neighborhood of each point of N, F is continuous (holomorphic in fact). A continuous extension of f must be unique on the closure of $U \setminus N$ in the subspace topology, $\overline{(U \setminus N)} \cap U$. Due to the identity theorem, the set N has empty interior, so $\overline{(U \setminus N)} \cap U = U$. Hence, F is the unique continuous extension of f to U.   □

Exercise 1.6.1: Let F be a meromorphic function on an open set $U \subset \mathbb{C}^n$. Show that if $p \in U$ is a pole (near p, $F = f/g$, and F does not extend through p), then there exists a sequence $\{p_k\}$ converging to p such that $F(p_k) \to \infty$. Namely, F is unbounded near p.

Exercise 1.6.2: Suppose that $U \subset \mathbb{C}^n$ is open and $N \subset U$ is a closed set such that for every $\zeta \in \mathbb{C}$, the set $\{z \in U : z_n = \zeta\} \cap N$ is countable. Suppose that $f : U \setminus N \to \mathbb{C}$ is holomorphic and locally bounded in U. Then f uniquely extends to a holomorphic function of U. Hint: Every countable closed subset of $\mathbb{C}$ has isolated points.

Exercise 1.6.3: Suppose $U = \{z \in \mathbb{D}^2 : z_1 \neq 0 \text{ and } z_2 \neq 0\}$. Compute the group of automorphisms $\mathrm{Aut}(U)$. Hint: See Exercise 1.5.3.

The set of zeros of a holomorphic function has a nice structure at most points.

Theorem 1.6.2. *Suppose $U \subset \mathbb{C}^n$ is a domain, $f \in \mathcal{O}(U)$, and f is not identically zero. Let $N = f^{-1}(0)$. Then there exists an open and dense (subspace topology) subset $N_{reg} \subset N$ such that at each $p \in N_{reg}$, after possibly reordering variables, N can be locally (that is, in some neighborhood) written as*

$$z_n = g(z_1, \ldots, z_{n-1})$$

for a holomorphic function g.

Proof. If N is locally a graph at p, then it is a graph for every point of N near p. So N_{reg} is open. If for every point $p_0 \in N$ and every neighborhood W of p_0, we show that $N \cap W$ has a regular point, then N_{reg} is dense. Replacing N with $N \cap W$, it thus suffices to show N_{reg} is nonempty.

Since f is not identically zero, then not all derivatives (of arbitrary order) of f vanish identically on N. If some first order derivative of f does not vanish identically on N, let $h = f$. Otherwise, suppose k is such that a derivative of f of order k does not vanish identically on N, and all derivatives of f of order less than k vanish identically on N. Let h be one of the derivatives of order $k - 1$. We obtain a function $h : U \to \mathbb{C}$, holomorphic, vanishing on N, and such that without loss of generality the z_n derivative does not vanish identically on N. Then there is some point $p \in N$ such that $\frac{\partial h}{\partial z_n}(p) \neq 0$. We apply the implicit function theorem at p to find g such that

$$h\big(z_1, \ldots, z_{n-1}, g(z_1, \ldots, z_{n-1})\big) = 0,$$

and $z_n = g(z_1, \ldots, z_{n-1})$ is the unique solution to $h = 0$ near p.

The zero set of h contains N, the zero set of f. We must show equality near p. That is, we need to show that near p, every zero of h is also a zero of f. Write $p = (p', p_n)$. Then the function

$$\xi \mapsto f(p', \xi)$$

has an isolated zero in a small disc Δ around p_n and is nonzero on the circle $\partial\Delta$. By Rouché's theorem, $\xi \mapsto f(z', \xi)$ must have a zero for all z' sufficiently close to p' (close enough to make $|f(p', \xi) - f(z', \xi)| < |f(p', \xi)|$ for all $\xi \in \partial\Delta$). Since $g(z')$ is the unique solution z_n to $h(z', z_n) = 0$ near p and the zero set of f is contained in the zero set of h, we are done. $\qquad\square$

The zero set N of a holomorphic function is an example of a so-called *subvariety* or an *analytic set*, although the general definition of a subvariety is more complicated. See chapter 6. Points where N is a graph of a holomorphic mapping are called *regular points*, and we write them as N_{reg} as above. In particular, since N is a graph of a single holomorphic function, they are called regular points of (complex) dimension $n - 1$, or (complex) codimension 1. The set of regular points is what is called an $(n-1)$-dimensional *complex submanifold*. It is also a real submanifold of real dimension $2n - 2$. The points on a subvariety that are not regular are called *singular points*.

To wit, one of important consequences of the theorem is that the zero set of a holomorphic function is always quite large when $n \geq 2$.

Example 1.6.3: For $U = \mathbb{C}^2$, let $f(z) = z_1^2 - z_2^2$ and consider $X = f^{-1}(0)$. As $\nabla f = (2z_1, -2z_2) \neq 0$ outside of the origin, we can solve for z_1 or z_2 and so all points of $X \setminus \{0\}$ are regular. In fact, $z_1 = z_2$ and $z_1 = -z_2$ are the two possibilities. In no neighborhood of the origin, however, is there a way to uniquely solve for either z_1 or z_2, since you always get two possible solutions: If you could solve $z_1 = g(z_2)$, then both $z_2 = g(z_2)$ and $-z_2 = g(z_2)$ must be true, a contradiction for any nonzero z_2. Similarly, we cannot solve for z_2. So the origin is a singular point.

To see that you may have needed to use derivatives of the function in the proof of the theorem, notice that the function $\varphi(z) = (z_1^2 - z_2^2)^2$ has the same zero set X, but both $\frac{\partial\varphi}{\partial z_1}$ and $\frac{\partial\varphi}{\partial z_2}$ vanish on X. Using $h = \frac{\partial\varphi}{\partial z_1}$ or $h = \frac{\partial\varphi}{\partial z_2}$ in the proof will work.

Similarly, $\psi(z) = (z_1 - z_2)^2(z_1 + z_2)$ has the same zero set X, and $h = \psi$ will work at regular points where $z_1 = -z_2$, but $h = \frac{\partial\psi}{\partial z_1}$ or $h = \frac{\partial\psi}{\partial z_2}$ must be used where $z_1 = z_2$.

Example 1.6.4: The theorem is not true in the nonholomorphic setting. The set where $x_1^2 + x_2^2 = 0$ in $\mathbb{R}^2$ is only the origin, clearly not a graph of any function of one variable. The first part of the theorem works, but the h you find is either $2x_1$ or $2x_2$, and its zero set is too big.

Exercise 1.6.4: *Find all the regular points of the subvariety $X = \{z \in \mathbb{C}^2 : z_1^2 = z_2^3\}$. Hint: The trick is showing that you've found all of them.*

Exercise 1.6.5: Suppose $U \subset \mathbb{C}^n$ is a domain and $f \in \mathcal{O}(U)$. Show that the complement of the zero set, $U \setminus f^{-1}(0)$, is connected.

Exercise 1.6.6: Suppose $U \subset \mathbb{C}^n$ is a domain, $n \geq 2$, and $f \in \mathcal{O}(U)$. Show that the zero set $f^{-1}(0)$ is not compact if it is nonempty. *Hint: A compact set has a point farthest from the origin.*

Remark 1.6.5. It is rather surprising that by a famous theorem of Whitney, any closed set whatsoever in $\mathbb{R}^n$ is the zero set of a C^∞-smooth function.

Let us now prove that a one-to-one holomorphic mapping is biholomorphic, a result definitely not true in the smooth setting: $x \mapsto x^3$ is smooth, one-to-one, onto map of $\mathbb{R}$ to $\mathbb{R}$, but the inverse is not differentiable.

Theorem 1.6.6. *Suppose $U \subset \mathbb{C}^n$ is an open set and $f \colon U \to \mathbb{C}^n$ is holomorphic and one-to-one. Then the Jacobian determinant is never equal to zero on U.*

In particular, if a holomorphic map $f \colon U \to V$ is one-to-one and onto for two open sets $U, V \subset \mathbb{C}^n$, then f is biholomorphic.

The function f is locally biholomorphic, in particular f^{-1} is holomorphic, on the set where the Jacobian determinant

$$J_f(z) = \det Df(z) = \det \left[\frac{\partial f_k}{\partial z_\ell}(z) \right]_{k\ell}$$

is not zero. This follows from the inverse function theorem, which is just a special case of the implicit function theorem. The trick to prove the theorem above is to prove that J_f is nowhere zero.

In one complex dimension, every holomorphic function f can, in the proper local holomorphic coordinates (and up to adding a constant), be written as z^d for $d = 0, 1, 2, \ldots$: Near a $z_0 \in \mathbb{C}$, there exists a constant c and a local biholomorphic g with $g(z_0) = 0$ such that $f(z) = c + \big(g(z)\big)^d$. So f is one-to-one precisely if $d = 1$. Such a simple result does not hold in several variables in general, but if the mapping is locally one-to-one, then the present theorem says that such a mapping can be locally written as the identity.

Proof of the theorem. We proceed by induction. We know the theorem for $n = 1$. Suppose $n > 1$ and suppose we know the theorem is true for dimension $n - 1$.

Suppose for contradiction that $J_f = 0$ somewhere. First suppose that J_f is not identically zero. Find a regular point q of the zero set of J_f. Write the zero set of J_f near q as

$$z_n = g(z_1, \ldots, z_{n-1})$$

for some holomorphic g. If we prove the theorem near q, we are done. Without loss of generality assume $q = 0$. The biholomorphic (near the origin) map

$$\Psi(z_1, \ldots, z_n) = \big(z_1, z_2, \ldots, z_{n-1}, z_n - g(z_1, \ldots, z_{n-1})\big)$$

takes the zero set of J_f to the set given by $z_n = 0$. By considering $f \circ \Psi^{-1}$ instead of f, we may assume that $J_f = 0$ on the set given by $z_n = 0$. We may also assume that $f(0) = 0$.

If J_f vanishes identically, then there is no need to do anything other than a translation. In either case, we may assume that $0 \in U$, $f(0) = 0$, and $J_f = 0$ when $z_n = 0$. Really, all we need is for the set where $J_f = 0$ to be a sufficiently large set.

We wish to show that all the derivatives of f in the $z_1, \ldots, z_{n-1}$ variables vanish whenever $z_n = 0$. This would clearly contradict f being one-to-one, as $f(z_1, \ldots, z_{n-1}, 0)$ would be constant. So for any point on $z_n = 0$, consider one of the components of f and one of the derivatives of that component. Without loss of generality, suppose the point is 0, and for contradiction suppose $\frac{\partial f_1}{\partial z_1}(0) \neq 0$. The map

$$G(z_1, \ldots, z_n) = \big(f_1(z), z_2, \ldots, z_n\big)$$

is biholomorphic on a small neighborhood of the origin. The function $f \circ G^{-1}$ is holomorphic and one-to-one on a small neighborhood. By the definition of G,

$$f \circ G^{-1}(w_1, \ldots, w_n) = \big(w_1, h(w)\big),$$

where h is a holomorphic mapping taking a neighborhood of the origin in $\mathbb{C}^n$ to $\mathbb{C}^{n-1}$. The mapping

$$\varphi(w_2, \ldots, w_n) = h(0, w_2, \ldots, w_n)$$

is a one-to-one holomorphic mapping of a neighborhood of the origin in $\mathbb{C}^{n-1}$ to $\mathbb{C}^{n-1}$. By the induction hypothesis, the Jacobian determinant of φ is nowhere zero.

If we differentiate $f \circ G^{-1}$, we notice $D(f \circ G^{-1}) = Df \circ D(G^{-1})$. So at the origin

$$\det D(f \circ G^{-1}) = \big(\det Df\big)\big(\det D(G^{-1})\big) = 0.$$

We obtain a contradiction, as at the origin

$$\det D(f \circ G^{-1}) = \det D\varphi \neq 0. \qquad \square$$

The theorem is no longer true if the dimensions of the domain and range of the mapping are not equal.

Exercise **1.6.7:** *Take the subvariety* $X = \{z \in \mathbb{C}^2 : z_1^2 = z_2^3\}$. *Find a one-to-one holomorphic mapping* $f : \mathbb{C} \to X$. *Note that the derivative of f vanishes at a certain point. So Theorem 1.6.6 has no analogue when the domain and range have different dimension.*

Exercise **1.6.8:** *Find a continuous function* $f : \mathbb{R} \to \mathbb{R}^2$ *that is one-to-one but such that the inverse* $f^{-1} : f(\mathbb{R}) \to \mathbb{R}$ *is not continuous.*

This is an appropriate place to state a well-known and as yet unsolved conjecture (and most likely ridiculously hard to solve): the *Jacobian conjecture*. This conjecture is a converse to the theorem above in a special case: *If $F \colon \mathbb{C}^n \to \mathbb{C}^n$ is a polynomial map (each component is a polynomial) and the Jacobian determinant J_F is never zero, then F is invertible with a polynomial inverse.* Clearly F would be locally one-to-one, but proving (or disproving) the existence of a global polynomial inverse is the content of the conjecture.

> **Exercise 1.6.9:** *Prove the Jacobian conjecture for $n = 1$. That is, prove that if $F \colon \mathbb{C} \to \mathbb{C}$ is a polynomial such that F' is never zero, then F has an inverse, which is a polynomial.*
>
> **Exercise 1.6.10:** *Let $F \colon \mathbb{C}^n \to \mathbb{C}^n$ be an injective polynomial map. Prove J_F is a nonzero constant.*
>
> **Exercise 1.6.11:** *Prove that the Jacobian conjecture is false if "polynomial" is replaced with "entire holomorphic," even for $n = 1$.*
>
> **Exercise 1.6.12:** *Prove that if a holomorphic $f \colon \mathbb{C} \to \mathbb{C}$ is injective, then it is onto, and therefore $f(z) = az + b$ for $a \neq 0$.*

We remark that while every injective holomorphic map $f \colon \mathbb{C} \to \mathbb{C}$ is onto, the same is not true in higher dimensions. In $\mathbb{C}^n$, $n \geq 2$, there exist so-called *Fatou–Bieberbach domains*, that is, proper subsets of $\mathbb{C}^n$ that are biholomorphic to $\mathbb{C}^n$.

2   Convexity and Pseudoconvexity

2.1   Domains of holomorphy & holomorphic extension

It turns out that not every domain in $\mathbb{C}^n$ is a natural domain for holomorphic functions.

Definition 2.1.1. Let $U \subset \mathbb{C}^n$ be a domain* (connected open set). The set U is a *domain of holomorphy* if there do not exist nonempty open sets V and W, with $V \subset U \cap W$, $W \not\subset U$, and W connected, such that for every $f \in \mathcal{O}(U)$ there exists an $F \in \mathcal{O}(W)$ with $f(z) = F(z)$ for all $z \in V$. See Figure 2.1.

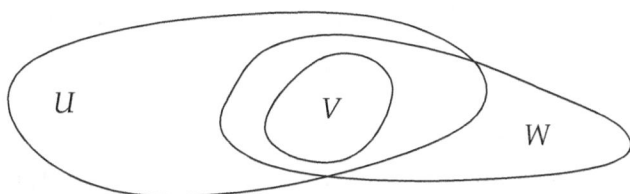

Figure 2.1: Definition of domain of holomorphy.

    The idea is that if a domain U is not a domain of holomorphy and V, W exist as in the definition, then f "extends across the boundary" somewhere.

Example 2.1.2: The unit ball $\mathbb{B}_n \subset \mathbb{C}^n$ is a domain of holomorphy. Proof: Consider $U = \mathbb{B}_n$, and suppose V, W as in the definition exist. As W is connected and open, it is path connected. There are points in W that are not in $\mathbb{B}_n$, so there is a path γ in W going from a point $q \in V$ to some $p \in \partial \mathbb{B}_n \cap W$, and assume $\gamma \setminus \{p\} \subset \mathbb{B}_n$. Without loss of generality (after composing with rotations, that is, unitary matrices), assume $p = (1, 0, 0, \ldots, 0)$. Consider $f(z) = \frac{1}{1-z_1}$. The function F equals f on the component of $\mathbb{B}_n \cap W$ that contains q. But that component contains p and so F blows up at p (so it cannot be holomorphic). The contradiction shows that no V and W exist.

Domain of holomorphy can make sense for disconnected sets (not domains), and some authors do define it so.

In one dimension, this notion has no real content: Every domain in $\mathbb{C}$ is a domain of holomorphy (exercise below).

Exercise 2.1.1 (Easy): *In $\mathbb{C}$, every domain is a domain of holomorphy.*

Exercise 2.1.2: *If $U_k \subset \mathbb{C}^n$ are domains of holomorphy (possibly an infinite set of domains), then the interior of $\bigcap_k U_k$ is either empty or every connected component is a domain of holomorphy.*

Exercise 2.1.3 (Easy): *Show that a polydisc in $\mathbb{C}^n$ is a domain of holomorphy.*

Exercise 2.1.4: *Suppose $U_k \subset \mathbb{C}^{n_k}$, $k = 1, \ldots, \ell$ are domains of holomorphy, show that $U_1 \times \cdots \times U_\ell$ is a domain of holomorphy. In particular, every cartesian product of domains in $\mathbb{C}$ is a domain of holomorphy.*

Exercise 2.1.5: *Suppose $U \subset \mathbb{C}^n$ is a domain of holomorphy and $f \in \mathcal{O}(U)$ is a function. Show that $U \setminus f^{-1}(0)$ is a domain of holomorphy.*

Exercise 2.1.6:

  a) *Given $p \in \partial \mathbb{B}_n$, find a function f holomorphic on $\mathbb{B}_n$, C^∞-smooth on $\overline{\mathbb{B}_n}$ (all real partial derivatives of all orders extend continuously to $\overline{\mathbb{B}_n}$), that does not extend past p as a holomorphic function. Hint: For the principal branch of $\sqrt{\cdot}$ the function $\xi \mapsto e^{-1/\sqrt{\xi}}$ is holomorphic for $\operatorname{Re} \xi > 0$ and extends to be continuous (even smooth) on all of $\operatorname{Re} \xi \geq 0$.*

  b) *Find a function f holomorphic on $\mathbb{B}_n$ that does not extend past any point of $\partial \mathbb{B}_n$.*

Various notions of convexity will play a big role later on. A set S is *geometrically convex* if $tx + (1-t)y \in S$ for all $x, y \in S$ and $t \in [0, 1]$. The exercise below says that every geometrically convex domain is a domain of holomorphy. Domains of holomorphy are often not geometrically convex (e.g. every domain in $\mathbb{C}$ is a domain of holomorphy), so classical convexity is not the correct notion, but it is in the right direction.

Exercise 2.1.7: *Show that a geometrically convex domain in $\mathbb{C}^n$ is a domain of holomorphy.*

In the following, when we say $f \in \mathcal{O}(U)$ extends holomorphically to V where $U \subset V$, we mean that there exists a function $F \in \mathcal{O}(V)$ such that $f = F$ on U.

Remark 2.1.3. The subtlety of the definition of a domain of holomorphy is that it does not necessarily talk about functions extending to a larger set, since we must take into account single-valuedness. For instance, let f be the principal branch of the logarithm defined on the slit plane $U = \mathbb{C} \setminus \{z \in \mathbb{C} : \operatorname{Im} z = 0, \operatorname{Re} z \leq 0\}$. We can locally define an extension from one side through the boundary of the domain, but we cannot define an extension on a open set that contains U. This one-dimensional example should be motivation for why we let V be a proper subset of $U \cap W$, and why W need not contain all of U. This one dimensional intuition can be extended to an actual example in $\mathbb{C}^n$, see Exercise 2.1.15.

In dimension two or higher, not every domain is a domain of holomorphy. We have the following theorem. The domain H in the theorem is called the *Hartogs figure*.

Theorem 2.1.4. *Let $(z, w) = (z_1, \ldots, z_m, w_1, \ldots, w_k) \in \mathbb{C}^m \times \mathbb{C}^k$ be the coordinates. For two numbers $0 < a, b < 1$, define the set $H \subset \mathbb{D}^{m+k}$ by*

$$H = \big\{ (z, w) \in \mathbb{D}^{m+k} : |z_\ell| > a \text{ for } \ell = 1, \ldots, m \big\}$$
$$\cup \big\{ (z, w) \in \mathbb{D}^{m+k} : |w_\ell| < b \text{ for } \ell = 1, \ldots, k \big\}.$$

If $f \in \mathcal{O}(H)$, then f extends holomorphically to $\mathbb{D}^{m+k}$.

In $\mathbb{C}^2$ if $m = 1$ and $k = 1$, see Figure 2.2 (the c will come up in the proof).

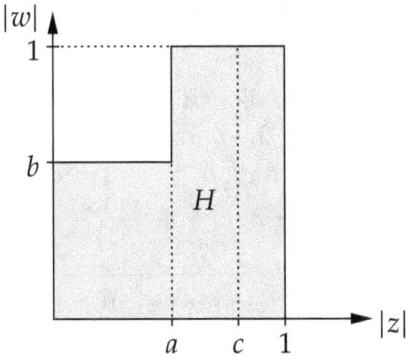

In diagrams, the Hartogs figure is often drawn as:

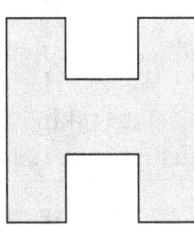

Figure 2.2: Hartogs figure.

Proof. Pick a $c \in (a, 1)$. Let

$$\Gamma = \big\{ z \in \mathbb{D}^m : |z_\ell| = c \text{ for } \ell = 1, \ldots, m \big\}.$$

The set Γ is the distinguished boundary of $c\mathbb{D}^m$, a polydisc centered at 0 of radius c in $\mathbb{C}^m$. Define

$$F(z, w) = \frac{1}{(2\pi i)^m} \int_\Gamma \frac{f(\xi, w)}{\xi - z} \, d\xi.$$

Clearly, F is well-defined on

$$c\mathbb{D}^m \times \mathbb{D}^k$$

as ξ only ranges through Γ and so as long as $w \in \mathbb{D}^k$ then $(\xi, w) \in H$.

The function F is holomorphic in w as we can differentiate underneath the integral and f is holomorphic in w on H. Furthermore, F is holomorphic in z as the kernel $\frac{1}{\xi-z}$ is holomorphic in z as long as $z \in c\mathbb{D}^m$.

For any fixed w with $|w_\ell| < b$ for all ℓ, the Cauchy integral formula says $F(z, w) = f(z, w)$ for all $z \in c\mathbb{D}^m$. Hence, $F = f$ on the open set $c\mathbb{D}^m \times b\mathbb{D}^k$, and so they are equal on $(c\mathbb{D}^m \times \mathbb{D}^k) \cap H$. Combining F and f, we obtain a holomorphic function on $\mathbb{D}^{m+k}$ that extends f. □

The theorem is used in many situations to extend holomorphic functions. We usually need to translate, scale, rotate (apply a unitary matrix), and even take more general biholomorphic mappings of H, to place it wherever we need it. The corresponding polydisc—or the image of the polydisc under the appropriate biholomorphic mapping if one was used—to which all holomorphic functions on H extend is denoted by $\widehat{H}$ and is called the *hull* of H.

Let us state a simple but useful case of the so-called *Hartogs phenomenon*. You have already proved a version of this result in Exercise 1.1.8, but let us prove it with the Hartogs figure.

Corollary 2.1.5. *Let $U \subset \mathbb{C}^n$, $n \geq 2$, be an open set and $p \in U$. Then every $f \in \mathcal{O}(U \setminus \{p\})$ extends holomorphically to U.*

Proof. Without loss of generality, by translating and scaling (those operations are after all holomorphic), we assume that $p = (0, \ldots, 0, \frac{3}{4})$ and the unit polydisc $\mathbb{D}^n$ is contained in U. We fit a Hartogs figure H in U by letting $m = n - 1$ and $k = 1$, writing $\mathbb{C}^n = \mathbb{C}^{n-1} \times \mathbb{C}^1$, and taking $a = b = \frac{1}{2}$. Then $H \subset U$, and $p \in \mathbb{D}^n \setminus H$. Theorem 2.1.4 says that f extends to be holomorphic through p. □

This result provides (yet) another reason why holomorphic functions in several variables have no isolated zeros (or poles). If a zero of f was isolated, then consider $1/f$ to obtain a contradiction. But the extension works in an even more surprising fashion. We could take out a very large set, for example, any geometrically convex compact subset:

Exercise 2.1.8: Suppose $U \subset \mathbb{C}^n$, $n \geq 2$, be an open set and $K \subset\subset U$ is a compact geometrically convex subset. If $f \in \mathcal{O}(U \setminus K)$, then f extends to be holomorphic in U. Hint: Find a nice point on ∂K and try extending a little bit. Then make sure your extension is single-valued.

Convexity of K is not needed; we only need that $U \setminus K$ is connected, but the proof is harder and we will get to it in section 4.3. The single-valuedness of the extension is the key point that makes the general proof harder.

Notice the surprising consequence of the exercise: Every holomorphic function on the shell

$$\mathbb{B}_n \setminus \overline{B_{1-\epsilon}(0)} = \{z \in \mathbb{C}^n : 1 - \epsilon < \|z\| < 1\}$$

for any $\epsilon > 0$ automatically extends to a holomorphic function of $\mathbb{B}_n$. In fact, we will show later that one can take this to the limit: A function only defined on a sphere that satisfies the Cauchy–Riemann equations on the sphere will also extend holomorphically to the interior. We need $n > 1$. The extension result decisively does not work in one dimension; consider $1/z$. You have already shown in an exercise that when $n \geq 2$, the zero sets of holomorphic functions is never compact, here is another reason why. If $n \geq 2$ and $f \in \mathcal{O}(\mathbb{B}_n)$ has a nonempty zero set, then the zero set must contain points arbitrarily close to the boundary. If the set of zeros were compact in $\mathbb{B}_n$, then we could try to extend the function $1/f$.

Exercise 2.1.9 (Hartogs triangle): *Let*

$$T = \left\{ (z_1, z_2) \in \mathbb{D}^2 : |z_2| < |z_1| \right\}.$$

Show that T is a domain of holomorphy. Then show that if

$$\widetilde{T} = T \cup B_\epsilon(0)$$

for an arbitrarily small $\epsilon > 0$, then $\widetilde{T}$ is not a domain of holomorphy. In fact, every function holomorphic on $\widetilde{T}$ extends to a holomorphic function of $\mathbb{D}^2$.

Exercise 2.1.10: Take the natural embedding of $\mathbb{R}^2 \subset \mathbb{C}^2$. Suppose $f \in \mathcal{O}(\mathbb{C}^2 \setminus \mathbb{R}^2)$. Show that f extends holomorphically to all of $\mathbb{C}^2$. Hint: Change coordinates before using Hartogs.

Exercise 2.1.11: Suppose

$$U = \left\{ (z, w) \in \mathbb{D}^2 : 1/2 < |z| \right\}.$$

Draw U. Let $\gamma = \left\{ z \in \mathbb{C} : |z| = 3/4 \right\}$ oriented positively. If $f \in \mathcal{O}(U)$, then show that the function

$$F(z, w) = \frac{1}{2\pi i} \int_\gamma \frac{f(\xi, w)}{\xi - z} \, d\xi$$

is well-defined in $\left((3/4)\mathbb{D} \right) \times \mathbb{D}$, holomorphic where defined, yet it is not necessarily true that $F = f$ on the intersections of their domains.

Exercise 2.1.12: Suppose $U \subset \mathbb{C}^n$ is an open set such that for every $z \in \mathbb{C}^n \setminus \{0\}$, there is a $\lambda \in \mathbb{C}$ such that $\lambda z \in U$. Let $f : U \to \mathbb{C}$ be holomorphic with $f(\lambda z) = f(z)$ whenever $z \in U$, $\lambda \in \mathbb{C}$ and $\lambda z \in U$.
  a) *(easy) Prove that f is constant.*
  b) *(hard) Relax the requirement on f to being meromorphic: $f = g/h$ for holomorphic g and h. Find a nonconstant example, and prove that such an f must be rational (that is, g and h must be polynomials).*

Exercise 2.1.13: Suppose

$$U = \left\{ z \in \mathbb{D}^3 : 1/2 < |z_1| \quad or \quad 1/2 < |z_2| \right\}.$$

Prove that every function $f \in \mathcal{O}(U)$ extends to $\mathbb{D}^3$. Compare to Exercise 2.1.11.

Exercise 2.1.14: *Let $U = \mathbb{C}^n \setminus \{z \in \mathbb{C}^n : z_1 = z_2 = 0\}$, $n \geq 2$. Show that every $f \in \mathcal{O}(U)$ extends holomorphically to $\mathbb{C}^n$.*

Exercise 2.1.15: *Construct an example domain $U \subset \mathbb{C}^2$ that is not a domain of holomorphy, but such that there is no domain $W \subset \mathbb{C}^2$ with $U \subset W$ such that every $f \in \mathcal{O}(U)$ extends to W. Hint: Extending the example from Remark 2.1.3 will almost give you a U, but it will be a domain of holomorphy, you need to modify it a little bit.*

Example 2.1.6: By Exercise 2.1.10, $U_1 = \mathbb{C}^2 \setminus \mathbb{R}^2$ is not a domain of holomorphy. On the other hand, $U_2 = \mathbb{C}^2 \setminus \{z \in \mathbb{C}^2 : z_2 = 0\}$ is a domain of holomorphy; the function $f(z) = \frac{1}{z_2}$ cannot extend. Therefore, U_1 and U_2 are rather different as far as complex variables are concerned, yet they are the same set if we ignore the complex structure. They are both a 4-dimensional real vector space minus a 2-dimensional real vector subspace. That is, U_1 is the set where either $\operatorname{Im} z_1 \neq 0$ or $\operatorname{Im} z_2 \neq 0$, while U_2 is the set where either $\operatorname{Re} z_2 \neq 0$ or $\operatorname{Im} z_2 \neq 0$.

The condition of being a domain of holomorphy, requires something more than just some real geometric condition on the set. Namely, we have shown that the image of a domain of holomorphy via an orthonormal real-linear mapping (so preserving distances, angles, straight lines, etc.) need not be a domain of holomorphy. In particular, when we want to "rotate" in complex analysis we use a complex linear mapping, a unitary matrix.

In fact, one does not need a whole Hartogs figure to extend a holomorphic function, a sequence of discs suffices. We will see another version of this theorem later, Theorem 2.5.2.

Theorem 2.1.7 (Kontinuitätssatz—Continuity principle, first version*). *Suppose $U \subset \mathbb{C}^n$ is open and there exists a sequence of closed analytic discs $\varphi_k \colon \overline{\mathbb{D}} \to \mathbb{C}^n$ converging (pointwise) to a closed analytic disc φ, such that $\varphi_k(\overline{\mathbb{D}}) \subset U$ and $\varphi(\partial \mathbb{D}) \subset U$. Then there exists an $s > 0$ such that for every $f \in \mathcal{O}(U)$ and for every $p \in \varphi(\mathbb{D})$, there is an $F \in \mathcal{O}\big(\Delta_s(p)\big)$ where $F = f$ on some open subset of $U \cap \Delta_s(p)$.*

In particular, a U that possesses such discs where $\varphi(\mathbb{D})$ does not lie entirely in U is not a domain of holomorphy. The continuity principle is illustrated in Figure 2.3, where analytic discs are drawn as lines and the boundaries as black dots. Note that the conclusion is that F continues analytically past any point in $\varphi(\mathbb{D})$. However, we do not necessarily get single-valued extension to a whole neighborhood of $\varphi(\mathbb{D})$ without a further hypothesis, see exercises below.[†]

Proof. Via Montel and considering slightly smaller discs (restrictions to discs of radii $1 - \epsilon$), we may assume that φ_k converge uniformly to φ on $\overline{\mathbb{D}}$. Fix some $f \in \mathcal{O}(U)$.

---

[*]Sometimes this (or similar) theorem is called Behnke–Sommer, although the first version of it (where the discs are complex lines) were proved by Hartogs.

[†]A counterexample can be found in S. Ivashkovich, *Discrete and Continuous Versions of the Continuity Principle*, The Journal of Geometric Analysis, **32** (2022), Paper No. 226.

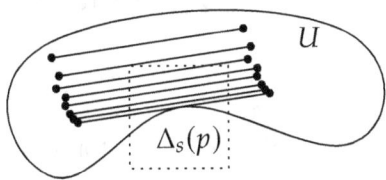

Figure 2.3: Continuity principle for extension of functions.

As $\varphi(\partial\mathbb{D})$ is compact, there exists an $r > 0$ such that for each $z \in \varphi(\partial\mathbb{D})$, $\Delta_r(z) \subset U$, meaning that the power series of f converges in $\Delta_r(z)$. Pick a positive $s < r$. As φ_k converge uniformly, for sufficiently high k,

$$\bigcup_{q \in \varphi_k(\partial\mathbb{D})} \overline{\Delta_s(q)}$$

is a compact subset of U and hence f is bounded by some M on this set. Cauchy estimates give that

$$\left| \frac{\partial^\alpha f}{\partial z^\alpha}(q) \right| \leq \frac{M\alpha!}{s^{|\alpha|}} \tag{2.1}$$

for all $q \in \varphi_k(\partial\mathbb{D})$. By the maximum principle, (2.1) holds for all $q \in \varphi_k(\overline{\mathbb{D}})$, and hence the power series for f at all $q \in \varphi_k(\overline{\mathbb{D}})$ converges in $\Delta_s(q)$. Thus we get an F defined by this power series on $\Delta_s(q)$ which agrees with f in a neighborhood of q. By considering a large enough k, and a slightly smaller s', then for every $p \in \varphi(\overline{\mathbb{D}})$ we can fit a $\Delta_{s'}(p) \subset \Delta_s(q)$ for some $q \in \varphi_k(\overline{\mathbb{D}})$ and where $q \in \Delta_{s'}(p)$. Since p must be in the closure of U by necessity, then $\Delta_{s'}(p)$ intersects U, and as it contains q, F agrees with f on some open subset of $\Delta_{s'}(p)$.

The s (and s') only depends on the distance between the boundary of U and $\varphi(\partial\mathbb{D})$, so it does not depend on f and moreover, the assumption to restrict to smaller discs in the beginning of the proof is valid. $\qquad\square$

Exercise 2.1.16: *Prove that given an analytic disc $\varphi \colon \mathbb{D} \to \mathbb{C}^n$ and a point $p \in \varphi(\mathbb{D})$, then for small enough $\epsilon > 0$, the set $\Delta_\epsilon(p) \cap \varphi(\mathbb{D})$ is connected. Hint: Pull back the coordinate functions from $\mathbb{C}^n$ to $\mathbb{D}$.*

Exercise 2.1.17: *Prove that if furthermore φ is injective in the proof, then there exists an entire neighborhood W of $\varphi(\overline{\mathbb{D}})$ and for every $f \in \mathcal{O}(U)$, there is an $F \in \mathcal{O}(W)$ such that $f = F$ on some open neighborhood of $\varphi(\partial\mathbb{D})$.*

Exercise 2.1.18: *Suppose that given an open $U \subset \mathbb{C}^n$ and there exists a collection of closed analytic discs $\Delta_\alpha \subset U$ such that $\bigcup_\alpha \partial\Delta_\alpha \subset\subset U$. Show that for every p in the closure of $\bigcup_\alpha \Delta_\alpha$ (closure in $\mathbb{C}^n$) there exists an analytic disc through p and a sequence of discs converging to it that satisfy the hypotheses of the theorem.*

2.2 \ Tangent vectors, the Hessian, and convexity

An exercise in the previous section showed that every convex domain is a domain of holomorphy. However, classical convexity is too strong. By Exercise 2.1.4, for any domains $U \subset \mathbb{C}$ and $V \subset \mathbb{C}$, the set $U \times V$ is a domain of holomorphy in $\mathbb{C}^2$. The domains U and V, and hence $U \times V$, can be spectacularly nonconvex. But we should not discard convexity completely. There is a notion of *pseudoconvexity*, which vaguely means "convexity in the complex directions" and is the correct notion to distinguish domains of holomorphy. Let us figure out what classical convexity means locally for a smooth boundary.

Definition 2.2.1. A set $M \subset \mathbb{R}^n$ is a C^k-smooth *hypersurface* if at each point $p \in M$, there exists a k-times continuously differentiable function $r \colon V \to \mathbb{R}$ with nonvanishing derivative, defined in a neighborhood V of p such that $M \cap V = \{x \in V : r(x) = 0\}$. The function r is called the *defining function* of M (at p).

    An open set (or domain) $U \subset \mathbb{R}^n$ with C^k-*smooth boundary* is a set where ∂U is a C^k-smooth hypersurface, and at every $p \in \partial U$ there is a defining function r such that $r < 0$ for points in U and $r > 0$ for points not in U. See Figure 2.4.

    By simply *smooth*, we mean C^∞-smooth, that is, the r is infinitely differentiable.

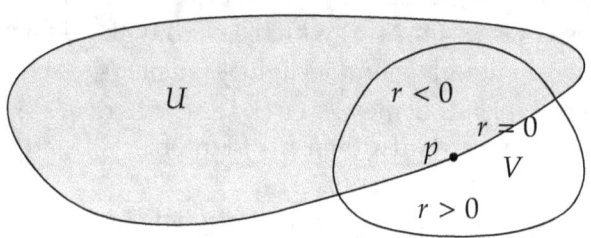

Figure 2.4: Local defining function for a domain.

    What we really defined is an *embedded hypersurface*. In particular, in this book the topology on the set M will be the subset topology. Furthermore, in this book we generally deal with smooth (that is, C^∞) functions and hypersurfaces. Dealing with C^k-smooth functions for finite k introduces technicalities that make certain theorems and arguments unnecessarily difficult.

    As the derivative of r is nonvanishing, a hypersurface M is locally the graph of one variable over the rest using the implicit function theorem. That is, M is a smooth hypersurface if it is locally a set defined by $x_k = \varphi(x_1, \ldots, x_{k-1}, x_{k+1}, \ldots, x_n)$ for some k and some smooth function φ.

    The definition of an open set with smooth boundary is not simply that the boundary is a smooth hypersurface, that is not enough. It says that one side of that hypersurface is in U and one side is not in U: As the derivative of r never vanishes, r

has different signs on different sides of $\{x \in V : r(x) = 0\}$. The verification of this fact is left to the reader. (Hint: Look at where the gradient points.) We can, in fact, find a single global defining function for every open set with smooth boundary, but we have no need of this.

The same definition works for $\mathbb{C}^n$, where we treat $\mathbb{C}^n$ as $\mathbb{R}^{2n}$. For example, the ball $\mathbb{B}_n$ is a domain with smooth boundary with defining function $r(z, \bar{z}) = \|z\|^2 - 1$. In $\mathbb{C}^n$ a hypersurface defined as above is a *real hypersurface,* to distinguish it from a complex hypersurface that would be the zero set of a holomorphic function, although we may leave out the word "real" if it is clear from context.

Definition 2.2.2. For a point $p \in \mathbb{R}^n$, the set of *tangent vectors* $T_p\mathbb{R}^n$ is given by

$$T_p\mathbb{R}^n = \operatorname{span}_{\mathbb{R}} \left\{ \frac{\partial}{\partial x_1}\Big|_p, \ldots, \frac{\partial}{\partial x_n}\Big|_p \right\}.$$

That is, a vector $X_p \in T_p\mathbb{R}^n$ is an object of the form

$$X_p = \sum_{k=1}^{n} a_k \frac{\partial}{\partial x_k}\Big|_p,$$

for real numbers a_k. For computations, X_p could be represented by an n-vector $a = (a_1, \ldots, a_n)$. However, if $p \neq q$, then $T_p\mathbb{R}^n$ and $T_q\mathbb{R}^n$ are distinct spaces. An object $\frac{\partial}{\partial x_k}\big|_p$ is a real linear functional* on the space of smooth functions: When applied to a smooth function g, it gives $\frac{\partial g}{\partial x_k}\big|_p$. Therefore, X_p is also such a functional. It is the directional derivative from calculus; it is computed as $X_p f = \nabla f|_p \cdot (a_1, \ldots, a_n)$.

> *Exercise* 2.2.1: *Let X be a real linear functional on the set of real polynomials[†] in n variables such that $X(fg) = (Xf)g(0) + f(0)(Xg)$. Show that X can be identified with an element of $T_0\mathbb{R}^n$.*

Definition 2.2.3. Let $M \subset \mathbb{R}^n$ be a smooth hypersurface, $p \in M$, and r is a defining function at p, then a vector $X_p \in T_p\mathbb{R}^n$ is *tangent* to M at p if

$$X_p r = 0, \qquad \text{or in other words} \qquad \sum_{k=1}^{n} a_k \frac{\partial r}{\partial x_k}\Big|_p = 0.$$

The set of such tangent vectors is denoted by T_pM, called the *tangent space* of M at p.

The space T_pM is an $(n-1)$-dimensional real vector space—it is a subspace of an n-dimensional $T_p\mathbb{R}^n$ given by a single linear equation. Recall from calculus that the gradient $\nabla r|_p$ is "normal" to M at p, and the tangent space is given by all the n-vectors a that are orthogonal to the normal, that is, $\nabla r|_p \cdot a = 0$.

---

*Linear real-valued function.

†This result works for smooth functions too by applying Taylor's theorem.

We cheated in the terminology, and assumed without justification that $T_p M$ depends only on M, not on r. Fortunately, the definition of $T_p M$ is independent of the choice of r by the next two exercises.

Exercise 2.2.2: *Suppose $M \subset \mathbb{R}^n$ is a smooth hypersurface and r is a smooth defining function for M at p.*

　　a) *Suppose φ is another smooth defining function of M on a neighborhood of p. Show that there exists a smooth nonvanishing function g such that $\varphi = gr$ (in a neighborhood of p).*

　　b) *Now suppose φ is an arbitrary smooth function that vanishes on M (not necessarily a defining function). Again show that $\varphi = gr$, but now g may possibly vanish.*
Hint: First suppose $r = x_n$ and find a g such that $\varphi = x_n g$. Then find a local change of variables to make M into the set given by $x_n = 0$. A useful calculus fact: If $f(0) = 0$ and f is smooth, then $s \int_0^1 f'(ts)\, dt = f(s)$, and $\int_0^1 f'(ts)\, dt$ is a smooth function of s.

Exercise 2.2.3: *Show that $T_p M$ is independent of the defining function: Prove that if r and $\tilde{r}$ are defining functions for M at p, then $\sum_k a_k \frac{\partial r}{\partial x_k}\big|_p = 0$ if and only if $\sum_k a_k \frac{\partial \tilde{r}}{\partial x_k}\big|_p = 0$.*

The tangent space $T_p M$ is the set of derivatives *along* M at p. If r is a defining function of M, and f and h are two smooth functions such that $f = h$ on M, then Exercise 2.2.2 says that

$$f - h = gr, \qquad \text{or} \qquad f = h + gr,$$

for some smooth g. Applying X_p we find

$$X_p f = X_p h + X_p(gr) = X_p h + (X_p g)r + g(X_p r) = X_p h + (X_p g)r.$$

So $X_p f = X_p h$ on M (where $r = 0$). In other words, $X_p f$ only depends on the values of f on M.

This brings up a natural question about what is a smooth function on M. By definition, a function f defined on M (or any other subset of $\mathbb{R}^n$ that is not open) is *smooth* if it is locally the restriction to M of a smooth function in an open neighborhood. This extension of f is not unique, so the above calculation shows that differentiating f via $T_p M$ is independent of how f is extended to a neighborhood.

Example 2.2.4: If $M \subset \mathbb{R}^n$ is given by $x_n = 0$, then $T_p M$ is given by derivatives of the form

$$X_p = \sum_{k=1}^{n-1} a_k \frac{\partial}{\partial x_k}\bigg|_p.$$

That is, derivatives along the first $n - 1$ variables only.

Definition 2.2.5. The disjoint union

$$TR^n = \bigsqcup_{p \in R^n} T_p R^n$$

is called the *tangent bundle*. There is a natural identification $R^n \times R^n \cong TR^n$:

$$(p, a) \in R^n \times R^n \quad \mapsto \quad \sum_{k=1}^{n} a_k \frac{\partial}{\partial x_k}\Big|_p \in TR^n.$$

The topology and smooth structure on TR^n comes from this identification. The wording "bundle" (a bundle of fibers) comes from the natural projection $\pi \colon TR^n \to R^n$, where fibers are $\pi^{-1}(p) = T_p R^n$.

A smooth *vector field* in TR^n is an object of the form

$$X = \sum_{k=1}^{n} a_k \frac{\partial}{\partial x_k},$$

where a_k are smooth functions. That is, X is a smooth function $X \colon V \subset R^n \to TR^n$ such that $X(p) \in T_p R^n$. Usually, we write X_p rather than $X(p)$. To be more fancy, say X is a *section* of TR^n.

Similarly, the tangent bundle of M is

$$TM = \bigsqcup_{p \in M} T_p M.$$

A vector field X in TM is a vector field such that $X_p \in T_p M$ for all $p \in M$.

Before we move on, we note how smooth maps transform tangent spaces. Given a smooth $f \colon U \subset R^n \to R^m$, the derivative at p is a linear mapping of the tangent spaces: $Df(p) \colon T_p R^n \to T_{f(p)} R^m$. If $X_p \in T_p R^n$, then $Df(p)X_p$ should be in $T_{f(p)} R^m$. The vector $Df(p)X_p$ is defined by how it acts on smooth functions φ of a neighborhood of $f(p)$ in R^m:

$$(Df(p)X_p)\varphi = X_p(\varphi \circ f).$$

It is the only reasonable way to put those three objects together. When the spaces are C^n and C^m, we denote this derivative as $D_R f$ to distinguish it from the holomorphic derivative. As far as calculus computations are concerned, the linear mapping $Df(p)$ is the Jacobian matrix acting on vectors in the standard basis of the tangent space as given above. This is why we use the same notation for the Jacobian matrix and the derivative acting on tangent spaces. To verify this claim, it is enough to see where the basis element $\frac{\partial}{\partial x_k}\big|_p$ goes, and the form of $Df(p)$ as a matrix follows by the chain rule. For instance, the derivative of the mapping $f(x_1, x_2) = (x_1 + 2x_2 + x_1^2, 3x_1 + 4x_2 + x_1 x_2)$ at the origin is given by the matrix $\begin{bmatrix} 1 & 2 \\ 3 & 4 \end{bmatrix}$, and so the vector $X_p = a\frac{\partial}{\partial x_1}\big|_0 + b\frac{\partial}{\partial x_2}\big|_0$ gets taken to $Df(0)X_0 = (a + 2b)\frac{\partial}{\partial y_1}\big|_0 + (3a + 4b)\frac{\partial}{\partial y_2}\big|_0$, where (y_1, y_2) are the coordinates on the target. You should check on some test function, such as $\varphi(y_1, y_2) = \alpha y_1 + \beta y_2$, that the definition above is satisfied.

Suppose that for a smooth map f and smooth hypersurfaces M and M' you have $f(M) \subset M'$. Then you get the same containment for the tangent spaces. Indeed, suppose that r is a defining function for M near p and r' is a defining function for M' near $f(p)$, and suppose that $X_p \in T_pM$. Then $r' \circ f$ is zero on M, and hence

$$(Df(p)X_p)r' = X_p(r' \circ f) = X_p(0) = 0.$$

If the map is a diffeomorphism (has an inverse), then $f(M)$ is a smooth hypersurface with defining function $r \circ f^{-1}$, the derivative is an invertible linear map, and we get that $Df(p)$ restricts to an isomorphism of T_pM and $T_{f(p)}f(M)$. That is, we proved the following proposition.

Proposition 2.2.6. *Suppose $U \subset \mathbb{R}^n$ is open, $M \subset U$ is a smooth hypersurface, $f: U \to \mathbb{R}^m$ is a smooth function, $M' \subset \mathbb{R}^m$ is a smooth hypersurface such that $f(M) \subset M'$, and $p \in M$. Then*

$$Df(p)(T_pM) \subset T_{f(p)}M'.$$

Moreover, if $m = n$ and f is a diffeomorphism (bijective onto some open set U' such that f^{-1} is smooth), then $f(M)$ is a smooth hypersurface and $Df(p)(T_pM) = T_{f(p)}f(M)$.

Now that we know what tangent vectors are and how they transform, let us define convexity for domains with smooth boundary.

Definition 2.2.7. Suppose $U \subset \mathbb{R}^n$ is an open set with smooth boundary, and r is a defining function for ∂U at $p \in \partial U$ such that $r < 0$ on U. If

$$\sum_{k=1,\ell=1}^{n} a_k a_\ell \frac{\partial^2 r}{\partial x_k \partial x_\ell}\bigg|_p \geq 0, \qquad \text{for all} \qquad X_p = \sum_{k=1}^{n} a_k \frac{\partial}{\partial x_k}\bigg|_p \in T_p\partial U,$$

then U is said to be *convex* at p. If the inequality above is strict for all nonzero $X_p \in T_p\partial U$, then U is said to be *strongly convex* at p.

A domain U is *convex* if it is convex at all $p \in \partial U$. If U is bounded*, we say U is *strongly convex* if it is strongly convex at all $p \in \partial U$.

The matrix

$$\left[\frac{\partial^2 r}{\partial x_k \partial x_\ell}\bigg|_p\right]_{k\ell}$$

is the *Hessian* of r at p. So, U is convex at $p \in \partial U$ if the Hessian of r at p as a bilinear form is positive semidefinite when restricted to $T_p\partial U$. More concretely, let H be the Hessian of r at p, and treat $a \in \mathbb{R}^n$ as a column vector. Then ∂U is convex at p whenever

$$a^t H a \geq 0, \qquad \text{for all } a \in \mathbb{R}^n \text{ such that } \nabla r|_p \cdot a = 0.$$

---

*Matters are a little more complicated with the "strong" terminology if U is unbounded, so sometimes *strictly convex* is used instead.

This bilinear form given by the Hessian is the second fundamental form from Riemannian geometry in mild disguise (or perhaps it is the other way around).

We cheated a little bit, since we have not proved that the notion of convexity is well-defined. In particular, there are many possible defining functions.

> **Exercise 2.2.4:** *Show that the definition of convexity is independent of the defining function. Hint: If $\tilde{r}$ is another defining function near p, then there is a smooth function $g > 0$ such that $\tilde{r} = gr$.*

Example 2.2.8: The unit disc in $\mathbb{R}^2$ is strongly convex. Proof: Let (x, y) be the coordinates and let $r(x, y) = x^2 + y^2 - 1$ be the defining function. The tangent space of the circle is one-dimensional, so we simply need to find a single nonzero tangent vector at each point. Consider the gradient $\nabla r = (2x, 2y)$ to check that

$$X = y\frac{\partial}{\partial x} - x\frac{\partial}{\partial y}$$

is tangent to the circle, that is, $Xr = X(x^2 + y^2 - 1) = (2x, 2y) \cdot (y, -x) = 0$ on the circle—by chance, $Xr = 0$ everywhere. The vector field X is nonzero on the circle, so at each point it gives a basis of the tangent space. See Figure 2.5.

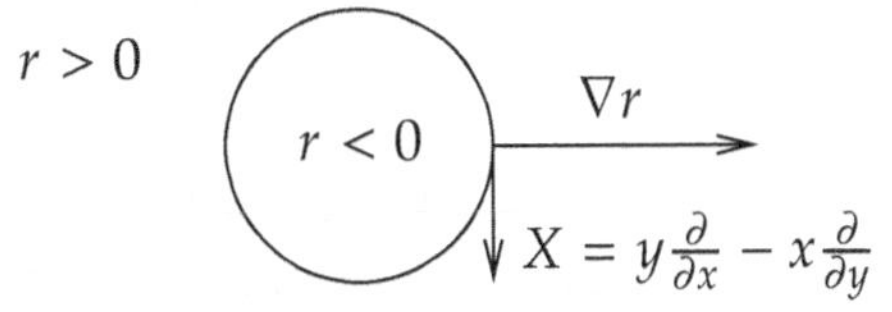

Figure 2.5: Tangent vector to a circle.

The Hessian matrix of r is

$$\begin{bmatrix} \frac{\partial^2 r}{\partial x^2} & \frac{\partial^2 r}{\partial x \partial y} \\ \frac{\partial^2 r}{\partial y \partial x} & \frac{\partial^2 r}{\partial y^2} \end{bmatrix} = \begin{bmatrix} 2 & 0 \\ 0 & 2 \end{bmatrix}.$$

Applying the vector $(y, -x)$ gets us

$$\begin{bmatrix} y & -x \end{bmatrix} \begin{bmatrix} 2 & 0 \\ 0 & 2 \end{bmatrix} \begin{bmatrix} y \\ -x \end{bmatrix} = 2y^2 + 2x^2 = 2 > 0.$$

So the domain given by $r < 0$ is strongly convex at all points.

In general, to construct a tangent vector field for a curve in $\mathbb{R}^2$, consider $r_y\frac{\partial}{\partial x} - r_x\frac{\partial}{\partial y}$. In higher dimensions, running through enough pairs of variables gets a basis of TM.

Exercise **2.2.5:** *Show that if an open set with smooth boundary is strongly convex at a point p, then it is strongly convex for all points in some neighborhood of p. Then find an example of an open set with smooth boundary that is convex at one point p, but not convex at points arbitrarily near p.*

Exercise **2.2.6:** *Show that the domain in $\mathbb{R}^2$ defined by $x^4 + y^4 < 1$ is convex, but not strongly convex. Find all the points where the domain is not strongly convex.*

Exercise **2.2.7:** *Show that the domain in $\mathbb{R}^3$ defined by $(x_1^2 + x_2^2)^2 < x_3$ is strongly convex at all points except the origin, where it is just convex (but not strongly).*

The right sort of changes of coordinates that preserve convexity are invertible real affine linear mappings. It is rather clear for geometric convexity, as these are precisely the maps that take lines to lines, but it takes a little bit of computation for convexity at a point of a smooth boundary (exercise below). A useful analogy to keep in mind (but not to go overboard with) is that holomorphic functions are sort of like affine functions. And so it will be with convexity being replaced with pseudoconvexity in just a little bit, and affine linear maps with holomorphic maps.

Exercise **2.2.8:** *Prove that translations and invertible linear maps (matrices) preserve convexity and strong convexity at a point for a domain with smooth boundary.*

In the following, we use the *big-oh notation*, although we use a perhaps less standard shorthand*. A smooth function is $O(\ell)$ at a point p (usually the origin), if all its derivatives of order $0, 1, \ldots, \ell - 1$ vanish at p. For example, if f is $O(3)$ at the origin, then $f(0) = 0$, and its first and second derivatives vanish at the origin.

For computations it is often useful to use a more convenient defining function, that is, it is convenient to write M as a graph.

Proposition 2.2.9. *Suppose $M \subset \mathbb{R}^n$ is a smooth hypersurface, and $p \in M$. Then after a rotation (orthogonal matrix) and translation, p is the origin, and near the origin, M is given by*

$$y = \varphi(x),$$

where $(x, y) \in \mathbb{R}^{n-1} \times \mathbb{R}$ are our coordinates and φ is a smooth function that is $O(2)$ at the origin, namely, $\varphi(0) = 0$ and $d\varphi(0) = 0$. Consequently,

$$T_0 M = \operatorname{span}_{\mathbb{R}} \left\{ \left. \frac{\partial}{\partial x_1} \right|_p, \ldots, \left. \frac{\partial}{\partial x_{n-1}} \right|_p \right\}.$$

If M is the boundary of an open set U with smooth boundary and $r < 0$ on U, then the rotation can be chosen such that $y > \varphi(x)$ for points in U. See Figure 2.6.

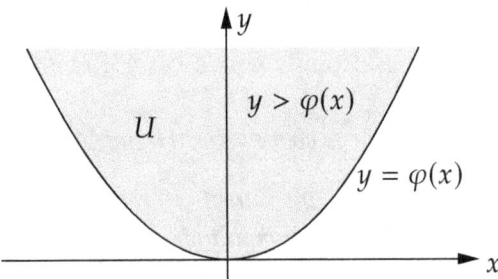

Figure 2.6: Defining a domain as a graph.

Proof. Let r be a defining function at p. Take $v = \nabla r|_p$. By translating p to the origin, and applying a rotation (an orthogonal matrix), we assume $v = (0, 0, \ldots, 0, v_n)$, where $v_n < 0$. Denote our coordinates by $(x, y) \in \mathbb{R}^{n-1} \times \mathbb{R}$. As $\nabla r|_0 = v$, then $\frac{\partial r}{\partial y}(0) \neq 0$. The implicit function theorem gives a smooth function φ such that $r\big(x, \varphi(x)\big) = 0$ for all x in a neighborhood of the origin, and $\{(x, y) : y = \varphi(x)\}$ are all the solutions to $r = 0$ near the origin.

We need to show that the derivative of φ at 0 vanishes. As $r\big(x, \varphi(x)\big) = 0$ for all x in a neighborhood of the origin, we differentiate. For every $k = 1, \ldots, n - 1$,

$$0 = \frac{\partial}{\partial x_k}\Big[r\big(x, \varphi(x)\big)\Big] = \left(\sum_{\ell=1}^{n-1} \frac{\partial r}{\partial x_\ell}\frac{\partial x_\ell}{\partial x_k}\right) + \frac{\partial r}{\partial y}\frac{\partial \varphi}{\partial x_k} = \frac{\partial r}{\partial x_k} + \frac{\partial r}{\partial y}\frac{\partial \varphi}{\partial x_k}.$$

At the origin, $\frac{\partial r}{\partial x_k}(0, 0) = 0$ and $\frac{\partial r}{\partial y}(0, 0) = v_n \neq 0$, and therefore $\frac{\partial \varphi}{\partial x_k}(0) = 0$. That $T_0 M$ is the span of the x_k derivatives follows at once from the fact that $\nabla r|_0 = (0, \ldots, 0, v_n)$.

To prove the final statement, note that $r < 0$ on U. It is enough to check that r is negative for $(0, y)$ if $y > 0$ is small, which follows as $\frac{\partial r}{\partial y}(0, 0) = v_n < 0$. □

The advantage of this representation is that the tangent space at p can be identified with the x coordinates for the purposes of computation. Considering x as a column vector, the Taylor expansion of a smooth function φ at the origin is

$$\varphi(x) = \varphi(0) + \nabla\varphi|_0 \cdot x + \frac{1}{2}x^t H x + E(x),$$

where $H = \left[\frac{\partial^2 \varphi}{\partial x_k \partial x_\ell}|_0\right]_{k\ell}$ is the Hessian matrix of φ at the origin, and E is $O(3)$, namely, $E(0) = 0$, and all first and second order derivatives of E vanish at 0. In the context of the lemma above, the φ is $O(2)$ at the origin, i.e. $\varphi(0) = 0$ and $\nabla\varphi|_0 = 0$. So we write the hypersurface M as

$$y = \frac{1}{2}x^t H x + E(x).$$

---

*The standard notation for $O(\ell)$ is $O(\|x\|^\ell)$ and it means that $\left|\frac{f(x)}{\|x\|^\ell}\right|$ is bounded as $x \to p$.

If M is the boundary ∂U of an open set U, then we pick the rotation so that $y > \frac{1}{2} x^t H x + E(x)$ on U. It is an easy exercise to show that U is convex at p if H is positive semidefinite, and U is strongly convex at p if H is positive definite.

Exercise 2.2.9: *Prove the statement above about H and convexity at p.*

Exercise 2.2.10: *Let r be a defining function at p for a smooth hypersurface $M \subset \mathbb{R}^n$. We say M is convex from both sides at p if both the set given by $r > 0$ and the set given by $r < 0$ are convex at p. Prove that if a hypersurface $M \subset \mathbb{R}^n$ is convex from both sides at all points, then it is locally just a hyperplane (the zero set of a real affine function).*

Exercise 2.2.11: *Suppose U is a domain with smooth boundary that is strongly convex at $p \in \partial U$. Then there exists a real affine change of variables (translation and an invertible linear map), such that after the change of variables, $p = 0$ and near 0, ∂U is given by $y = x^t x + E(x)$ where $E(x)$ is $O(3)$ and $y > x^t x + E(x)$ on U.*

Recall that U is *geometrically convex* if for every $p, q \in U$ the line between p and q is in U, that is, $tp + (1 - t)q \in U$ for all $t \in [0, 1]$. Geometric convexity is a *global* condition; it considers the entire U. The notion of convexity for a smooth boundary is *local* in that you only need to know ∂U in a small neighborhood. For domains with smooth boundaries, the two notions are equivalent. Proving one direction is easy.

Exercise 2.2.12: *Suppose a domain $U \subset \mathbb{R}^n$ with smooth boundary is geometrically convex. Show that U is convex.*

The other direction is considerably more complicated, and we will not worry about it here. Proving a global condition from a local one is often trickier, but also often more interesting. Similar difficulties will be present once we move back to several complex variables and try to relate pseudoconvexity with domains of holomorphy.

2.3 \ Holomorphic vectors, Levi form, pseudoconvexity

As $\mathbb{C}^n$ is identified with $\mathbb{R}^{2n}$ using $z = x + iy$, we have $T_p\mathbb{C}^n = T_p\mathbb{R}^{2n}$. If we take the complex span instead of the real span, we get the *complexified tangent space*[*]

$$\mathbb{C}T_p\mathbb{C}^n = \operatorname{span}_{\mathbb{C}} \left\{ \frac{\partial}{\partial x_1}\Big|_p, \frac{\partial}{\partial y_1}\Big|_p, \ldots, \frac{\partial}{\partial x_n}\Big|_p, \frac{\partial}{\partial y_n}\Big|_p \right\}.$$

We simply replace all the real coefficients with complex ones. The space $\mathbb{C}T_p\mathbb{C}^n$ is a $2n$-dimensional complex vector space. Both $\frac{\partial}{\partial z_k}\big|_p$ and $\frac{\partial}{\partial \bar{z}_k}\big|_p$ are in $\mathbb{C}T_p\mathbb{C}^n$, and

$$\mathbb{C}T_p\mathbb{C}^n = \operatorname{span}_{\mathbb{C}} \left\{ \frac{\partial}{\partial z_1}\Big|_p, \frac{\partial}{\partial \bar{z}_1}\Big|_p, \ldots, \frac{\partial}{\partial z_n}\Big|_p, \frac{\partial}{\partial \bar{z}_n}\Big|_p \right\}.$$

---

[*]Abstractly, a real vector space X can be complexified as $X \otimes_{\mathbb{R}} \mathbb{C}$.

Define

$$T_p^{(1,0)}\mathbb{C}^n \overset{\text{def}}{=} \operatorname{span}_{\mathbb{C}}\left\{\frac{\partial}{\partial z_1}\Big|_p, \ldots, \frac{\partial}{\partial z_n}\Big|_p\right\} \quad \text{and} \quad T_p^{(0,1)}\mathbb{C}^n \overset{\text{def}}{=} \operatorname{span}_{\mathbb{C}}\left\{\frac{\partial}{\partial \bar{z}_1}\Big|_p, \ldots, \frac{\partial}{\partial \bar{z}_n}\Big|_p\right\}.$$

The vectors in $T_p^{(1,0)}\mathbb{C}^n$ are the *holomorphic vectors* and vectors in $T_p^{(0,1)}\mathbb{C}^n$ are the *antiholomorphic vectors*. We decompose the full tangent space as the direct sum

$$\mathbb{C}T_p\mathbb{C}^n = T_p^{(1,0)}\mathbb{C}^n \oplus T_p^{(0,1)}\mathbb{C}^n.$$

A holomorphic function is one that vanishes on $T_p^{(0,1)}\mathbb{C}^n$.

Let us see what holomorphic mappings do to these spaces when we treat holomorphic mappings as smooth mappings. Given a smooth mapping f from (an open subset of) $\mathbb{C}^n$ to $\mathbb{C}^m$, its derivative at $p \in \mathbb{C}^n$ is a real-linear mapping $D_{\mathbb{R}}f(p)\colon T_p\mathbb{C}^n \to T_{f(p)}\mathbb{C}^m$. Given the basis above, this mapping is represented by the standard real Jacobian matrix, that is, a real $2m \times 2n$ matrix that we wrote before as $D_{\mathbb{R}}f(p)$. As a basis for $T_p\mathbb{C}^n$ is a basis for $\mathbb{C}T_p\mathbb{C}^n$, the mapping $D_{\mathbb{R}}f(p)\colon T_p\mathbb{C}^n \to T_{f(p)}\mathbb{C}^m$ naturally uniquely extends to

$$D_{\mathbb{C}}f(p)\colon \mathbb{C}T_p\mathbb{C}^n \to \mathbb{C}T_{f(p)}\mathbb{C}^m.$$

Proposition 2.3.1. *Let $f\colon U \subset \mathbb{C}^n \to \mathbb{C}^m$ be a holomorphic mapping with $p \in U$. Then*

$$D_{\mathbb{C}}f(p)\left(T_p^{(1,0)}\mathbb{C}^n\right) \subset T_{f(p)}^{(1,0)}\mathbb{C}^m \quad \text{and} \quad D_{\mathbb{C}}f(p)\left(T_p^{(0,1)}\mathbb{C}^n\right) \subset T_{f(p)}^{(0,1)}\mathbb{C}^m.$$

If f is a biholomorphism, then $D_{\mathbb{C}}f(p)$ restricted to $T_p^{(1,0)}\mathbb{C}^n$ is a vector space isomorphism. Similarly for $T_p^{(0,1)}\mathbb{C}^n$.

> **Exercise 2.3.1:** *Prove the proposition. Hint: Start with $D_{\mathbb{R}}f(p)$ as a real $2m \times 2n$ matrix to show it extends (it is the same matrix if you think of it as a matrix and use the same basis vectors). Think of $\mathbb{C}^n$ and $\mathbb{C}^m$ in terms of the zs and the $\bar{z}$s and think of f as a mapping*
>
> $$(z, \bar{z}) \mapsto \left(f(z), \bar{f}(\bar{z})\right).$$
>
> *Write the derivative as a matrix in terms of the zs and the $\bar{z}$s and fs and $\bar{f}$s and the result will follow. That is just changing the basis.*
>
> **Exercise 2.3.2:** *Prove a converse to the proposition. If $f\colon U \subset \mathbb{C}^n \to \mathbb{C}^m$ is a smooth mapping such that $D_{\mathbb{C}}f(p)\left(T_p^{(1,0)}\mathbb{C}^n\right) \subset T_{f(p)}^{(1,0)}\mathbb{C}^m$ at every $p \in U$, then f is holomorphic.*

For holomorphic mappings and holomorphic vectors, when we say "derivative of f," we mean the holomorphic part of the derivative, which we write as

$$Df(p)\colon T_p^{(1,0)}\mathbb{C}^n \to T_{f(p)}^{(1,0)}\mathbb{C}^m, \qquad Df(p) = D_{\mathbb{C}}f(p)\big|_{T_p^{(1,0)}\mathbb{C}^n}.$$

That is, we restrict $D_{\mathbb{C}}f(p)$ to $T_p^{(1,0)}\mathbb{C}^n$. Let z be the coordinates on $\mathbb{C}^n$ and w the coordinates on $\mathbb{C}^m$. In the bases $\left\{\frac{\partial}{\partial z_1}\big|_p, \ldots, \frac{\partial}{\partial z_n}\big|_p\right\}$ and $\left\{\frac{\partial}{\partial w_1}\big|_{f(p)}, \ldots, \frac{\partial}{\partial w_m}\big|_{f(p)}\right\}$, the holomorphic derivative $Df(p)$ is represented by the $m \times n$ Jacobian matrix

$$\left[\frac{\partial f_k}{\partial z_\ell}\Big|_p\right]_{k\ell},$$

which we have seen in section 1.3 and for which we also used the notation $Df(p)$.

As before, define the tangent bundles

$$\mathbb{C}T\mathbb{C}^n, \quad T^{(1,0)}\mathbb{C}^n, \quad \text{and} \quad T^{(0,1)}\mathbb{C}^n$$

by taking the disjoint unions. One can also define vector fields in these bundles.

Let us describe $\mathbb{C}T_pM$ for a smooth real hypersurface $M \subset \mathbb{C}^n$. Let r be a real-valued defining function of M at p. A vector $X_p \in \mathbb{C}T_p\mathbb{C}^n$ is in $\mathbb{C}T_pM$ whenever $X_p r = 0$. That is,

$$X_p = \sum_{k=1}^n \left(a_k \frac{\partial}{\partial z_k}\Big|_p + b_k \frac{\partial}{\partial \bar{z}_k}\Big|_p\right) \in \mathbb{C}T_pM \quad \text{whenever} \quad \sum_{k=1}^n \left(a_k \frac{\partial r}{\partial z_k}\Big|_p + b_k \frac{\partial r}{\partial \bar{z}_k}\Big|_p\right) = 0.$$

Therefore, $\mathbb{C}T_pM$ is a $(2n-1)$-dimensional complex vector space. We decompose $\mathbb{C}T_pM$ as

$$\mathbb{C}T_pM = T_p^{(1,0)}M \oplus T_p^{(0,1)}M \oplus B_p,$$

where

$$T_p^{(1,0)}M \stackrel{\text{def}}{=} (\mathbb{C}T_pM) \cap \left(T_p^{(1,0)}\mathbb{C}^n\right) \quad \text{and} \quad T_p^{(0,1)}M \stackrel{\text{def}}{=} (\mathbb{C}T_pM) \cap \left(T_p^{(0,1)}\mathbb{C}^n\right).$$

The B_p is the "leftover" and must be included for the dimensions to work out.*

Exercise 2.3.3: *Prove that there is another way of getting at these spaces. Consider a smooth hypersurface M and $p \in M$. Let J be the linear map of $T_p\mathbb{C}^n$ to itself that corresponds to multiplication by i (the derivative of the actual multiplication by i). Write $T_p^cM = J(T_pM) \cap T_pM$ (the subspace fixed by J, sometimes called the* **complex tangent space** *despite being a real vector space). The map J restricts to an endomorphism of T_p^cM and thus it naturally induces an endomorphism of $\mathbb{C}T_p^cM$. Then $T_p^{(1,0)}M$ and $T_p^{(0,1)}M$ are the eigenspaces of J, which has eigenvalues $\pm i$.*

Make sure to notice what sort of vector spaces these are. The space T_pM is a real vector space; $\mathbb{C}T_pM$, $T_p^{(1,0)}M$, $T_p^{(0,1)}M$, and B_p are complex vector spaces. To see that these give vector bundles, we must first show that their dimensions do not vary

---

*The B_p is sometimes colloquially called the "bad direction."

from point to point. The easiest way to see this fact is to write down convenient local coordinates. First, let us note that a biholomorphic map preserves the tangent holomorphic and antiholomorphic vectors. That is, we get the following analogue of Proposition 2.2.6. Note that a biholomorphic map is a diffeomorphism.

Proposition 2.3.2. *Suppose $M \subset \mathbb{C}^n$ is a smooth real hypersurface, $p \in M$, and $U \subset \mathbb{C}^n$ is open with $M \subset U$, and suppose $M' \subset \mathbb{C}^m$ a smooth real hypersurface. Let $f : U \to \mathbb{C}^m$ be holomorphic such that $f(M) \subset M'$. Let $D_{\mathbb{C}} f(p)$ be the complexified real derivative as before. Then*

$$D_{\mathbb{C}} f(p)\left(T_p^{(1,0)} M\right) \subset T_{f(p)}^{(1,0)} M', \qquad D_{\mathbb{C}} f(p)\left(T_p^{(0,1)} M\right) \subset T_{f(p)}^{(0,1)} M'.$$

Moreover, if $m = n$ and f is a biholomorphism, then $f(M)$ is a smooth real hypersurface, $D_{\mathbb{C}} f(p)$ is invertible, $D_{\mathbb{C}} f(p)\left(T_p^{(1,0)} M\right) = T_{f(p)}^{(1,0)} f(M)$ and $D_{\mathbb{C}} f(p)\left(T_p^{(0,1)} M\right) = T_{f(p)}^{(0,1)} f(M)$. That is, the spaces are isomorphic as complex vector spaces.

The proposition is local; if U is only a neighborhood of p, replace M with $M \cap U$.

Proof. Apply Proposition 2.3.1 and Proposition 2.2.6. That is,

$$D_{\mathbb{C}} f(p)\left(T_p^{(1,0)} \mathbb{C}^n\right) \subset T_{f(p)}^{(1,0)} \mathbb{C}^m, \quad D_{\mathbb{C}} f(p)\left(T_p^{(0,1)} \mathbb{C}^n\right) \subset T_{f(p)}^{(0,1)} \mathbb{C}^m, \quad \text{and}$$

$$D_{\mathbb{C}} f(p)\left(\mathbb{C} T_p M\right) \subset \mathbb{C} T_{f(p)} M'.$$

Then $D_{\mathbb{C}} f(p)$ must take $T_p^{(1,0)} M$ to $T_{f(p)}^{(1,0)} M'$ and $T_p^{(0,1)} M$ to $T_{f(p)}^{(0,1)} M'$. The "Moreover" follows from the "Moreover" of Proposition 2.2.6. $\qquad\square$

We again wish to write a hypersurface as a graph. In this context, the right sort of transformations are biholomorphic transformations. Translations are biholomorphic, and the rotation we will want to use is applying a unitary matrix to $\mathbb{C}^n$.

Proposition 2.3.3. *Let $M \subset \mathbb{C}^n$ be a smooth real hypersurface, $p \in M$. After a translation and a rotation by a unitary matrix, p is the origin, and near the origin, M is written in variables $(z, w) \in \mathbb{C}^{n-1} \times \mathbb{C}$ as*

$$\operatorname{Im} w = \varphi(z, \bar{z}, \operatorname{Re} w),$$

with $\varphi(0) = 0$ and $d\varphi(0) = 0$. Consequently,

$$T_0^{(1,0)} M = \operatorname{span}_{\mathbb{C}} \left\{ \frac{\partial}{\partial z_1}\Big|_0, \dots, \frac{\partial}{\partial z_{n-1}}\Big|_0 \right\}, \qquad T_0^{(0,1)} M = \operatorname{span}_{\mathbb{C}} \left\{ \frac{\partial}{\partial \bar{z}_1}\Big|_0, \dots, \frac{\partial}{\partial \bar{z}_{n-1}}\Big|_0 \right\},$$

$$B_0 = \operatorname{span}_{\mathbb{C}} \left\{ \frac{\partial}{\partial (\operatorname{Re} w)}\Big|_0 \right\}.$$

In particular, $\dim_{\mathbb{C}} T_p^{(1,0)} M = \dim_{\mathbb{C}} T_p^{(0,1)} M = n - 1$ and $\dim_{\mathbb{C}} B_p = 1$.

If M is the boundary of a open set U with smooth boundary, the rotation can be chosen so that $\operatorname{Im} w > \varphi(z, \bar{z}, \operatorname{Re} w)$ on U.

Remark the notation $\varphi(z, \bar{z}, \operatorname{Re} w)$, where we are using the $z, \bar{z}$ notation for the z directions, but since φ does not depend on $\operatorname{Im} w$, we cannot do the same with the w.

Proof. Apply a translation to put $p = 0$ and in the same manner as in Proposition 2.2.9 apply a unitary matrix to make sure that ∇r is in the direction $-\frac{\partial}{\partial(\operatorname{Im} w)}\big|_0$. That $\varphi(0) = 0$ and $d\varphi(0) = 0$ follows as before. A translation and a unitary matrix are holomorphic and, in fact, biholomorphic, so via Proposition 2.3.2 the tangent spaces are all transformed correctly. The rest of the proposition follows at once as $\frac{\partial}{\partial(\operatorname{Im} w)}\big|_0$ is the normal vector to M at the origin.   □

Remark 2.3.4. When M is of dimension less than $2n - 1$ (not a hypersurface anymore), the conclusion of the proposition on the dimensions does not hold. That is, we still have $\dim_{\mathbb{C}} T_p^{(1,0)} M = \dim_{\mathbb{C}} T_p^{(0,1)} M$, but this number need not be constant from point to point. Fortunately, the boundaries of domains with smooth boundaries are by definition hypersurfaces and this complication does not arise.

Definition 2.3.5. Suppose $U \subset \mathbb{C}^n$ is an open set with smooth boundary, and r is a defining function for ∂U at $p \in \partial U$ such that $r < 0$ on U. If

$$\sum_{k=1,\ell=1}^{n} \bar{a}_k a_\ell \frac{\partial^2 r}{\partial \bar{z}_k \partial z_\ell}\bigg|_p \geq 0 \qquad \text{for all} \qquad X_p = \sum_{k=1}^{n} a_k \frac{\partial}{\partial z_k}\bigg|_p \quad \in \quad T_p^{(1,0)} \partial U,$$

then U is *pseudoconvex* at p (or *Levi pseudoconvex*). If the inequality above is strict for all nonzero $X_p \in T_p^{(1,0)} \partial U$, then U is *strongly pseudoconvex* at p. If U is pseudoconvex, but not strongly pseudoconvex, at p, then U is *weakly pseudoconvex*.

A domain U is *pseudoconvex* if it is pseudoconvex at all $p \in \partial U$. For a bounded* U, we say U is *strongly pseudoconvex* if it is strongly pseudoconvex at all $p \in \partial U$.

For $X_p \in T_p^{(1,0)} \partial U$, the Hermitian quadratic form

$$\mathcal{L}(X_p, X_p) = \sum_{k=1,\ell=1}^{n} \bar{a}_k a_\ell \frac{\partial^2 r}{\partial \bar{z}_k \partial z_\ell}\bigg|_p$$

is called the *Levi form* at p. So U is pseudoconvex (resp. strongly pseudoconvex) at $p \in \partial U$ if the Levi form is positive semidefinite (resp. positive definite) at p. The Levi form can be defined for any real hypersurface M, although one has to decide which side of M is "the inside."

The matrix

$$\left[\frac{\partial^2 r}{\partial \bar{z}_k \partial z_\ell}\bigg|_p\right]_{k\ell}$$

is called the *complex Hessian* of r at p.[†] So, U is pseudoconvex at $p \in \partial U$ if the complex Hessian of r at p as a Hermitian form is positive (semi)definite when restricted to

---

*The definition for unbounded domains is not consistent in the literature. Sometimes *strictly pseudoconvex* is used.

[†]People sometimes call the complex Hessian the "Levi form of r," which is incorrect. The Levi form is something defined for a boundary or a hypersurface acting only on its tangent vectors.

tangent vectors in $T_p^{(1,0)} \partial U$. For example, the unit ball $\mathbb{B}_n$ is strongly pseudoconvex as can be seen by computing the Levi form directly from $r(z, \bar{z}) = \|z\|^2 - 1$, that is, the complex Hessian of r is the identity matrix.

We remark that the complex Hessian is not the full (real) Hessian. Let us write down the full Hessian, using the basis of $\frac{\partial}{\partial z}$s and $\frac{\partial}{\partial \bar{z}}$s. It is the Hermitian matrix

$$
\begin{bmatrix}
\frac{\partial^2 r}{\partial \bar{z}_1 \partial z_1} & \cdots & \frac{\partial^2 r}{\partial \bar{z}_1 \partial z_n} & \frac{\partial^2 r}{\partial \bar{z}_1 \partial \bar{z}_1} & \cdots & \frac{\partial^2 r}{\partial \bar{z}_1 \partial \bar{z}_n} \\
\vdots & \ddots & \vdots & \vdots & \ddots & \vdots \\
\frac{\partial^2 r}{\partial \bar{z}_n \partial z_1} & \cdots & \frac{\partial^2 r}{\partial \bar{z}_n \partial z_n} & \frac{\partial^2 r}{\partial \bar{z}_n \partial \bar{z}_1} & \cdots & \frac{\partial^2 r}{\partial \bar{z}_n \partial \bar{z}_n} \\
\frac{\partial^2 r}{\partial z_1 \partial z_1} & \cdots & \frac{\partial^2 r}{\partial z_1 \partial z_n} & \frac{\partial^2 r}{\partial z_1 \partial \bar{z}_1} & \cdots & \frac{\partial^2 r}{\partial z_1 \partial \bar{z}_n} \\
\vdots & \ddots & \vdots & \vdots & \ddots & \vdots \\
\frac{\partial^2 r}{\partial z_n \partial z_1} & \cdots & \frac{\partial^2 r}{\partial z_n \partial z_n} & \frac{\partial^2 r}{\partial z_n \partial \bar{z}_1} & \cdots & \frac{\partial^2 r}{\partial z_n \partial \bar{z}_n}
\end{bmatrix}.
$$

To make it a Hermitian form, note that when multiplying on the left by X_p we are also taking the conjugate so the rows for the zs and the $\bar{z}$s are flipped.* Note that it is Hermitian only for a real-valued r (see an exercise below). The complex Hessian is the upper left, or the transpose of the lower right, block—if you write the full Hessian as $\begin{bmatrix} L & \bar{Z} \\ Z & L^t \end{bmatrix}$, then L is the complex Hessian. Note that L is a smaller matrix and we apply it only to a subspace of the complexified tangent space.

We illustrate the change of basis in one dimension, and leave higher dimensions to the student. Let $z = x + iy$ be in $\mathbb{C}$, and denote by T the change of basis matrix:

$$
T = \begin{bmatrix} 1/2 & 1/2 \\ -i/2 & i/2 \end{bmatrix}, \qquad T^* \begin{bmatrix} \frac{\partial^2 r}{\partial x \partial x} & \frac{\partial^2 r}{\partial x \partial y} \\ \frac{\partial^2 r}{\partial y \partial x} & \frac{\partial^2 r}{\partial y \partial y} \end{bmatrix} T = \begin{bmatrix} \frac{\partial^2 r}{\partial \bar{z} \partial z} & \frac{\partial^2 r}{\partial \bar{z} \partial \bar{z}} \\ \frac{\partial^2 r}{\partial z \partial z} & \frac{\partial^2 r}{\partial z \partial \bar{z}} \end{bmatrix},
$$

where $T^* = \bar{T}^t$ is the conjugate transpose. By Sylvester's law of inertia from linear algebra, star-congruence preserves the inertia (the number of positive, negative, and zero eigenvalues). So the inertia of the full Hessian in terms of xs and ys is the same as for the full Hessian in terms of zs and $\bar{z}$s. The relationship between the eigenvalues of the full Hessian and the complex Hessian is not as straightforward as may at first seem, but there is a relationship there nonetheless.

Exercise 2.3.4 (Easy): *If r is real-valued, then both the complex Hessian of r and the full Hessian in terms of zs and $\bar{z}$ are Hermitian matrices.*

Exercise 2.3.5: *Consider one dimension, $z = x + iy$, and the real Hessian in terms of x and y:*

$$
H = \begin{bmatrix} \frac{\partial^2 r}{\partial x \partial x} & \frac{\partial^2 r}{\partial x \partial y} \\ \frac{\partial^2 r}{\partial y \partial x} & \frac{\partial^2 r}{\partial y \partial y} \end{bmatrix}.
$$

*It is common to also write it not flipped, in which case it will be a symmetric matrix.

Prove that the complex Hessian L (a number now) is ¹/₄ of the trace of H. Thus, if H is positive definite, then L > 0, and if H is negative definite, then L < 0. Then show by example that if H has mixed eigenvalues (positive and negative), then L can be positive, negative, or zero.

Exercise 2.3.6: *For every dimension, find the change of variables T^*HT to go from the real Hessian in terms of x and y to the Hessian in terms of z and z̄. Hint: If you figure it out for n = 2, it will be easy to do in general.*

Exercise 2.3.7: *Prove in every dimension that if the real Hessian (in terms of x and y) is positive (semi)definite, then the complex Hessian is positive (semi)definite. Hint: A Hermitian matrix L is positive definite if $v^*Lv > 0$ for all nonzero vectors v and semidefinite if $v^*Lv \geq 0$ for all v.*

Let us also see how a complex linear change of variables acts on the Hessian. A complex linear mapping A as an $n \times n$ complex matrix transforms the tangent space in the basis of $\frac{\partial}{\partial z}$s and $\frac{\partial}{\partial \bar{z}}$s via the derivative $D_{\mathbb{C}}A$ written as a $2n \times 2n$ matrix. A direct computation shows $D_{\mathbb{C}}A = A \oplus \bar{A} = \begin{bmatrix} A & 0 \\ 0 & \bar{A} \end{bmatrix}$. Write the full Hessian as $\begin{bmatrix} L & \bar{Z} \\ Z & L^t \end{bmatrix}$, where L is the complex Hessian. The complex linear change of variables A transforms the full Hessian as

$$\begin{bmatrix} A & 0 \\ 0 & \bar{A} \end{bmatrix}^* \begin{bmatrix} L & \bar{Z} \\ Z & L^t \end{bmatrix} \begin{bmatrix} A & 0 \\ 0 & \bar{A} \end{bmatrix} = \begin{bmatrix} A^*LA & \overline{A^tZA} \\ A^tZA & (A^*LA)^t \end{bmatrix},$$

Again by Sylvester's law of inertia, L and A^*LA have the same inertia, that is, the number of positive, negative, and zero eigenvalues.

The Levi form itself does depend on the defining function, but the signs of the eigenvalues do not. It is common to say "the Levi form" without mentioning a specific defining function even though that is not completely correct. The proof of the following proposition is left as an exercise.

Proposition 2.3.6. *If $U \subset \mathbb{C}^n$ is an open set with smooth boundary and $p \in \partial U$, then the inertia of the Levi form at p does not depend on the defining function. Consequently, the notion of pseudoconvexity and strong pseudoconvexity is independent of the defining function.*

Exercise 2.3.8: *Prove Proposition 2.3.6.*

Exercise 2.3.9: *Show that a convex domain with smooth boundary is pseudoconvex, and show that (a bounded) strongly convex domain with smooth boundary is strongly pseudoconvex.*

Exercise 2.3.10: *Show that if an open set with smooth boundary is strongly pseudoconvex at a point, it is strongly pseudoconvex at all nearby points.*

We are generally interested what happens under a holomorphic change of co-ordinates, that is, a biholomorphic mapping. And as far as pseudoconvexity is concerned we are interested in local changes of coordinates as pseudoconvexity is a local property. Before proving that pseudoconvexity is a biholomorphic invariant, let us note where the Levi form appears in the graph coordinates from Proposition 2.3.3, that is, when our boundary (the hypersurface) is given near the origin by

$$\operatorname{Im} w = \varphi(z, \bar{z}, \operatorname{Re} w),$$

where φ is $O(2)$. Let $r(z, \bar{z}, w, \bar{w}) = \varphi(z, \bar{z}, \operatorname{Re} w) - \operatorname{Im} w$ be our defining function. The complex Hessian of r (an $n \times n$ matrix) has the form

$$\begin{bmatrix} L & * \\ * & * \end{bmatrix} \qquad \text{where} \quad L = \left[\frac{\partial^2 \varphi}{\partial \bar{z}_k \partial z_\ell} \Big|_0 \right]_{k\ell}.$$

Note that L is an $(n-1) \times (n-1)$ matrix. The vectors in $T_0^{(1,0)} \partial U$ are the span of $\left\{ \frac{\partial}{\partial z_1} \big|_0, \ldots, \frac{\partial}{\partial z_{n-1}} \big|_0 \right\}$. That is, as an n-vector, a vector in $T_0^{(1,0)} \partial U$ is represented by $(a, 0) \in \mathbb{C}^n$ for some $a \in \mathbb{C}^{n-1}$. The Levi form at the origin is then $a^* L a$, in other words, it is given by the $(n-1) \times (n-1)$ matrix L. If this matrix L is positive semidefinite, then ∂U is pseudoconvex at 0.

Example 2.3.7: Let us change variables to write the ball $\mathbb{B}_n$ in different local holomorphic coordinates where the Levi form is displayed nicely. The sphere $\partial \mathbb{B}_n$ is defined in the variables $Z = (Z_1, \ldots, Z_n) \in \mathbb{C}^n$ by $\|Z\| = 1$. We change variables to $(z_1, \ldots, z_{n-1}, w)$ where

$$z_k = \frac{Z_k}{1 - Z_n} \quad \text{for all } k = 1, \ldots, n-1, \qquad w = i \frac{1 + Z_n}{1 - Z_n}.$$

This change of variables is a biholomorphic map from the set where $Z_n \neq 1$ to the set where $w \neq -i$ (exercise). For us, it suffices that the map is invertible near $(0, \ldots, 0, -1)$, which follows by computing the derivative. Notice that the last component is the inverse of the Cayley transform (which takes the disc to the upper half-plane).

We claim that the mapping takes the unit sphere given by $\|Z\| = 1$ (without the point $(0, \ldots, 0, 1)$), to the set defined by

$$\operatorname{Im} w = |z_1|^2 + \cdots + |z_{n-1}|^2,$$

and that it takes $(0, \ldots, 0, -1)$ to the origin (this part is trivial). Let us check:

$$
\begin{aligned}
|z_1|^2 + \cdots + |z_{n-1}|^2 - \operatorname{Im} w &= \left| \frac{Z_1}{1 - Z_n} \right|^2 + \cdots + \left| \frac{Z_{n-1}}{1 - Z_n} \right|^2 - \frac{i \frac{1+Z_n}{1-Z_n} - \overline{i \frac{1+Z_n}{1-Z_n}}}{2i} \\
&= \frac{|Z_1|^2}{|1 - Z_n|^2} + \cdots + \frac{|Z_{n-1}|^2}{|1 - Z_n|^2} - \frac{1 + Z_n}{2(1 - Z_n)} - \frac{1 + \bar{Z}_n}{2(1 - \bar{Z}_n)} \\
&= \frac{|Z_1|^2 + \cdots + |Z_{n-1}|^2 + |Z_n|^2 - 1}{|1 - Z_n|^2}.
\end{aligned}
$$

Therefore, $|Z_1|^2 + \cdots + |Z_n|^2 = 1$ if and only if $\operatorname{Im} w = |z_1|^2 + \cdots + |z_{n-1}|^2$. As the map takes the point $(0, \ldots, 0, -1)$ to the origin, we can think of the set given by

$$\operatorname{Im} w = |z_1|^2 + \cdots + |z_{n-1}|^2$$

as the sphere in local holomorphic coordinates at $(0, \ldots, 0, -1)$ (by symmetry of the sphere we could have done this at any point by rotation). In the coordinates (z, w), the ball (the inside of the sphere) is the set given by

$$\operatorname{Im} w > |z_1|^2 + \cdots + |z_{n-1}|^2.$$

In these coordinates, the Levi form is just the identity matrix at the origin, and so the domain is strongly pseudoconvex at the origin. We will prove that (strong) pseudo-convexity is a biholomorphic invariant, and so the ball is strongly pseudoconvex.

Not the entire sphere gets transformed, the points where $Z_n = 1$ get "sent to infinity." The hypersurface $\operatorname{Im} w = |z_1|^2 + \cdots + |z_{n-1}|^2$ is sometimes called the *Lewy hypersurface*, and in the literature some even say it *is* the sphere*. Pretending z is just one real direction, see Figure 2.7. As an aside, the hypersurface $\operatorname{Im} w = |z_1|^2 + \cdots + |z_{n-1}|^2$ is also called the *Heisenberg group*. The group in this case is defined on the parameters $(z, \operatorname{Re} w)$ of this hypersurface with the group law $(z, \operatorname{Re} w)(z', \operatorname{Re} w') = (z + z', \operatorname{Re} w + \operatorname{Re} w' + 2\operatorname{Im}\langle z, z'\rangle)$.

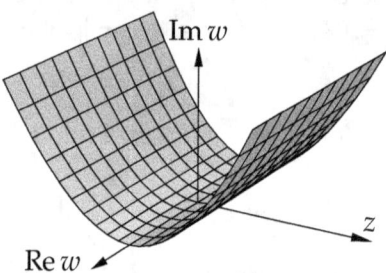

Figure 2.7: Lewy hypersurface.

Exercise 2.3.11: Prove the assertion in the example about the mapping being biholomorphic on the sets described above.

Let us see how the Hessian of r changes under a biholomorphic change of coordinates. Let $f \colon V \to V'$ be a biholomorphic map between two domains in $\mathbb{C}^n$, and let $r \colon V' \to \mathbb{R}$ be a smooth function with nonvanishing derivative. Let us compute the Hessian of $r \circ f \colon V \to \mathbb{R}$. We first compute what happens to the

---

*That is not, in fact, completely incorrect. If we think of the sphere in the complex projective space, we are simply looking at the sphere in a different coordinate patch.

nonmixed derivatives. As we have to apply chain rule twice, to keep better track of things, we write where the derivatives are being evaluated inside the computation, as they are, after all, functions. For clarity, let z be the coordinates in V and ζ the coordinates in V'. That is, r is a function of ζ and $\bar{\zeta}$, f is a function of z, and $\bar{f}$ is a function of $\bar{z}$. So $r \circ f$ is a function of z and $\bar{z}$.

$$
\begin{aligned}
\frac{\partial^2 (r \circ f)}{\partial z_k \partial z_\ell} &= \frac{\partial}{\partial z_k} \sum_{m=1}^{n} \left(\frac{\partial r}{\partial \zeta_m}\bigg|_{(f(z),\bar{f}(\bar{z}))} \frac{\partial f_m}{\partial z_\ell}\bigg|_z + \frac{\partial r}{\partial \bar{\zeta}_m}\bigg|_{(f(z),\bar{f}(\bar{z}))} \underbrace{\frac{\partial \bar{f}_m}{\partial z_\ell}\bigg|_{\bar{z}}}_{0} \right) \\
&= \sum_{m,v=1}^{n} \left(\frac{\partial^2 r}{\partial \zeta_v \partial \zeta_m}\bigg|_{(f(z),\bar{f}(\bar{z}))} \frac{\partial f_v}{\partial z_k}\bigg|_z \frac{\partial f_m}{\partial z_\ell}\bigg|_z + \frac{\partial^2 r}{\partial \bar{\zeta}_v \partial \zeta_m}\bigg|_{(f(z),\bar{f}(\bar{z}))} \underbrace{\frac{\partial \bar{f}_v}{\partial z_k}\bigg|_{\bar{z}}}_{0} \frac{\partial f_m}{\partial z_\ell}\bigg|_z \right) \\
&\quad + \sum_{m=1}^{n} \frac{\partial r}{\partial \zeta_m}\bigg|_{(f(z),\bar{f}(\bar{z}))} \frac{\partial^2 f_m}{\partial z_k \partial z_\ell}\bigg|_z \\
&= \sum_{m,v=1}^{n} \frac{\partial^2 r}{\partial \zeta_v \partial \zeta_m} \frac{\partial f_v}{\partial z_k} \frac{\partial f_m}{\partial z_\ell} + \sum_{m=1}^{n} \frac{\partial r}{\partial \zeta_m} \frac{\partial^2 f_m}{\partial z_k \partial z_\ell}.
\end{aligned}
$$

The matrix $\left[\frac{\partial^2 (r \circ f)}{\partial z_k \partial z_\ell} \right]$ can have different eigenvalues than the matrix $\left[\frac{\partial^2 r}{\partial \zeta_k \partial \zeta_\ell} \right]$. If r has nonvanishing gradient, then using the second term, we can (locally) choose f in such a way as to make the matrix $\left[\frac{\partial^2 (r \circ f)}{\partial z_k \partial z_\ell} \right]$ be the zero matrix (or anything else) at a certain point as we can choose the second derivatives of f arbitrarily at that point. See the exercise below. Nothing about the matrix $\left[\frac{\partial^2 r}{\partial \zeta_k \partial \zeta_\ell} \right]$ is preserved under a biholomorphic map. And that is precisely why it does not appear in the definition of pseudoconvexity. The story for $\left[\frac{\partial^2 (r \circ f)}{\partial \bar{z}_k \partial \bar{z}_\ell} \right]$ and $\left[\frac{\partial^2 r}{\partial \bar{\zeta}_k \partial \bar{\zeta}_\ell} \right]$ is exactly the same.

> *Exercise 2.3.12:* Given a real function r with nonvanishing gradient at $p \in \mathbb{C}^n$. Find a local change of coordinates f at p (so f ought to be a holomorphic mapping with an invertible derivative at p) such that $\left[\frac{\partial^2 (r \circ f)}{\partial z_k \partial z_\ell}\bigg|_p \right]$ and $\left[\frac{\partial^2 (r \circ f)}{\partial \bar{z}_k \partial \bar{z}_\ell}\bigg|_p \right]$ are just the zero matrices.

Let us look at the mixed derivatives:

$$
\begin{aligned}
\frac{\partial^2 (r \circ f)}{\partial \bar{z}_k \partial z_\ell} &= \frac{\partial}{\partial \bar{z}_k} \sum_{m=1}^{n} \left(\frac{\partial r}{\partial \zeta_m}\bigg|_{(f(z),\bar{f}(\bar{z}))} \frac{\partial f_m}{\partial z_\ell}\bigg|_z \right) \\
&= \sum_{m,v=1}^{n} \frac{\partial^2 r}{\partial \bar{\zeta}_v \partial \zeta_m}\bigg|_{(f(z),\bar{f}(\bar{z}))} \frac{\partial \bar{f}_v}{\partial \bar{z}_k}\bigg|_{\bar{z}} \frac{\partial f_m}{\partial z_\ell}\bigg|_z + \sum_{m=1}^{n} \frac{\partial r}{\partial \zeta_m}\bigg|_{(f(z),\bar{f}(\bar{z}))} \underbrace{\frac{\partial^2 f_m}{\partial \bar{z}_k \partial z_\ell}\bigg|_z}_{0} \\
&= \sum_{m,v=1}^{n} \frac{\partial^2 r}{\partial \bar{\zeta}_v \partial \zeta_m} \frac{\partial \bar{f}_v}{\partial \bar{z}_k} \frac{\partial f_m}{\partial z_\ell}.
\end{aligned}
$$

The complex Hessian of $r \circ f$ is the complex Hessian L of r conjugated as D^*LD, where D is the holomorphic derivative matrix of f at z and D^* is its conjugate transpose. Sylvester's law of inertia says that the number of positive, negative, and zero eigenvalues of D^*LD is the same as that for L. The eigenvalues may change, but their signs do not. We are only considering L and D^*LD on a subspace. In linear algebra language, consider an invertible D, a subspace T, and its image DT. Then the inertia of L restricted to DT is the same as the inertia of D^*LD restricted to T.

Let M be a smooth real hypersurface given by $r = 0$, then $f^{-1}(M)$ is a smooth real hypersurface given by $r \circ f = 0$. The holomorphic derivative $D = Df(p)$ takes $T_p^{(1,0)} f^{-1}(M)$ isomorphically to $T_{f(p)}^{(1,0)} M$. So L is positive (semi)definite on $T_{f(p)}^{(1,0)} M$ if and only if D^*LD is positive (semi)definite on $T_p^{(1,0)} f^{-1}(M)$. We have almost proved the following theorem. In short, pseudoconvexity is a biholomorphic invariant.

Theorem 2.3.8. *Suppose $U, U' \subset \mathbb{C}^n$ are open sets with smooth boundary, $p \in \partial U$, $V \subset \mathbb{C}^n$ a neighborhood of p, $q \in \partial U'$, $V' \subset \mathbb{C}^n$ a neighborhood of q, and $f : V \to V'$ a biholomorphic map with $f(p) = q$, such that $f(U \cap V) = U' \cap V'$. See Figure 2.8.*

Then the inertia of the Levi form of U at p is the same as the inertia of the Levi form of U' at q. In particular, U is pseudoconvex at p if and only if U' is pseudoconvex at q. Similarly, U is strongly pseudoconvex at p if and only if U' is strongly pseudoconvex at q.

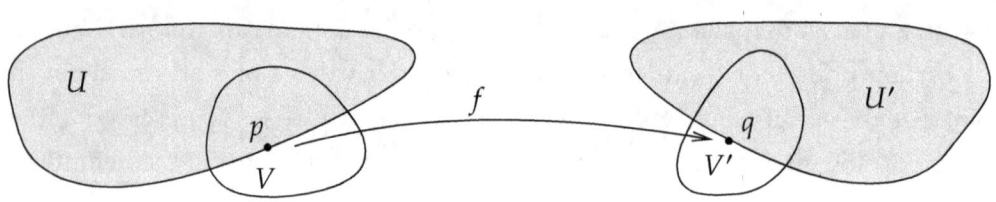

Figure 2.8: Local boundary biholomorphism.

To finish proving the theorem, the only thing left is to observe that if $f(U \cap V) = U' \cap V'$, then $f(\partial U \cap V) = \partial U' \cap V'$, and to note that if r is a defining function for U' at q, then $r \circ f$ is a defining function for U at p.

Exercise 2.3.13: Find an example of a bounded domain in $\mathbb{C}^n$, $n \geq 2$, with smooth boundary that is not convex, but that is pseudoconvex.

So while the Levi form is not invariant under holomorphic changes of coordinates, its inertia is. Putting this together with the other observations we made, we find the normal form for the quadratic part of the defining equation for a smooth real hypersurface under biholomorphic transformations. It is possible to do better than the following lemma, but it is not always possible to get rid of the dependence on $\mathrm{Re}\, w$ in the higher order terms.

Lemma 2.3.9. *Let $M \subset \mathbb{C}^n$ be a smooth real hypersurface and $p \in M$. Then there exists a local biholomorphic change of coordinates taking p to 0 and M to the hypersurface given by*

$$\operatorname{Im} w = \sum_{k=1}^{\alpha} |z_k|^2 - \sum_{k=\alpha+1}^{\alpha+\beta} |z_k|^2 + E(z, \bar{z}, \operatorname{Re} w),$$

where E is $O(3)$ at the origin. If M is a boundary, then the coordinates are chosen so that the domain is given by replacing $=$ with $>$ as usual and where α is the number of positive eigenvalues of the Levi form at p, β is the number of negative eigenvalues, and $\alpha + \beta \leq n - 1$.

Recall that $O(\ell)$ at the origin means a function that together with its derivatives up to and including order $\ell - 1$ vanish at the origin.

Proof. Change coordinates so that M is given by $\operatorname{Im} w = \varphi(z, \bar{z}, \operatorname{Re} w)$, where φ is $O(2)$. Apply Taylor's theorem to φ up to the second order:

$$\varphi(z, \bar{z}, \operatorname{Re} w) = q(z, \bar{z}) + (\operatorname{Re} w)(Lz + \overline{Lz}) + a(\operatorname{Re} w)^2 + O(3),$$

where q is quadratic, $L \colon \mathbb{C}^{n-1} \to \mathbb{C}$ is linear, and $a \in \mathbb{R}$. If $L \neq 0$, do a linear change of coordinates in the z only to make $Lz = z_1$. So assume $Lz = \epsilon z_1$ where $\epsilon = 0$ or $\epsilon = 1$.

Change coordinates by leaving z unchanged and letting $w = w' + bw'^2 + cw'z_1$. Ignore $q(z, \bar{z})$ for a moment, as this change of coordinates does not affect it. Also, only work up to second order.

$$- \operatorname{Im} w + \epsilon(\operatorname{Re} w)(z_1 + \bar{z}_1) + a(\operatorname{Re} w)^2$$

$$= -\frac{w - \bar{w}}{2i} + \epsilon \frac{w + \bar{w}}{2}(z_1 + \bar{z}_1) + a\left(\frac{w + \bar{w}}{2}\right)^2$$

$$= -\frac{w' + bw'^2 + cw'z_1 - \bar{w}' - \bar{b}\bar{w}'^2 - \bar{c}\bar{w}'\bar{z}_1}{2i}$$

$$+ \epsilon \frac{w' + bw'^2 + cw'z_1 + \bar{w}' + \bar{b}\bar{w}'^2 + \bar{c}\bar{w}'\bar{z}_1}{2}(z_1 + \bar{z}_1)$$

$$+ a\frac{(w' + bw'^2 + cw'z_1 + \bar{w}' + \bar{b}\bar{w}'^2 + \bar{c}\bar{w}'\bar{z}_1)^2}{4}$$

$$= -\frac{w' - \bar{w}'}{2i} + \frac{((\epsilon i - c)w' + \epsilon i \bar{w}')z_1 + ((\epsilon i + \bar{c})\bar{w}' + \epsilon i w')\bar{z}_1}{2i}$$

$$+ \frac{(ia - 2b)w'^2 + (ia + 2\bar{b})\bar{w}'^2 + 2iaw'\bar{w}'}{4i} + O(3).$$

We cannot quite get rid of all the quadratic terms in φ, but we choose b and c to make the second order terms not depend on $\operatorname{Re} w'$. Set $b = ia$ and $c = 2i\epsilon$, and add $q(z, \bar{z}) + O(3)$ into the mix to get

$$- \operatorname{Im} w + \varphi(z, \bar{z}, \operatorname{Re} w) = - \operatorname{Im} w + q(z, \bar{z}) + \epsilon(\operatorname{Re} w)(z_1 + \bar{z}_1) + a(\operatorname{Re} w)^2 + O(3)$$

$$= -\frac{w' - \bar{w}'}{2i} + q(z, \bar{z}) - \epsilon i \frac{w' - \bar{w}'}{2i}(z_1 - \bar{z}_1) + a\left(\frac{w' - \bar{w}'}{2i}\right)^2 + O(3)$$

$$= - \operatorname{Im} w' + q(z, \bar{z}) - \epsilon i(\operatorname{Im} w')(z_1 - \bar{z}_1) + a(\operatorname{Im} w')^2 + O(3).$$

The right-hand side is the defining function in the (z, w') coordinates. As M is no longer written as a graph of $\operatorname{Im} w'$ over the rest, apply the implicit function theorem to solve for $\operatorname{Im} w'$ and write the hypersurface as a graph again. The expression for $\operatorname{Im} w'$ is $O(2)$, and so $-i\epsilon(\operatorname{Im} w')(z_1 - \bar{z}_1) + a(\operatorname{Im} w')^2$ is $O(3)$. If we write M as a graph,

$$\operatorname{Im} w' = q(z, \bar{z}) + E(z, \bar{z}, \operatorname{Re} w'),$$

then E is $O(3)$.

Write the quadratic polynomial q as

$$q(z, \bar{z}) = \sum_{k,\ell=1}^{n-1} a_{k\ell} z_k z_\ell + b_{k\ell} \bar{z}_k \bar{z}_\ell + c_{k\ell} \bar{z}_k z_\ell. \tag{2.2}$$

The $a_{k\ell}$ and $b_{k\ell}$ are not uniquely determined, but we can pick the matrices $[a_{k\ell}]$ and $[b_{k\ell}]$ to be symmetric to make them uniquely determined. As q is real-valued, it is left as an exercise to show that $a_{k\ell} = \overline{b_{k\ell}}$ and $c_{k\ell} = \overline{c_{\ell k}}$. That is, the matrix $[b_{k\ell}]$ is the complex conjugate of $[a_{k\ell}]$ and $[c_{k\ell}]$ is Hermitian.

We make another change of coordinates. Fix the zs again, and set

$$w' = w'' + 2i \sum_{k,\ell=1}^{n-1} a_{k\ell} z_k z_\ell. \tag{2.3}$$

In particular,

$$\operatorname{Im} w' = \operatorname{Im} w'' + \operatorname{Im}\left(2i \sum_{k,\ell=1}^{n-1} a_{k\ell} z_k z_\ell\right) = \operatorname{Im} w'' + \sum_{k,\ell=1}^{n-1} \left(a_{k\ell} z_k z_\ell + b_{k\ell} \bar{z}_k \bar{z}_\ell\right),$$

as $\overline{a_{k\ell}} = b_{k\ell}$. Plugging (2.3) into $\operatorname{Im} w' = q(z, \bar{z}) + E(z, \bar{z}, \operatorname{Re} w')$ and solving for $\operatorname{Im} w''$ cancels the holomorphic and antiholomorphic terms in q, and leaves E as $O(3)$. After this change of coordinates we may assume that q is a Hermitian form,

$$q(z, \bar{z}) = \sum_{k,\ell=1}^{n-1} c_{k\ell} z_k \bar{z}_\ell.$$

As q is real-valued, the matrix $C = [c_{k\ell}]$ is Hermitian. In linear algebra notation, $q(z, \bar{z}) = z^* C z$, where we think of z as a column vector. If T is a linear transformation on the z variables, say $z' = Tz$, we obtain $z'^* C z' = (Tz)^* C T z = z^* (T^* C T) z$. Thus, we normalize C up to $*$-congruence. A Hermitian matrix is $*$-congruent to a diagonal matrix with only 1s, -1s, and 0s on the diagonal, again by Sylvester's law of inertia. Writing out what that means is precisely the conclusion of the proposition. If M is a boundary, we make sure the interior of the domain is given by $\operatorname{Im} w > \varphi(z, \bar{z}, \operatorname{Re} w)$ by possibly replacing w with $-w$, which reverses the signs of the eigenvalues. $\square$

Exercise **2.3.14:** *Prove the assertions in the proof. First, that if q is a quadratic as in (2.2), then the matrices $[a_{k\ell}]$ and $[b_{k\ell}]$ can be chosen to be symmetric, in which case all the coefficients are uniquely determined. Second, that if q is real valued, then $a_{k\ell} = \overline{b_{k\ell}}$ and $c_{k\ell} = \overline{c_{\ell k}}$ for all k and ℓ.*

Lemma 2.3.10 (Narasimhan's lemma[*]). *Let $U \subset \mathbb{C}^n$ be an open set with smooth boundary that is strongly pseudoconvex at $p \in \partial U$. Then there exists a local biholomorphic change of coordinates fixing p such that in these new coordinates, U is strongly convex at p and hence strongly convex at all points near p.*

Exercise **2.3.15:** *Prove Narasimhan's lemma. Hint: See the proof of Lemma 2.3.9.*

Exercise **2.3.16:** *Prove that an open $U \subset \mathbb{C}^n$ with smooth boundary is pseudoconvex at p if and only if there exist local holomorphic coordinates at p such that U is convex at p.*

To make use of convexity, the domain needs to be convex at all points (all points near p), so Narasimhan's lemma only works at points of strong pseudoconvexity. For weakly pseudoconvex points the situation is far more complicated. While it is possible to use weak pseudoconvexity at p to make the domain convex at p, the same change of variables does not necessarily make the domain convex at nearby points. In particular, it is not always possible for a domain that is weakly pseudoconvex at all points to be made convex in a neighborhood. What makes the lemma work is that if U is strongly (pseudo)convex at p, it will also be so at nearby points.

Let us prove the easy direction of the famous *Levi problem*. The Levi problem was a long-standing problem[†] in several complex variables to classify domains of holomorphy in $\mathbb{C}^n$. The answer is that a domain is a domain of holomorphy if and only if it is pseudoconvex. Just as the problem of trying to show that the classical geometric convexity is the same as convexity as we have defined it, the Levi problem has an easier direction and a harder direction. The easier direction is to show that a domain of holomorphy is pseudoconvex, and the harder direction is to show that a pseudoconvex domain is a domain of holomorphy. See Hörmander's book [H] for the proof of the hard direction.

Theorem 2.3.11 (Tomato can principle). *Suppose $U \subset \mathbb{C}^n$ is an open set with smooth boundary and the Levi form has a negative eigenvalue at $p \in \partial U$. Then every holomorphic function on U extends to a neighborhood of p. In particular, U is not a domain of holomorphy.*

---

[*]A statement essentially of Narasimhan's lemma was already used by Helmut Knesser in 1936.
[†]E. E. Levi stated the problem in 1911, but it was not fully solved until the 1950s, by Oka and others.

Pseudoconvex at p means that all eigenvalues of the Levi form are nonnegative. The theorem says that a domain of holomorphy must be pseudoconvex. The theorem's name comes from the proof, and sometimes other theorems using a similar proof of a "tomato can" of analytic discs are called tomato can principles. The general statement of proof of the principle is that "an analytic function holomorphic in a neighborhood of the sides and the bottom of a tomato can extends to the inside." And the theorem we gave as the principle states that "if the Levi form at p has a negative eigenvalue, we can fit a tomato can from inside the domain over p."

Proof. We change variables so that $p = 0$, and near p, U is given by

$$\operatorname{Im} w > -|z_1|^2 + \sum_{k=2}^{n-1} \epsilon_k |z_k|^2 + E(z_1, z', \bar{z}_1, \bar{z}', \operatorname{Re} w),$$

where $z' = (z_2, \ldots, z_{n-1})$, $\epsilon_k = -1, 0, 1$, and E is $O(3)$. We embed an analytic disc via the map $\xi \in \overline{\mathbb{D}} \overset{\varphi}{\mapsto} (\lambda\xi, 0, 0, \ldots, 0)$ for some small $\lambda > 0$. Clearly $\varphi(0) = 0 \in \partial U$. For $\xi \neq 0$ near the origin

$$-\lambda^2 |\xi|^2 + \sum_{k=2}^{n-1} \epsilon_k |0|^2 + E(\lambda\xi, 0, \lambda\bar{\xi}, 0, 0) = -\lambda^2 |\xi|^2 + E(\lambda\xi, 0, \lambda\bar{\xi}, 0, 0) < 0,$$

because the function above has a strict maximum at $\xi = 0$ by the second derivative test. Therefore, for $\xi \neq 0$ near the origin, $\varphi(\xi) \in U$. By picking λ small enough, $\varphi(\overline{\mathbb{D}} \setminus \{0\}) \subset U$.

We can "wiggle the disc a little" and find discs entirely in U. In particular, for all small enough $s > 0$, the closed disc given by

$$\xi \in \overline{\mathbb{D}} \overset{\varphi_s}{\mapsto} (\lambda\xi, 0, 0, \ldots, 0, is)$$

(that is, for slightly positive $\operatorname{Im} w$) is entirely inside U. Fix such a small $s > 0$. Suppose $\epsilon > 0$ is small and $\epsilon < s$. Define the Hartogs figure

$$H = \big\{(z, w) : \lambda - \epsilon < |z_1| < \lambda + \epsilon, |z_k| < \epsilon \text{ for } k = 2, \ldots, n-1, \text{ and } |w - is| < s + \epsilon\big\}$$
$$\cup \big\{(z, w) : |z_1| < \lambda + \epsilon, |z_k| < \epsilon \text{ for } k = 2, \ldots, n-1, \text{ and } |w - is| < \epsilon\big\}.$$

The set where $|z_1| = \lambda$, $z' = 0$, and $|w - is| \leq s$ is inside U, so an ϵ-neighborhood of that is in U. For $w = is$ the whole disc where $|z_1| \leq \lambda$ and $z' = 0$ is in U, so an ϵ-neighborhood of that is in U. Thus, for small enough $\epsilon > 0$, $H \subset U$. We are really just taking a Hartogs figure in the z_1, w variables, and then "fattening it up" to the z' variables. In Figure 2.9, we picture the Hartogs figure in the $|z_1|$ and $|w - is|$ variables. The boundary ∂U and U are only pictured diagrammatically. Also, we make a "picture" the analytic discs giving the "tomato can." In the picture, the U is below its boundary ∂U, unlike usually.

The origin is in the hull of H, and so every function holomorphic in U, and so in H, extends through the origin. Hence U is not a domain of holomorphy. □

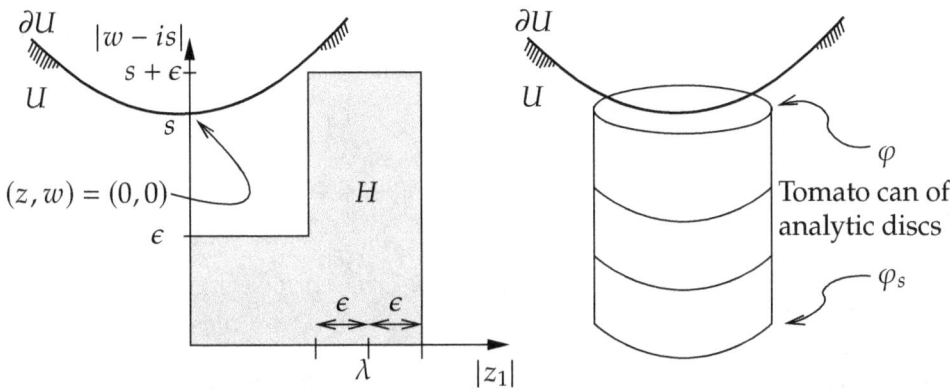

Figure 2.9: Tomato can principle.

Another, perhaps a little less concrete, way to finish the proof that does not use a Hartogs figure is to apply the first version of Kontinuitätssatz (Theorem 2.1.7) with the sequence of discs $\{\varphi_{1/k}\}$.

Exercise 2.3.17: For the following domains in $U \subset \mathbb{C}^2$, find all the points in ∂U where U is weakly pseudoconvex, all the points where it is strongly pseudoconvex, and all the points where it is not pseudoconvex. Is U pseudoconvex?
  a) $\operatorname{Im} w > |z|^4$
  b) $\operatorname{Im} w > |z|^2(\operatorname{Re} w)$
  c) $\operatorname{Im} w > (\operatorname{Re} z)(\operatorname{Re} w)$

Exercise 2.3.18: Let $U \subset \mathbb{C}^n$ be an open set with smooth boundary that is strongly pseudoconvex at $p \in \partial U$. Show that p is a so-called **peak point**: There exists a neighborhood W of p and a holomorphic $f : W \to \mathbb{C}$ such that $f(p) = 1$ and $|f(z)| < 1$ for all $z \in W \cap \overline{U} \setminus \{p\}$.

Exercise 2.3.19: Suppose $U \subset \mathbb{C}^n$ is an open set with smooth boundary. Suppose for $p \in \partial U$, there is a neighborhood W of p and a holomorphic $f : W \to \mathbb{C}$ such that $df(p) \neq 0$, $f(p) = 0$, but f is never zero on $W \cap U$. Show that U is pseudoconvex at p. Hint: You may need the holomorphic implicit function theorem (Theorem 1.3.8). Note: The result does not require the df to not vanish, but it is harder to prove without that hypothesis.

A hyperplane is the "degenerate" case of normal convexity, that is, a hyperplane is convex from both sides. There is also a flat case of pseudoconvexity. A smooth real hypersurface $M \subset \mathbb{C}^n$ is *Levi-flat* if the Levi form vanishes at every point of M. The zero matrix is positive semidefinite and negative semidefinite, so both sides of M are pseudoconvex. Conversely, the only hypersurface pseudoconvex from both sides is a Levi-flat one.

Exercise 2.3.20: Suppose $U = V \times \mathbb{C}^{n-1} \subset \mathbb{C}^n$, where $V \subset \mathbb{C}$ is an open set with smooth boundary. Show that U has a smooth Levi-flat boundary.

Exercise 2.3.21: Prove that a real hyperplane is Levi-flat.

Exercise 2.3.22: Let $U \subset \mathbb{C}^n$ be open, $f \in \mathscr{O}(U)$, and $M = \{z \in U : \operatorname{Im} f(z) = 0\}$. Show that if $df(p) \neq 0$ for some $p \in M$, then near p, M is a Levi-flat hypersurface.

Exercise 2.3.23: Suppose $M \subset \mathbb{C}^n$ is a smooth Levi-flat hypersurface, $p \in M$, and a complex line L is tangent to M at p. Prove that p is not an isolated point of $L \cap M$.

Exercise 2.3.24: Suppose $U \subset \mathbb{C}^n$ is an open set with smooth boundary and ∂U is Levi-flat. Show that U is unbounded. Hint: If U were bounded, consider the point on ∂U farthest from the origin.

2.4 \ From harmonic to plurisubharmonic functions

We start with a quick review of harmonic and subharmonic functions in $\mathbb{C}$. For a more detailed treatment, see a one-variable book such as [L].

Definition 2.4.1. Let $U \subset \mathbb{R}^n$ be open. A C^2-smooth $f \colon U \to \mathbb{R}$ is *harmonic* if[*]

$$\nabla^2 f = \frac{\partial^2 f}{\partial x_1^2} + \cdots + \frac{\partial^2 f}{\partial x_n^2} = 0 \quad \text{on } U.$$

A function $f \colon U \to \mathbb{R} \cup \{-\infty\}$ is *subharmonic* if it is upper-semicontinuous[†] and for every ball $B_r(a)$ with $\overline{B_r(a)} \subset U$, and every function g continuous on $\overline{B_r(a)}$ and harmonic on $B_r(a)$ such that $f(x) \leq g(x)$ for $x \in \partial B_r(a)$, we have

$$f(x) \leq g(x), \quad \text{for all } x \in B_r(a).$$

In other words, a subharmonic function is a function that is less than every harmonic function on every ball whenever that holds on the boundary. When $n = 1$ in the definition of a subharmonic function, it is the same as the standard definition of a convex function of one real variable, where affine linear functions play the role of harmonic functions: A function of one real variable is *convex* if for every interval it is less than the affine linear function with the same end points. A function of one real variable is harmonic if the second derivative vanishes, and it is therefore affine linear. In one real dimension it is also easier to picture. The function f is convex if on every interval $[\alpha, \beta]$, $f \leq g$ for every affine linear g bigger than f at the endpoints α and β. In particular, we can take the g that is equal to f at the endpoints. See Figure 2.10. The picture is analogous for subharmonic functions for $n > 1$, but it is harder to draw.

---

[*]The operator ∇^2, sometimes also written Δ, is the *Laplacian*. It is the trace of the Hessian matrix.
[†]Recall f is *upper-semicontinuous* if $\limsup_{t \to x} f(t) \leq f(x)$ for all $x \in U$.

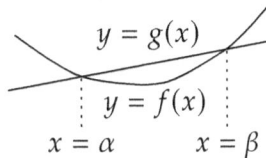

Figure 2.10: Convex function.

We will consider harmonic and subharmonic functions in $\mathbb{C} \cong \mathbb{R}^2$. Let us go through some basic results on harmonic and subharmonic functions in $\mathbb{C}$ that you have probably seen in detail in your one-variable class. Consequently, we leave some of these results as exercises. In this section (and not just here) we often write $f(z)$ for a function even if it is not holomorphic.

Exercise 2.4.1: An upper-semicontinuous function achieves a maximum on compact sets. You may assume the function to be extended-real-valued.

Exercise 2.4.2: Let $U \subset \mathbb{C}$ be open. Show that for a C^2 function $f: U \to \mathbb{R}$,

$$\frac{\partial^2}{\partial \bar{z} \partial z} f = \frac{1}{4} \nabla^2 f.$$

Use this to show that f is harmonic if and only if it is (locally) the real or imaginary part of a holomorphic function. Hint: The key is to find an antiderivative of a holomorphic function.

Exercise 2.4.3: Prove the identity theorem. Let $U \subset \mathbb{C}$ be a domain and $f: U \to \mathbb{R}$ harmonic such that $f = 0$ on a nonempty open subset of U. Then $f \equiv 0$.

Via Exercise 2.4.2, harmonic functions are locally real parts of holomorphic functions, and hence they are infinitely differentiable. In fact, on a simply connected domain in $\mathbb{C}$, any harmonic function is the real part of a holomorphic function.

It is useful to find a harmonic function given boundary values. This problem is called the *Dirichlet problem*, and it is solvable for many (though not all) domains. The proof of the following special case is contained in the exercises following the theorem. The *Poisson kernel* for the unit disc $\mathbb{D} \subset \mathbb{C}$ is

$$P_r(\theta) = \frac{1}{2\pi} \frac{1 - r^2}{1 + r^2 - 2r \cos \theta} = \frac{1}{2\pi} \operatorname{Re}\left(\frac{1 + re^{i\theta}}{1 - re^{i\theta}}\right), \qquad \text{for } 0 \leq r < 1.$$

Theorem 2.4.2. *Let $u: \partial \mathbb{D} \to \mathbb{R}$ be a continuous function. Define $Pu: \overline{\mathbb{D}} \to \mathbb{R}$ by*

$$Pu(re^{i\theta}) = \int_{-\pi}^{\pi} u(e^{it}) P_r(\theta - t) \, dt \quad \text{if } r < 1 \qquad \text{and} \qquad Pu(e^{i\theta}) = u(e^{i\theta}).$$

Then Pu is harmonic in $\mathbb{D}$ and continuous on $\overline{\mathbb{D}}$.

In the proof, it is useful to consider how the graph of P_r as a function of θ looks for a fixed r. That is, P_r acts like an approximate identity; integrating against $P_r(\theta - t)$ gives a weighted average of u with the values near $e^{i\theta}$ getting weighted more and more as $r \to 1$. See Figure 2.11.

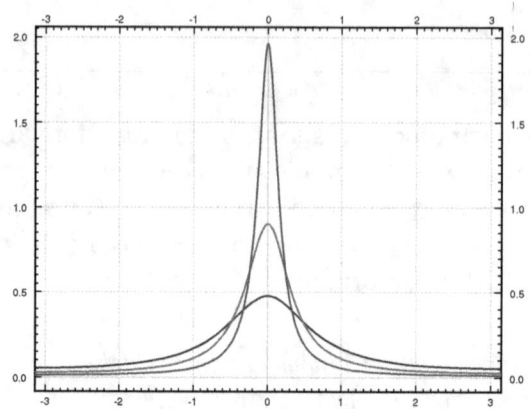

Figure 2.11: Graph of P_r for $r = 0.5$, $r = 0.7$, and $r = 0.85$ on $[-\pi, \pi]$.

Exercise 2.4.4:
  a) *Prove $P_r(\theta) > 0$ for all $0 \le r < 1$ and all θ.*
  b) *Prove $\int_{-\pi}^{\pi} P_r(\theta)\, d\theta = 1$.*
  c) *Prove for any given $\delta > 0$, $\sup\{P_r(\theta) : \delta \le |\theta| \le \pi\} \to 0$ as $r \to 1$.*

Exercise 2.4.5: *Prove Theorem 2.4.2 using the following guideline:*
  a) *Poisson kernel is harmonic as a function of $z = re^{i\theta} \in \mathbb{D}$, and hence Pu is harmonic.*
  b) *P acts like an approximate identity: Prove that $Pu(re^{i\theta}) \to u(e^{i\theta})$ uniformly as $r \to 1$. Hint: Split the integral to $[-\delta, \delta]$ and the rest and use the previous exercise.*
  c) *Prove that $Pu(z)$ tends to $u(z_0)$ as $z \in \mathbb{D} \to z_0 \in \partial\mathbb{D}$.*

Exercise 2.4.6: *State and prove a version of Theorem 2.4.2 for an arbitrary disc $\Delta_r(a)$.*

Exercise 2.4.7: *Prove that the Dirichlet problem is not solvable in the punctured disc $\mathbb{D} \setminus \{0\}$. Hint: Let $u = 0$ on $\partial\mathbb{D}$ and $u(0) = 1$. The solution would be less than $-\epsilon \log|z|$ for every $\epsilon > 0$.*

The Poisson kernel is a reproducing kernel for holomorphic (and antiholomorphic) functions, as (the real and imaginary parts of) holomorphic functions are harmonic. Poisson kernel exists for higher dimensions as well. Solving the Dirichlet problem using the Poisson kernel leads to the following result.

Proposition 2.4.3 (Mean-value property and sub-mean-value property). *Let $U \subset \mathbb{C}$ be an open set.*

(i) *A continuous function $f : U \to \mathbb{R}$ is harmonic if and only if*

$$f(a) = \frac{1}{2\pi} \int_0^{2\pi} f(a + re^{i\theta}) \, d\theta \qquad \text{whenever} \quad \overline{\Delta_r(a)} \subset U.$$

(ii) *An upper-semicontinuous function $f : U \to \mathbb{R} \cup \{-\infty\}$ is subharmonic if and only if*

$$f(a) \leq \frac{1}{2\pi} \int_0^{2\pi} f(a + re^{i\theta}) \, d\theta \qquad \text{whenever} \quad \overline{\Delta_r(a)} \subset U.$$

For the sub-mean-value property you may have to use the Lebesgue integral to integrate an upper-semicontinuous function, and to use the version of the Poisson integral above, you need to approximate by continuous functions on the boundary in the right way. On first reading, feel free to think of continuous subharmonic functions and not too much will be lost.

Exercise 2.4.8: Fill in the details of the proof of Proposition 2.4.3. Hint 1: (i) follows from (ii) if you can solve the Dirichlet problem in a disc. Hint 2: Prove the reverse direction of (ii) by contrapositive.

Exercise 2.4.9: Suppose $U \subset \mathbb{C}$ is open and $f : U \to \mathbb{R} \cup \{-\infty\}$ is subharmonic. Prove

$$\limsup_{w \to z} f(w) = f(z) \qquad \text{for all } z \in U.$$

Exercise 2.4.10: Suppose $U \subset \mathbb{C}$ is open and $g : U \to \mathbb{R}$ is harmonic. Then $f : U \to \mathbb{R} \cup \{-\infty\}$ is subharmonic if and only if $f - g$ is subharmonic.

Proposition 2.4.4 (Maximum principle). *Suppose $U \subset \mathbb{C}$ is a domain and $f : U \to \mathbb{R} \cup \{-\infty\}$ is subharmonic. If f attains a maximum in U, then f is constant.*

Proof. Suppose f attains a maximum at $a \in U$. If $\overline{\Delta_r(a)} \subset U$, then

$$f(a) \leq \frac{1}{2\pi} \int_0^{2\pi} f(a + re^{i\theta}) \, d\theta \leq f(a).$$

Hence, $f = f(a)$ almost everywhere on $\partial \Delta_r(a)$. By upper-semicontinuity, $f = f(a)$ everywhere on $\partial \Delta_r(a)$. This is true for all r with $\overline{\Delta_r(a)} \subset U$, so $f = f(a)$ on $\Delta_r(a)$, and so the set where $f = f(a)$ is open. The set where an upper-semicontinuous function attains a maximum is closed. So $f = f(a)$ on U as U is connected. $\square$

A very useful fact we will use over and over without mentioning much is that subharmonicity (like harmonicity) is a local property, even if it does not seem so from the definition. The proof is left as an exercise.

Proposition 2.4.5. *Given an open $U \subset \mathbb{C}$, an upper-semicontinuous $f : U \to \mathbb{R} \cup \{-\infty\}$ is subharmonic if for every $p \in U$ there is an $R_p > 0$ where $\Delta_{R_p}(p) \subset U$ and such that the estimate in Proposition 2.4.3 part (ii) holds for all r with $0 < r < R_p$.*

In particular, a function $f : U \to \mathbb{R} \cup \{-\infty\}$ is subharmonic if and only if for every $p \in U$ there exists a neighborhood W of p, $W \subset U$, such that $f|_W$ is subharmonic.

> *Exercise 2.4.11: Prove Proposition 2.4.5. Hint: Analyze your proof of Proposition 2.4.3.*
>
> *Exercise 2.4.12: Suppose $U \subset \mathbb{C}$ is a bounded open set, $f : \overline{U} \to \mathbb{R} \cup \{-\infty\}$ is upper-semicontinuous such that $f|_U$ is subharmonic, $g : \overline{U} \to \mathbb{R}$ is continuous such that $g|_U$ is harmonic and $f(z) \leq g(z)$ for all $z \in \partial U$. Prove that $f(z) \leq g(z)$ for all $z \in U$.*
>
> *Exercise 2.4.13: Let g be a function harmonic on a disc $\Delta \subset \mathbb{C}$ and continuous on $\overline{\Delta}$. Prove that for every $\epsilon > 0$ there exists a function g_ϵ, harmonic in a neighborhood of $\overline{\Delta}$, such that $g(z) \leq g_\epsilon(z) \leq g(z) + \epsilon$ for all $z \in \overline{\Delta}$. In particular, to test subharmonicity, we only need to consider those g that are harmonic a bit past the boundary of the disc.*

Proposition 2.4.6. *Suppose $U \subset \mathbb{C}$ is an open set and $f : U \to \mathbb{R}$ is a C^2 function. The function f is subharmonic if and only if $\nabla^2 f \geq 0$.*

In analogy to convex functions, a C^2-smooth function f of one real variable is convex if and only if $f''(x) \geq 0$ for all x.

Proof. Suppose f is a C^2-smooth function on a subset of $\mathbb{C} \cong \mathbb{R}^2$ with $\nabla^2 f \geq 0$. We wish to show that f is subharmonic. Take a disc Δ such that $\overline{\Delta} \subset U$. Consider a function g continuous on $\overline{\Delta}$, harmonic on Δ, and such that $f \leq g$ on the boundary $\partial \Delta$. Because $\nabla^2(f - g) = \nabla^2 f \geq 0$, we assume $g = 0$ and $f \leq 0$ on the boundary $\partial \Delta$. We need to show that $f \leq 0$ on Δ.

Suppose $\nabla^2 f > 0$ at all points on Δ. The Laplacian $\nabla^2 f$ is the trace of the Hessian matrix, that is, the sum of the eigenvalues. Thus f has no maximum in Δ, since at a maximum both eigenvalues of the Hessian matrix would be nonpositive. Therefore, $f \leq 0$ on all of $\overline{\Delta}$.

Next suppose only that $\nabla^2 f \geq 0$. Let M be the maximum of $x^2 + y^2$ on $\overline{\Delta}$. Take $f_n(x, y) = f(x, y) + \frac{1}{n}(x^2 + y^2) - \frac{1}{n}M$. Clearly $\nabla^2 f_n > 0$ everywhere on Δ and $f_n \leq 0$ on the boundary, so $f_n \leq 0$ on all of $\overline{\Delta}$. As $f_n \to f$, we obtain that $f \leq 0$ on all of $\overline{\Delta}$.

The other direction is left as an exercise. $\qquad\qquad\qquad\qquad\qquad\qquad\qquad \square$

> *Exercise 2.4.14: Finish the proof of the proposition above.*

Proposition 2.4.7. *Suppose $U \subset \mathbb{C}$ is an open set and $f_\alpha : U \to \mathbb{R} \cup \{-\infty\}$ is a family of subharmonic functions. Let*

$$\varphi(z) = \sup_\alpha f_\alpha(z).$$

If the family is finite, then φ is subharmonic. If the family is infinite, $\varphi(z) \neq \infty$ for all z, and φ is upper-semicontinuous, then φ is subharmonic.

Proof. Suppose $\overline{\Delta_r(a)} \subset U$. For all α,

$$\frac{1}{2\pi} \int_0^{2\pi} \varphi(a + re^{i\theta}) \, d\theta \geq \frac{1}{2\pi} \int_0^{2\pi} f_\alpha(a + re^{i\theta}) \, d\theta \geq f_\alpha(a).$$

Taking the supremum on the right over α obtains the results. $\square$

Exercise 2.4.15: Prove that if $\varphi : \mathbb{R} \to \mathbb{R}$ is a monotonically increasing convex function, $U \subset \mathbb{C}$ is an open set, and $f : U \to \mathbb{R}$ is subharmonic, then $\varphi \circ f$ is subharmonic.

Exercise 2.4.16: Let $U \subset \mathbb{C}$ be open, $\{f_n\}$ a sequence of subharmonic functions uniformly bounded above on compact subsets, and $\{c_n\}$ a sequence of positive real numbers such that $\sum_{n=1}^\infty c_n < \infty$. Prove that $f = \sum_{n=1}^\infty c_n f_n$ is subharmonic. Make sure to prove the function is upper-semicontinuous.

Exercise 2.4.17: Suppose $U \subset \mathbb{C}$ is a bounded open set, and $\{p_n\}$ a sequence of points in U. For $z \in U$, define $f(z) = \sum_{n=1}^\infty 2^{-n} \log|z - p_n|$, possibly taking on the value $-\infty$.
a) Show that f is a subharmonic function in U.
b) If $U = \mathbb{D}$ and $p_n = 1/n$, show that f is discontinuous at 0 (the natural topology on $\mathbb{R} \cup \{-\infty\}$).
c) If $\{p_n\}$ is dense in U, show that f is discontinuous on a dense set. Hint: Prove that $f^{-1}(-\infty)$ is a small (but dense) set. Another hint: Integrate the partial sums, and use polar coordinates.

There are too many harmonic functions in $\mathbb{C}^n \cong \mathbb{R}^{2n}$. The real and imaginary parts of holomorphic functions in $\mathbb{C}^n$ form a smaller set when $n > 1$. Notice that when a holomorphic function is restricted to a complex line, we obtain a holomorphic function of one variable. So the real and imaginary parts of a holomorphic function had better be harmonic on every complex line. It turns out, this is precisely the right class of functions.

Definition 2.4.8. Let $U \subset \mathbb{C}^n$ be open. A C^2-smooth $f : U \to \mathbb{R}$ is *pluriharmonic* if for every $a, b \in \mathbb{C}^n$, the function of one variable

$$\xi \mapsto f(a + b\xi)$$

is harmonic where defined (on $\{\xi \in \mathbb{C} : a + b\xi \in U\}$). That is, f is harmonic on every complex line.

A function $f: U \to \mathbb{R} \cup \{-\infty\}$ is *plurisubharmonic*, sometimes *plush* or *psh* for short, if it is upper-semicontinuous and for every $a, b \in \mathbb{C}^n$, the function of one variable

$$\xi \mapsto f(a + b\xi)$$

is subharmonic where defined.

A harmonic function of one complex variable is in some sense a generalization of an affine linear function of one real variable. Similarly, as far as several complex variables are concerned, a pluriharmonic function is the right generalization to $\mathbb{C}^n$ of an affine linear function on $\mathbb{R}^n$. In the same way, plurisubharmonic functions are the correct complex variable generalizations of convex functions. A convex function of one real variable is like a subharmonic function, and a convex function of several real variables is a function that is convex when restricted to any real line.

Many properties of harmonic and subharmonic functions in $\mathbb{C}$ have immediate generalizations to pluriharmonic and plurisubharmonic functions in $\mathbb{C}^n$. We emphasize three such immediate generalizations, the maximum principle, the fact that the property of plurisubharmonicity is local, and the fact that functions are pluriharmonic if and only if they are (locally) the real and imaginary parts of holomorphic functions. We will leave these as exercises.

Exercise 2.4.18: Let $U \subset \mathbb{C}^n$ be open. Prove that a C^2-smooth $f: U \to \mathbb{R}$ is pluriharmonic if and only if

$$\frac{\partial^2 f}{\partial \bar{z}_k \partial z_\ell} = 0 \quad \text{on } U \text{ for all } k, \ell = 1, \ldots, n.$$

Exercise 2.4.19: Show that a pluriharmonic function is harmonic. On the other hand, find an example of a harmonic function that is not pluriharmonic.

Exercise 2.4.20: Let $U \subset \mathbb{C}^n$ be open. Show that $f: U \to \mathbb{R}$ is pluriharmonic if and only if it is locally the real or imaginary part of a holomorphic function. Hint: Using a previous exercise, $\frac{\partial f}{\partial z_\ell}$ is holomorphic for all ℓ. Assume that U is simply connected, $p \in U$, and $f(p) = 0$. Consider the line integral from p to a nearby $q \in U$:

$$F(q) = \int_p^q \sum_{\ell=1}^n \frac{\partial f}{\partial z_\ell}(\zeta) \, d\zeta_\ell.$$

Prove that it is path independent, compute derivatives of F, and find out what $F + \bar{F} - f$ is.

Exercise 2.4.21: Prove the maximum principle: If $U \subset \mathbb{C}^n$ is a domain and $f: U \to \mathbb{R} \cup \{-\infty\}$ is plurisubharmonic and achieves a maximum at $p \in U$, then f is constant.

Exercise 2.4.22: Show that plurisubharmonicity is a local property, that is, f is plurisubharmonic if and only if f is plurisubharmonic in some neighborhood of each point.

Proposition 2.4.9. *Let $U \subset \mathbb{C}^n$ be open. A C^2-smooth $f \colon U \to \mathbb{R}$ is plurisubharmonic if and only if the complex Hessian matrix*

$$\left[\frac{\partial^2 f}{\partial \bar{z}_k \partial z_\ell} \right]_{k\ell}$$

is positive semidefinite at every point.

Proof. First suppose the complex Hessian has a negative eigenvalue at some $p \in U$. After a translation assume $p = 0$. As f is real-valued, the complex Hessian $\left[\frac{\partial^2 f}{\partial \bar{z}_k \partial z_\ell} \big|_0 \right]_{k\ell}$ is Hermitian. A complex linear change of coordinates acts on the complex Hessian by $*$-congruence, and therefore we can diagonalize, using Sylvester's Law of Inertia again. So assume that $\left[\frac{\partial^2 f}{\partial \bar{z}_k \partial z_\ell} \big|_0 \right]_{k\ell}$ is diagonal. If the complex Hessian has a negative eigenvalue, then one of the diagonal entries is negative. Without loss of generality suppose $\frac{\partial^2 f}{\partial \bar{z}_1 \partial z_1} \big|_0 < 0$. The function $z_1 \mapsto f(z_1, 0, \dots, 0)$ has a negative Laplacian and therefore is not subharmonic, and thus f itself is not plurisubharmonic.

For the other direction, suppose the complex Hessian is positive semidefinite at all points. After an affine change of coordinates assume that an arbitrary complex line $\xi \mapsto a + b\xi$ is setting all but the first variable to zero, that is, $a = 0$ and $b = (1, 0, \dots, 0)$. As the complex Hessian is positive semidefinite, $\frac{\partial^2 f}{\partial \bar{z}_1 \partial z_1} \geq 0$ for all points $(z_1, 0, \dots, 0)$. We proved above that $\nabla^2 g \geq 0$ implies g is subharmonic, and we are done. $\qquad \square$

Exercise 2.4.23: Suppose $U \subset \mathbb{C}^n$ is open and $f \colon U \to \mathbb{C}$ is holomorphic.
  a) Show $\log|f(z)|$ is pluriharmonic on $U \setminus f^{-1}(0)$ and plurisubharmonic on U.
  b) Show $|f(z)|^\eta$ is plurisubharmonic for all $\eta > 0$.

Exercise 2.4.24: Show that the set of plurisubharmonic functions on an open set $U \subset \mathbb{C}^n$ is a cone in the sense that if $a, b > 0$ are constants and $f, g \colon U \to \mathbb{R} \cup \{-\infty\}$ are plurisubharmonic, then $af + bg$ is plurisubharmonic.

Theorem 2.4.10. *Suppose $U \subset \mathbb{C}^n$ is an open set and $f \colon U \to \mathbb{R} \cup \{-\infty\}$ is plurisubharmonic. Let $U_\epsilon \subset U$ be the set of points further than $\epsilon > 0$ away from ∂U. For every $\epsilon > 0$, there exists a smooth plurisubharmonic function $f_\epsilon \colon U_\epsilon \to \mathbb{R}$ such that $f_\epsilon(z) \geq f(z)$, and*

$$f(z) = \lim_{\epsilon \to 0} f_\epsilon(z) \qquad \text{for all } z \in U.$$

That is, f is a (pointwise) limit of smooth plurisubharmonic functions. The idea of the proof is important and useful in many other contexts.

Proof. We smooth f out by convolving with so-called *mollifiers*, or *approximate delta functions*. Many different mollifiers work, but we use a specific one for concreteness.

For $\epsilon > 0$, define

$$g(z) = \begin{cases} Ce^{-1/(1-\|z\|^2)} & \text{if } \|z\| < 1, \\ 0 & \text{if } \|z\| \geq 1, \end{cases} \quad \text{and} \quad g_\epsilon(z) = \frac{1}{\epsilon^{2n}} g(z/\epsilon).$$

It is left as an exercise that g, and so g_ϵ, is smooth. The function g has compact support as it is only nonzero inside the unit ball. The support of g_ϵ is the ϵ-ball. Both are nonnegative. Choose C so that

$$\int_{\mathbb{C}^n} g \, dV = 1, \quad \text{and therefore} \quad \int_{\mathbb{C}^n} g_\epsilon \, dV = 1.$$

Here dV is the volume measure. The function g only depends on $\|z\|$. To get an idea of how these functions work, see Figure 2.12.

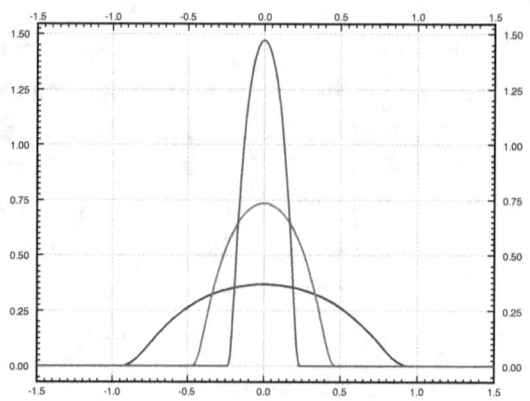

Figure 2.12: Graphs of $e^{-1/(1-x^2)}$, $\frac{1}{0.5}e^{-1/(1-(x/0.5)^2)}$, and $\frac{1}{0.25}e^{-1/(1-(x/0.25)^2)}$.

Compare the graphs to the graphs of the Poisson kernel as a function of θ, which is also a type of mollifier. The idea of integrating against the right approximate delta function with the desired properties is similar to the solution of the Dirichlet problem using the Poisson kernel.

The function f is bounded above on compact sets as it is upper semicontinuous. If f is not bounded below, replace f with $\max\{f, \frac{-1}{\epsilon}\}$, which is still plurisubharmonic. Thus, without loss of generality, assume that f is locally bounded. For $z \in U_\epsilon$, define f_ϵ as the convolution with g_ϵ:

$$f_\epsilon(z) = (f * g_\epsilon)(z) = \int_{\mathbb{C}^n} f(w)g_\epsilon(z - w) \, dV(w) = \int_{\mathbb{C}^n} f(z - w)g_\epsilon(w) \, dV(w).$$

The two forms of the integral follow easily via change of variables. We are perhaps abusing notation a bit as f is only defined on U, but it is not a problem as long as $z \in U_\epsilon$, because g_ϵ is then zero when f is undefined. By differentiating the first form under the integral, we find that f_ϵ is smooth.

Let us show that f_ϵ is plurisubharmonic. Restrict to a line $\xi \mapsto a + b\xi$. We wish to prove the sub-mean-value property using a circle of radius r around $\xi = 0$:

$$\frac{1}{2\pi} \int_0^{2\pi} f_\epsilon(a + bre^{i\theta}) \, d\theta = \frac{1}{2\pi} \int_0^{2\pi} \int_{\mathbb{C}^n} f(a + bre^{i\theta} - w) g_\epsilon(w) \, dV(w) \, d\theta$$

$$= \int_{\mathbb{C}^n} \left(\frac{1}{2\pi} \int_0^{2\pi} f(a - w + bre^{i\theta}) \, d\theta \right) g_\epsilon(w) \, dV(w)$$

$$\geq \int_{\mathbb{C}^n} f(a - w) g_\epsilon(w) \, dV(w) = f_\epsilon(a).$$

For the inequality, we used $g_\epsilon \geq 0$. So f_ϵ is plurisubharmonic.

Let us show that $f_\epsilon(z) \geq f(z)$ for all $z \in U_\epsilon$. The function $g_\epsilon(w)$ only depends on $|w_1|, \ldots, |w_n|$, in fact, $g_\epsilon(w_1, \ldots, w_n) = g_\epsilon(|w_1|, \ldots, |w_n|)$. Without loss of generality, we consider $z = 0$ and we use polar coordinates for the integral.

$$f_\epsilon(0) = \int_{\mathbb{C}^n} f(-w) g_\epsilon(|w_1|, \ldots, |w_n|) \, dV(w)$$

$$= \int_0^\epsilon \cdots \int_0^\epsilon \left(\int_0^{2\pi} \cdots \int_0^{2\pi} f(-r_1 e^{i\theta_1}, \ldots, -r_n e^{i\theta_n}) \, d\theta_1 \cdots d\theta_n \right)$$
$$g_\epsilon(r_1, \ldots, r_n) \, r_1 \cdots r_n \, dr_1 \cdots dr_n$$

$$\geq \int_0^\epsilon \cdots \int_0^\epsilon \left(\int_0^{2\pi} \cdots \int_0^{2\pi} (2\pi) f(0, -r_2 e^{i\theta_2}, \ldots, -r_n e^{i\theta_n}) \, d\theta_2 \cdots d\theta_n \right)$$
$$g_\epsilon(r_1, \ldots, r_n) \, r_1 \cdots r_n \, dr_1 \cdots dr_n$$

$$\geq f(0) \int_0^\epsilon \cdots \int_0^\epsilon (2\pi)^n g_\epsilon(r_1, \ldots, r_n) \, r_1 \cdots r_n \, dr_1 \cdots dr_n$$

$$= f(0) \int_{\mathbb{C}^n} g_\epsilon(w) \, dV(w) = f(0).$$

The second equality above follows as g_ϵ is zero outside the polydisc of radius ϵ. For the inequalities, we again needed that $g_\epsilon \geq 0$. The penultimate equality follows from the fact that $2\pi = \int_0^{2\pi} d\theta$.

Finally, for a fixed z, we show $\lim_{\epsilon \to 0} f_\epsilon(z) = f(z)$. For subharmonic, and so for plurisubharmonic, functions, $\limsup_{\zeta \to z} f(\zeta) = f(z)$, see Exercise 2.4.9. So given $\delta > 0$, find an $\epsilon > 0$ such that $f(\zeta) - f(z) \leq \delta$ for all $\zeta \in B_\epsilon(z)$.

$$f_\epsilon(z) - f(z) = \int_{B_\epsilon(0)} f(z - w) g_\epsilon(w) \, dV(w) - f(z) \int_{B_\epsilon(0)} g_\epsilon(w) \, dV(w)$$

$$= \int_{B_\epsilon(0)} \big(f(z - w) - f(z) \big) g_\epsilon(w) \, dV(w)$$

$$\leq \delta \int_{B_\epsilon(0)} g_\epsilon(w) \, dV(w) = \delta.$$

Again we used that $g_\epsilon \geq 0$. We find $0 \leq f_\epsilon(z) - f(z) \leq \delta$, and so $f_\epsilon(z) \to f(z)$. $\qquad \square$

Exercise **2.4.25:** *Show that g in the proof above is smooth on all of $\mathbb{C}^n$.*

Exercise **2.4.26:**
  a) *Show that for a subharmonic function f, $\int_0^{2\pi} f(a + re^{i\theta})\, d\theta$ is a monotone function of r. Hint: Try a C^2 function first and use Green's theorem.*
  b) *Use this fact to show that the $f_\epsilon(z)$ from Theorem 2.4.10 is monotone increasing in ϵ (so $f_\epsilon(z)$ decreases to $f(z)$ as ϵ decreases).*

As smooth plurisubharmonic functions have a local characterization in terms of derivatives, we obtain the following useful corollary, whose proof is an exercise.

Corollary 2.4.11. *Let $U \subset \mathbb{C}^n$ and $V \subset \mathbb{C}^m$ be open. Prove that if $g : U \to V$ is holomorphic and $f : V \to \mathbb{R} \cup \{-\infty\}$ is plurisubharmonic, then $f \circ g$ is plurisubharmonic.*

Exercise **2.4.27:** *Prove Corollary 2.4.11. Hint: Prove it first for C^2 functions, then use the approximation. Monotone convergence is useful.*

Exercise **2.4.28:** *With the computation from Theorem 2.4.10 show that if f is pluriharmonic, then $f_\epsilon = f$ (where it makes sense), obtaining another proof that a pluriharmonic f is C^∞.*

Exercise **2.4.29:** *Let the f in Theorem 2.4.10 be continuous and suppose $K \subset\subset U$. For small enough $\epsilon > 0$, $K \subset U_\epsilon$. Show that f_ϵ converges uniformly to f on K.*

Exercise **2.4.30:** *Let the f in Theorem 2.4.10 be C^k-smooth for some $k \geq 0$. Show that all derivatives of f_ϵ up to order k converge uniformly on compact sets to the corresponding derivatives of f. See also previous exercise.*

We turn to the theorem of Radó, a result complementary to the Riemann extension theorem. Here the function to be extended is continuous and vanishes on the set you wish to extend across, but you know nothing about this set. We prove the one-variable result and leave its straightforward extension to several variables as an exercise.

Theorem 2.4.12 (Radó). *Let $U \subset \mathbb{C}^n$ be open and $f : U \to \mathbb{C}$ a continuous function that is holomorphic on the set*
$$U' = \{z \in U : f(z) \neq 0\}.$$
Then $f \in \mathcal{O}(U)$.

Proof. First assume $n = 1$. The conclusion is local, so it is enough to prove it for a small disc Δ such that f is continuous on the closure $\overline{\Delta}$. Let $\Delta' \subset \Delta$ be the set where f is nonzero. If Δ' is empty, we are done as f is identically zero and hence holomorphic.

Let u be the real part of f. On Δ', u is a harmonic function. Let Pu be the Poisson integral of u on Δ. Hence Pu equals u on $\partial\Delta$, and Pu is harmonic in all of Δ. Consider the function $Pu(z) - u(z)$ on $\overline{\Delta}$. The function is zero on $\partial\Delta$ and it is harmonic on

Δ'. By rescaling f, we assume $|f(z)| < 1$ for all $z \in \overline{\Delta}$. The function $z \mapsto \log|f(z)|$ is harmonic on Δ', it is $-\infty$ when $f(z) = 0$, and hence it is upper-semicontinuous on $\overline{\Delta}$. Applying the sub-mean-value property near points where f vanishes and the fact that subharmonicity is local, we find that $\log|f(z)|$ is subharmonic on Δ. As $|f(z)| < 1$, we find that $\log|f(z)|$ is negative on $\overline{\Delta}$. So for every $t > 0$, the function $z \mapsto t \log|f(z)|$ is subharmonic and negative and the function $z \mapsto -t \log|f(z)|$ is superharmonic (minus a subharmonic function) and positive. See Figure 2.13. It is immediate that for all $t > 0$ and $z \in \partial\Delta$, we have

$$t \log|f(z)| \le Pu(z) - u(z) \le -t \log|f(z)|. \tag{2.4}$$

The functions $z \mapsto t \log|f(z)| - (Pu(z) - u(z))$ and $z \mapsto t \log|f(z)| - (u(z) - Pu(z))$ are harmonic on Δ' and $-\infty$ whenever $f(z) = 0$. Thus both are upper-semicontinuous on $\overline{\Delta}$ and subharmonic on Δ. The maximum principle shows that (2.4) holds for all $z \in \overline{\Delta}$ and all $t > 0$.

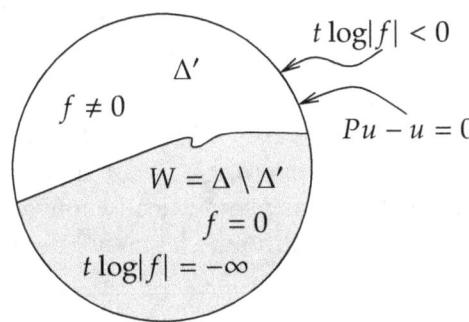

Figure 2.13: Proof of Radó's theorem.

Taking the limit $t \to 0$ shows that $Pu = u$ on Δ'. Let $W = \Delta \setminus \overline{\Delta'}$. On W, $u = 0$ and so $Pu - u$ is harmonic on W and continuous on $\overline{W}$. Furthermore, $Pu - u = 0$ on $\overline{\Delta'} \cup \partial\Delta$, and so $Pu - u = 0$ on ∂W. By the maximum principle, $Pu = u$ on W and therefore on all of $\overline{\Delta}$. Similarly, if v is the imaginary part of f, then $Pv = v$ on $\overline{\Delta}$. In other words, u and v are harmonic on Δ. As Δ is simply connected, let $\tilde{v}$ be the harmonic conjugate of u that equals v at some point of Δ'. As f is holomorphic on Δ', the harmonic functions $\tilde{v}$ and v are equal on the nonempty open subset Δ' of Δ and so they are equal everywhere. Consequently, $f = u + iv$ is holomorphic on Δ.

The extension of the proof to several variables is left as an exercise. $\qquad\square$

Exercise **2.4.31:** *Use the one-variable result to extend the theorem to several variables.*

2.5 \ Hartogs pseudoconvexity

By Corollary 2.4.11, plurisubharmonicity is preserved under holomorphic mappings. In particular, if $\varphi \colon \overline{\mathbb{D}} \to \mathbb{C}^n$ is an analytic disc and f is plurisubharmonic in a neighborhood of $\varphi(\overline{\mathbb{D}})$, then $f \circ \varphi$ is subharmonic on $\mathbb{D}$. As subharmonic functions satisfy the maximum principle, we find that $f(z) \leq \sup_{w \in \varphi(\partial \mathbb{D})} f(w)$ for all $z \in \varphi(\mathbb{D})$. Let us give a general definition for this type of situation.

Definition 2.5.1. Let $\mathcal{F}$ be a class of (extended*)-real-valued functions defined on an open $U \subset \mathbb{R}^n$. If $K \subset U$, define $\widehat{K}$, the *hull* of K with respect to $\mathcal{F}$, as the set

$$\widehat{K} \stackrel{\text{def}}{=} \left\{ x \in U : f(x) \leq \sup_{y \in K} f(y) \text{ for all } f \in \mathcal{F} \right\}.$$

An open set U is *convex with respect to $\mathcal{F}$* if for every $K \subset\subset U$, the hull $\widehat{K} \subset\subset U$.[†]

Clearly $K \subset \widehat{K}$. The key is to show that $\widehat{K}$ is not "too large" for U. Keep in mind that the functions in $\mathcal{F}$ are defined on U, so $\widehat{K}$ depends on U not just on K. An easy mistake is to consider functions defined on a larger set, obtaining a smaller $\mathcal{F}$ and hence a larger $\widehat{K}$. Sometimes it is useful to write $\widehat{K}_{\mathcal{F}}$ to denote the dependence on $\mathcal{F}$, especially when talking about several different hulls.

For example, if $U = \mathbb{R}$ and $\mathcal{F}$ is the set of real-valued smooth $f \colon \mathbb{R} \to \mathbb{R}$ with $f''(x) \geq 0$, then $\widehat{\{a,b\}} = [a,b]$ for any $a, b \in \mathbb{R}$. In general, if $\mathcal{F}$ is the set of convex functions, then a domain $U \subset \mathbb{R}^n$ is geometrically convex if and only if it is convex with respect to convex functions[‡], although let us not define what that means except for smooth functions in exercises below.

> *Exercise* 2.5.1: *Suppose $U \subset \mathbb{R}^n$ is a domain.*
> a) *Show that U is geometrically convex if and only if it is convex with respect to the affine linear functions.*
> b) *Suppose U has smooth boundary. Show that U is convex if and only if it is convex with respect to the smooth convex functions on U, that is, with respect to smooth functions with positive semidefinite Hessian.*
>
> *Exercise* 2.5.2: *Show that every open set $U \subset \mathbb{R}^n$ is convex with respect to real polynomials.*

Theorem 2.5.2 (Kontinuitätssatz—Continuity principle, second version). *Suppose an open set $U \subset \mathbb{C}^n$ is convex with respect to plurisubharmonic functions, then given any collection of closed analytic discs $\Delta_\alpha \subset U$ such that $\bigcup_\alpha \partial \Delta_\alpha \subset\subset U$, we have $\bigcup_\alpha \Delta_\alpha \subset\subset U$.*

---

*By extended reals we mean $\mathbb{R} \cup \{-\infty, \infty\}$.

[†]Recall that $\subset\subset$ means a relatively compact subset; the closure in the relative (subspace) topology is compact.

[‡]The technicality is, of course, that we must define convex functions on not-necessarily-convex sets, and that is not completely straightforward.

Various similar theorems are named the *continuity principle*—we have now seen two. What they have in common is a family of analytic discs whose boundaries stay inside a domain, and where the conclusion has to do with extension of holomorphic functions, with domains of holomorphy, or with some sort of convexity as above.

Proof. Let f be a plurisubharmonic function on U. If $\varphi_\alpha \colon \overline{\mathbb{D}} \to U$ is the holomorphic (in $\mathbb{D}$) mapping giving the closed analytic disc, then $f \circ \varphi_\alpha$ is subharmonic. By the maximum principle, f on Δ_α must be less than or equal to the supremum of f on $\partial\Delta_\alpha$, so $\overline{\Delta_\alpha}$ is in the hull of $\partial\Delta_\alpha$. In other words, $\bigcup_\alpha \Delta_\alpha$ is in the hull of $\bigcup_\alpha \partial\Delta_\alpha$ and therefore $\bigcup_\alpha \Delta_\alpha \subset\subset U$ by convexity. $\qquad\square$

Let us illustrate the failure of the continuity principle. If you have discs (denoted by straight line segments) that approach the boundary as in Figure 2.14, then the domain is not convex with respect to plurisubharmonic functions. In the diagram, the boundaries of the discs are denoted by the dark dots at the end of the segments. In fact, for standard geometric convexity, we can prove a continuity principle where we do replace discs with line segments, see the exercises below.

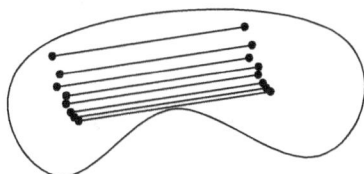

Figure 2.14: Failure of the continuity principle.

Exercise 2.5.3: Suppose $U \subset \mathbb{C}^n$ is a domain and $K \subset\subset U$ is a nonempty compact subset. Prove that $U \setminus K$ is not convex with respect to plurisubharmonic functions.

Exercise 2.5.4: Suppose $U \subset \mathbb{C}^n$ is a domain with smooth boundary, $p \in \partial U$, and Δ is an affine linear analytic disc with $p \in \Delta$, but $\Delta \setminus \{p\} \subset U$. Prove that U is not convex with respect to the plurisubharmonic functions.

Exercise 2.5.5: Prove the corresponding Kontinuitätssatz, and its converse, for geometric convexity: Prove that a domain $U \subset \mathbb{R}^n$ is geometrically convex if and only if whenever $[x_\alpha, y_\alpha] \subset U$ is a collection of straight line segments such that $\bigcup_\alpha \{x_\alpha, y_\alpha\} \subset\subset U$ implies $\bigcup_\alpha [x_\alpha, y_\alpha] \subset\subset U$.

We now define another version of pseudoconvexity, this time only in terms of the interior of the domain.

Definition 2.5.3. Let $U \subset \mathbb{C}^n$ be open. An $f : U \to \mathbb{R}$ is an *exhaustion function* for U if

$$\{z \in U : f(z) < r\} \subset\subset U \qquad \text{for every } r \in \mathbb{R}.$$

A domain $U \subset \mathbb{C}^n$ is *Hartogs pseudoconvex* if there exists a continuous plurisubharmonic exhaustion function. The set $\{z \in U : f(z) < r\}$ is called the *sublevel set* of f, or the *r-sublevel set*.

Example 2.5.4: The unit ball $\mathbb{B}_n$ is Hartogs pseudoconvex. The continuous function

$$z \mapsto -\log(1 - \|z\|^2)$$

is an exhaustion function, and it is easy to check directly that it is plurisubharmonic.

Example 2.5.5: The entire $\mathbb{C}^n$ is Hartogs pseudoconvex as $\|z\|^2$ is a continuous plurisubharmonic exhaustion function. Also, because $\|z\|^2$ is plurisubharmonic, then given any $K \subset\subset \mathbb{C}^n$, the hull $\widehat{K}$ with respect to plurisubharmonic functions must be bounded. In other words, $\mathbb{C}^n$ is convex with respect to plurisubharmonic functions.

Theorem 2.5.6. *Suppose $U \subsetneq \mathbb{C}^n$ is a domain. The following are equivalent:*

(i) $-\log \rho(z)$ *is plurisubharmonic, where $\rho(z)$ is the distance from z to ∂U.*

(ii) U *is Hartogs pseudoconvex.*

(iii) U *is convex with respect to plurisubharmonic functions defined on U.*

(iv) *The conclusion of Kontinuitätssatz (second version) holds: for any collection of closed analytic discs $\Delta_\alpha \subset U$ such that $\bigcup_\alpha \partial \Delta_\alpha \subset\subset U$, we have $\bigcup_\alpha \Delta_\alpha \subset\subset U$.*

Proof. (i) $\Rightarrow$ (ii): If U is bounded, the function $-\log \rho(z)$ is clearly a continuous exhaustion function. If U is unbounded, take $z \mapsto \max\{-\log \rho(z), \|z\|^2\}$.

(ii) $\Rightarrow$ (iii): Suppose f is a continuous plurisubharmonic exhaustion function. If $K \subset\subset U$, then for some r we have $K \subset \{z \in U : f(z) < r\} \subset\subset U$. But then by definition of the hull $\widehat{K}$ we have $\widehat{K} \subset \{z \in U : f(z) < r\} \subset\subset U$.

(iii) $\Rightarrow$ (iv): That is simply the statement of Kontinuitätssatz (second version).

(iv) $\Rightarrow$ (i): As long as $U \neq \mathbb{C}^n$, the function $-\log \rho(z)$ is real-valued and continuous. For $c \in \mathbb{C}^n$ with $\|c\| = 1$, let $\rho_c(z)$ be the supremum of the radii of the affine discs centered at z in the direction c that lie in U. That is,

$$\rho_c(z) = \sup\{\lambda > 0 : z + \zeta c \in U \text{ for all } \zeta \in \lambda \mathbb{D}\}.$$

As $\rho(z) = \inf_c \rho_c(z)$,

$$-\log \rho(z) = \sup_{\|c\|=1} \left(-\log \rho_c(z)\right).$$

If we prove that for all $a, b \in \mathbb{C}^n$ and $c \in \mathbb{C}^n$ with $\|c\| = 1$, the function $\xi \mapsto -\log \rho_c(a + b\xi)$ is subharmonic, then $\xi \mapsto -\log \rho(a + b\xi)$ is subharmonic, and we are done. See Figure 2.15 for the setup.

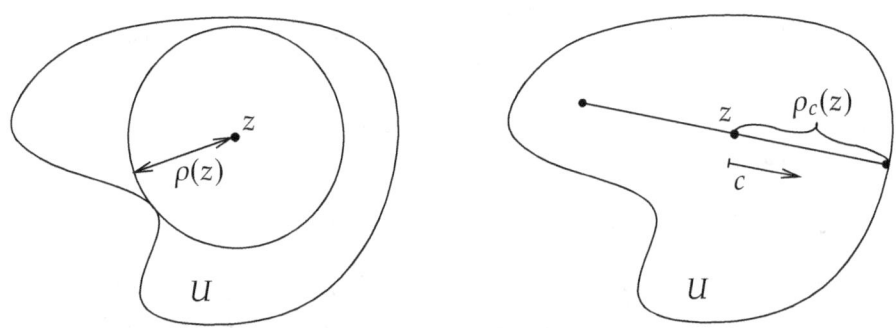

Figure 2.15: Largest disc in the direction of c. The disc is drawn as a line.

Suppose $\Delta \subset \mathbb{C}$ is a disc such that $a + b\xi \in U$ for all $\xi \in \overline{\Delta}$. If u is a harmonic function on Δ continuous on $\overline{\Delta}$ such that $-\log \rho_c(a + b\xi) \leq u(\xi)$ on $\partial\Delta$, we must show that the inequality holds on Δ. By Exercise 2.4.13, we may assume u is harmonic on a neighborhood of $\overline{\Delta}$ and so let $u = \operatorname{Re} f$ for a holomorphic function f. Suppose $\xi \in \partial\Delta$ for a moment. We have $-\log \rho_c(a + b\xi) \leq \operatorname{Re} f(\xi)$, or in other words

$$\rho_c(a + b\xi) \geq e^{-\operatorname{Re} f(\xi)} = \left|e^{-f(\xi)}\right|.$$

Using $\zeta = te^{-f(\xi)}$ in the definition of $\rho_c(a + b\xi)$, the statement above is equivalent to

$$(a + b\xi) + te^{-f(\xi)}c \in U \quad \text{for all } t \in \mathbb{D}.$$

This statement holds whenever $\xi \in \partial\Delta$. We must prove that it also holds for all $\xi \in \Delta$.

The function $\varphi_t(\xi) = (a + b\xi) + te^{-f(\xi)}c$ gives a closed analytic disc with boundary inside U. We have a family of analytic discs, parametrized by t, whose boundaries are in U for all t with $|t| < 1$. For $t = 0$ the entire disc is inside U. As $\varphi_t(\xi)$ is continuous in both t and ξ and $\overline{\Delta}$ is compact, $\varphi_t(\Delta) \subset U$ for t in some neighborhood of 0. Take $0 < t_0 < 1$ such that $\varphi_t(\Delta) \subset U$ for all t with $|t| < t_0$. Then

$$\bigcup_{|t|<t_0} \varphi_t(\partial\Delta) \subset \bigcup_{|t|\leq t_0} \varphi_t(\partial\Delta) \subset\subset U,$$

because continuous functions take compact sets to compact sets. The hypothesis (iv) implies

$$\bigcup_{|t|<t_0} \varphi_t(\Delta) \subset\subset U.$$

By continuity again, $\bigcup_{|t|\leq t_0} \varphi_t(\Delta) \subset\subset U$, and so $\bigcup_{|t|<t_0+\epsilon} \varphi_t(\Delta) \subset\subset U$ for some $\epsilon > 0$. Consequently $\varphi_t(\Delta) \subset U$ for all t with $|t| < 1$. Thus $(a + b\xi) + te^{-f(\xi)}c \in U$ for all $\xi \in \Delta$ and all $|t| < 1$. This implies $\rho_c(a + b\xi) \geq e^{-\operatorname{Re} f(\xi)}$ for all $\xi \in \Delta$, which in turn implies $-\log \rho_c(a + b\xi) \leq \operatorname{Re} f(\xi) = u(\xi)$ for all $\xi \in \Delta$. Therefore, $-\log \rho_c(a + b\xi)$ is subharmonic. $\qquad\square$

Exercise 2.5.6: Show that if $U_1 \subset \mathbb{C}^n$ and $U_2 \subset \mathbb{C}^n$ are Hartogs pseudoconvex domains, then so are all the topological components of $U_1 \cap U_2$.

Exercise 2.5.7: Show that if $U \subset \mathbb{C}^n$ and $V \subset \mathbb{C}^m$ are Hartogs pseudoconvex domains, then so is $U \times V$.

Exercise 2.5.8: Show that every domain $U \subset \mathbb{C}$ is Hartogs pseudoconvex.

Exercise 2.5.9: Consider the union $U = \bigcup_k U_k$ of a nested sequence of Hartogs pseudoconvex domains, $U_{k-1} \subset U_k \subset \mathbb{C}^n$. Show that U is Hartogs pseudoconvex.

Exercise 2.5.10: Let $\mathbb{R}^2 \subset \mathbb{C}^2$ be naturally embedded (that is, it is the set where z_1 and z_2 are real). Show that the set $\mathbb{C}^2 \setminus \mathbb{R}^2$ is not Hartogs pseudoconvex.

Exercise 2.5.11: Let $U \subset \mathbb{C}^n$ be a domain and $f \in \mathcal{O}(U)$. Prove that $U' = \{z \in U : f(z) \neq 0\}$ is a Hartogs pseudoconvex domain. Hint: See also Exercise 1.6.5.

Exercise 2.5.12: Suppose $U, V \subset \mathbb{C}^n$ are biholomorphic domains. Prove that U is Hartogs pseudoconvex if and only if V is Hartogs pseudoconvex.

Exercise 2.5.13: Let $U = \{z \in \mathbb{C}^2 : |z_1| > |z_2|\}$.
  a) Prove that U is a Hartogs pseudoconvex domain.
  b) Find a closed analytic disc Δ in $\mathbb{C}^2$ such that $0 \in \Delta$ ($0 \notin U$) and $\Delta \setminus \{0\} \subset U$ (in particular $\partial \Delta \subset U$).
  c) What do you think would happen if you tried to move Δ a little bit to avoid the intersection with the complement? Think about the continuity principle (second version). Compare with Exercise 2.5.4.

Exercise 2.5.14: Let $U \subset \mathbb{C}^n$ be a domain and $f : \overline{U} \to \mathbb{R}$ be continuous, plurisubharmonic, negative on U, and $f = 0$ on ∂U. Prove that U is Hartogs pseudoconvex.

The statement corresponding to Exercise 2.5.9 on nested unions for domains of holomorphy is the *Behnke–Stein theorem*, which follows using this exercise and the solution of the Levi problem. Behnke–Stein is easier to prove without the solution to the Levi problem, see Exercise 2.6.13, and is, in fact, generally used in the solution of the Levi problem.

Exercise 2.5.12 says that (Hartogs) pseudoconvexity is a biholomorphic invariant. That is a good indication that we are looking at a correct notion. It also allows us to change variables to more convenient ones when proving a specific domain is (Hartogs) pseudoconvex.

It is not immediately clear from the definition, but Hartogs pseudoconvexity is also a local property of the boundary.

Lemma 2.5.7. *A domain $U \subset \mathbb{C}^n$ is Hartogs pseudoconvex if and only if for every point $p \in \partial U$ there exists a neighborhood W of p such that $W \cap U$ is Hartogs pseudoconvex.*

Proof. One direction is trivial, so consider the other. Suppose $p \in \partial U$, and let W be such that $U \cap W$ is Hartogs pseudoconvex. Intersection of Hartogs pseudoconvex domains is Hartogs pseudoconvex, see Exercise 2.5.6, so assume $W = B_r(p)$. Let $B = B_{r/2}(p)$. If $q \in B \cap U$, the distance from q to the boundary of $W \cap U$ is the same as the distance to ∂U. The setup is illustrated in Figure 2.16.

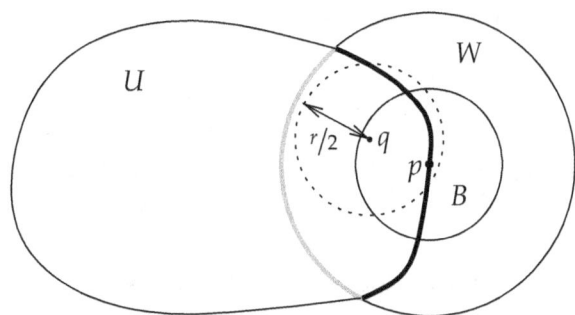

Figure 2.16: Local Hartogs pseudoconvexity.

The part of the boundary ∂U in W is marked by a thick black line, the part of the boundary of $\partial(W \cap U)$ that is the boundary of W is marked by a thick gray line. A point $q \in B$ is marked and a ball of radius $r/2$ around q is dotted. No point of distance $r/2$ from q is in ∂W, and the distance of q to ∂U is at most $r/2$ as $p \in \partial U$ and p is the center of B. Let $\operatorname{dist}(x, y)$ denote the euclidean distance function*. Then for $z \in B \cap U$

$$- \log \operatorname{dist}(z, \partial U) = - \log \operatorname{dist}(z, \partial(U \cap W)).$$

The right-hand side is plurisubharmonic as $U \cap W$ is Hartogs pseudoconvex. Such a ball B exists around every $p \in \partial U$, so near the boundary, $- \log \operatorname{dist}(z, \partial U)$ is plurisubharmonic.

If U is bounded, then ∂U is compact. So there is some $\epsilon > 0$ such that $- \log \operatorname{dist}(z, \partial U)$ is plurisubharmonic if $\operatorname{dist}(z, \partial U) < 2\epsilon$. The function

$$\varphi(z) = \max\left\{- \log \operatorname{dist}(z, \partial U), - \log \epsilon\right\}$$

is a continuous plurisubharmonic exhaustion function. The proof for unbounded U is left as an exercise. □

Exercise 2.5.15: Finish the proof of the lemma for unbounded domains. See Exercise 2.5.9.

It may seem that we defined a totally different concept, but it turns out that Levi and Hartogs pseudoconvexity are one and the same on domains where both concepts make sense. As a consequence of the following theorem we say simply "pseudoconvex" and there is no ambiguity.

---

*If x and/or y are sets of points, we take the infimum of the euclidean distance over all the points.

Theorem 2.5.8. *Let $U \subset \mathbb{C}^n$ be a domain with smooth boundary. Then U is Hartogs pseudoconvex if and only if U is Levi pseudoconvex.*

Proof. Suppose $U \subset \mathbb{C}^n$ is a domain with smooth boundary that is not Levi pseudoconvex at $p \in \partial U$. As in Theorem 2.3.11, change coordinates so that $p = 0$ and U is defined by

$$\operatorname{Im} z_n > -|z_1|^2 + \sum_{k=2}^{n-1} \epsilon_k |z_k|^2 + O(3).$$

For a small fixed $\lambda > 0$, the closed analytic discs defined by $\xi \in \overline{\mathbb{D}} \mapsto (\lambda\xi, 0, \cdots, 0, is)$ are in U for all small enough $s > 0$. The origin is a limit point of the insides of the discs, but not a limit point of their boundaries. Kontinuitätssatz (second version) is not satisfied, and U is not convex with respect to the plurisubharmonic functions. Therefore, U is not Hartogs pseudoconvex.

Next suppose U is Levi pseudoconvex. Take any $p \in \partial U$. After translation and rotation by a unitary, assume $p = 0$ and write a defining function r as

$$r(z) = \varphi(z', \bar{z}', \operatorname{Re} z_n) - \operatorname{Im} z_n,$$

where $z' = (z_1, \ldots, z_{n-1})$ and $\varphi \in O(2)$. Levi pseudoconvexity says

$$\sum_{k,\ell=1}^{n} \bar{a}_k a_\ell \frac{\partial^2 r}{\partial \bar{z}_k \partial z_\ell}\Big|_q \geq 0 \quad \text{whenever} \quad \sum_{k=1}^{n} a_k \frac{\partial r}{\partial z_k}\Big|_q = 0, \tag{2.5}$$

for all $q \in \partial U$ near 0. Let s be a small real constant, and let $\tilde{q} = (q_1, \ldots, q_{n-1}, q_n + is)$. As no derivatives of r depend on $\operatorname{Im} z_n$, we have $\frac{\partial r}{\partial z_\ell}\big|_{\tilde{q}} = \frac{\partial r}{\partial z_\ell}\big|_q$ and $\frac{\partial^2 r}{\partial \bar{z}_k \partial z_\ell}\big|_{\tilde{q}} = \frac{\partial^2 r}{\partial \bar{z}_k \partial z_\ell}\big|_q$ for all k and ℓ. So condition (2.5) holds for all $q \in U$ near 0. We will use r to manufacture a plurisubharmonic exhaustion function. We need a positive semidefinite complex Hessian, and r already has what we need in all but one direction.

Let $\nabla_z r|_q = \left(\frac{\partial r}{\partial z_1}\big|_q, \ldots, \frac{\partial r}{\partial z_n}\big|_q\right)$ denote the gradient of r in the holomorphic directions only. Given $q \in U$ near 0, decompose an arbitrary $c \in \mathbb{C}^n$ as $c = a + b$, where a and b are orthogonal and b is a scalar multiple of $\overline{\nabla_z r|_q}$. That is, $a = (a_1, \ldots, a_n)$ satisfies

$$\sum_{k=1}^{n} a_k \frac{\partial r}{\partial z_k}\Big|_q = \langle a, \overline{\nabla_z r|_q} \rangle = 0.$$

By the equality part of Cauchy–Schwarz,

$$\left|\sum_{k=1}^{n} c_k \frac{\partial r}{\partial z_k}\Big|_q\right| = \left|\sum_{k=1}^{n} b_k \frac{\partial r}{\partial z_k}\Big|_q\right| = \left|\langle b, \overline{\nabla_z r|_q} \rangle\right| = \|b\| \, \|\nabla_z r|_q\|.$$

As $\nabla_z r|_0 = (0, \ldots, 0, -1/2i)$, for q sufficiently near 0, we have $\|\nabla_z r|_q\| \geq 1/3$, and

$$\|b\| = \frac{1}{\|\nabla_z r|_q\|}\left|\sum_{k=1}^{n} c_k \frac{\partial r}{\partial z_k}\Big|_q\right| \leq 3\left|\sum_{k=1}^{n} c_k \frac{\partial r}{\partial z_k}\Big|_q\right|. \tag{2.6}$$

As $c = a + b$ is the orthogonal decomposition, $\|c\| \geq \|b\|$. The complex Hessian matrix of r is continuous, and so let $M \geq 0$ be an upper bound on its operator norm for q near the origin. Note that $\bar{c}_k c_\ell = (\bar{a}_k + \bar{b}_k)(a_\ell + b_\ell) = \bar{a}_k a_\ell + \bar{b}_k (a_\ell + b_\ell) + (\bar{a}_k + \bar{b}_k) b_\ell - \bar{b}_k b_\ell = \bar{a}_k a_\ell + \bar{b}_k c_\ell + \bar{c}_k b_\ell - \bar{b}_k b_\ell$. Using Cauchy–Schwarz,

$$
\sum_{k=1,\ell=1}^{n} \bar{c}_k c_\ell \frac{\partial^2 r}{\partial \bar{z}_k \partial z_\ell}\bigg|_q = \sum_{k=1,\ell=1}^{n} \bar{a}_k a_\ell \frac{\partial^2 r}{\partial \bar{z}_k \partial z_\ell}\bigg|_q + \sum_{k=1,\ell=1}^{n} \bar{b}_k c_\ell \frac{\partial^2 r}{\partial \bar{z}_k \partial z_\ell}\bigg|_q
$$

$$
+ \sum_{k=1,\ell=1}^{n} \bar{c}_k b_\ell \frac{\partial^2 r}{\partial \bar{z}_k \partial z_\ell}\bigg|_q - \sum_{k=1,\ell=1}^{n} \bar{b}_k b_\ell \frac{\partial^2 r}{\partial \bar{z}_k \partial z_\ell}\bigg|_q \tag{2.7}
$$

$$
\geq \sum_{k=1,\ell=1}^{n} \bar{a}_k a_\ell \frac{\partial^2 r}{\partial \bar{z}_k \partial z_\ell}\bigg|_q - M\|b\|\|c\| - M\|c\|\|b\| - M\|b\|^2
$$

$$
\geq -3M\|c\|\|b\|.
$$

Putting (2.7) and (2.6) together, for $q \in U$ near the origin,

$$
\sum_{k=1,\ell=1}^{n} \bar{c}_k c_\ell \frac{\partial^2 r}{\partial \bar{z}_k \partial z_\ell}\bigg|_q \geq -3M\|c\|\|b\| \geq -3^2 M\|c\| \left| \sum_{k=1}^{n} c_k \frac{\partial r}{\partial z_k}\bigg|_q \right|.
$$

For $z \in U$ sufficiently close to 0, define

$$
f(z) = -\log(-r(z)) + A\|z\|^2,
$$

where $A > 0$ is some constant we will choose later. The log is there to make f blow up as we approach the boundary. The $A\|z\|^2$ is there to add a constant diagonal matrix to the complex Hessian of f, which we hope is enough to make it positive semidefinite at all z near 0. Compute:

$$
\frac{\partial^2 f}{\partial \bar{z}_k \partial z_\ell} = \frac{1}{r^2} \frac{\partial r}{\partial \bar{z}_k} \frac{\partial r}{\partial z_\ell} - \frac{1}{r} \frac{\partial^2 r}{\partial \bar{z}_k \partial z_\ell} + A\delta_k^\ell,
$$

where δ_k^ℓ is the Kronecker delta*. Apply the complex Hessian of f to c at $q \in U$ near the origin (recall that r is negative on U and so for $q \in U$, $-r = |r|$):

$$
\sum_{k=1,\ell=1}^{n} \bar{c}_k c_\ell \frac{\partial^2 f}{\partial \bar{z}_k \partial z_\ell}\bigg|_q = \frac{1}{r^2} \left| \sum_{\ell=1}^{n} c_\ell \frac{\partial r}{\partial z_\ell}\bigg|_q \right|^2 + \frac{1}{|r|} \sum_{k=1,\ell=1}^{n} \bar{c}_k c_\ell \frac{\partial^2 r}{\partial \bar{z}_k \partial z_\ell}\bigg|_q + A\|c\|^2
$$

$$
\geq \frac{1}{r^2} \left| \sum_{\ell=1}^{n} c_\ell \frac{\partial r}{\partial z_\ell}\bigg|_q \right|^2 - \frac{3^2 M}{|r|} \|c\| \left| \sum_{k=1}^{n} c_k \frac{\partial r}{\partial z_k}\bigg|_q \right| + A\|c\|^2.
$$

Now comes a somewhat funky trick. As a quadratic polynomial in $\|c\|$, the right-hand side of the inequality is always nonnegative if $A > 0$ and if the discriminant is negative

---

*Recall $\delta_k^\ell = 0$ if $k \neq \ell$ and $\delta_k^\ell = 1$ if $k = \ell$.

or zero. Let us see if we can make the discriminant zero:

$$0 = \left(\frac{3^2 M}{|r|} \left| \sum_{k=1}^{n} c_k \frac{\partial r}{\partial z_k} \right|_q \right)^2 - 4A \frac{1}{r^2} \left| \sum_{\ell=1}^{n} c_\ell \frac{\partial r}{\partial z_\ell} \right|_q^2 .$$

All the nonconstant terms go away and $A = \frac{3^4 M^2}{4}$ makes the discriminant zero. Any larger A would also work by making the discriminant negative. Thus for that A,

$$\sum_{k=1,\ell=1}^{n} \bar{c}_k c_\ell \frac{\partial^2 f}{\partial \bar{z}_k \partial z_\ell} \bigg|_q \geq 0.$$

In other words, the complex Hessian of f is positive semidefinite at all points $q \in U$ near 0. The function $f(z)$ goes to infinity as z approaches ∂U. So for every $t \in \mathbb{R}$, the t-sublevel set (the set where $f(z) < t$) is a positive distance away from ∂U near 0.

   We have constructed a local continuous plurisubharmonic exhaustion function for U near p. If we intersect with a small ball B centered at p, then $U \cap B$ is Hartogs pseudoconvex. This is true at all $p \in \partial U$, so U is Hartogs pseudoconvex.   □

Exercise 2.5.16: Show that any defining function works in the construction above. Suppose U is Levi pseudoconvex, $p \in \partial U$, r is the defining function from the proof, and let $\tilde{r} = \varphi r$ for some smooth φ defined near p, $\varphi(p) \neq 0$. Show that for $z \in U$ near p, the function $f(z) = -\log(-\tilde{r}(z)) + A\|z\|^2$ is plurisubharmonic if A is sufficiently large.

2.6 \ Holomorphic convexity

Definition 2.6.1. Let $U \subset \mathbb{C}^n$ be a domain. For $K \subset U$, define the *holomorphic hull*

$$\widehat{K}_U \stackrel{\text{def}}{=} \left\{ z \in U : |f(z)| \leq \sup_{w \in K} |f(w)| \text{ for all } f \in \mathcal{O}(U) \right\}.$$

A domain U is *holomorphically convex* if whenever $K \subset\subset U$, then $\widehat{K}_U \subset\subset U$. In other words, U is holomorphically convex if it is convex with respect to moduli of holomorphic functions on U.*

   It is a simple exercise (see below) to show that a holomorphically convex domain is Hartogs pseudoconvex. We will prove that holomorphic convexity is equivalent to being a domain of holomorphy. That a Hartogs pseudoconvex domain is holomorphically convex is the Levi problem for Hartogs pseudoconvex domains and is considerably more difficult. The thing is, there are lots of plurisubharmonic functions, and they are easy to construct; we can even construct them locally, and then piece

---

*Sometimes simply $\widehat{K}$ is used, but we use $\widehat{K}_U$ to emphasize the dependence on U.

them together by taking maxima. There are far fewer holomorphic functions, and we cannot just construct them locally and expect the pieces to somehow fit together. As the result is so fundamental, let us state it as a theorem.

Theorem 2.6.2 (Solution of the Levi problem). *A domain $U \subset \mathbb{C}^n$ is holomorphically convex if and only if it is Hartogs pseudoconvex.*

Proof. The forward direction follows from an exercise below, which is sometimes called Oka's lemma. We skip the proof of the backward direction in order to save some hundred pages or so. See Hörmander's book [H] for the proof. □

Exercise **2.6.1** (Oka's lemma)*: Prove that if a domain $U \subset \mathbb{C}^n$ is holomorphically convex, then it is Hartogs pseudoconvex. See Exercise 2.4.23.*

Exercise **2.6.2***: Prove that every domain $U \subset \mathbb{C}$ is holomorphically convex by giving a topological description of $\widehat{K}_U$ for every compact $K \subset\subset U$. Hint: Runge may be useful.*

Exercise **2.6.3***: Suppose $f : \mathbb{C}^n \to \mathbb{C}$ is holomorphic and U is a topological component of $\{z \in \mathbb{C}^n : |f(z)| < 1\}$. Prove that U is a holomorphically convex domain.*

Exercise **2.6.4***: Compute the hull $\widehat{K}_{\mathbb{D}^n}$ of the set $K = \{z \in \mathbb{D}^n : |z_\ell| = \lambda_\ell \text{ for } \ell = 1, \ldots, n\}$, where $0 \leq \lambda_\ell < 1$. Prove that the unit polydisc is holomorphically convex.*

Exercise **2.6.5***: Prove that a geometrically convex domain $U \subset \mathbb{C}^n$ is holomorphically convex.*

Exercise **2.6.6***: Prove the Hartogs figure (see Theorem 2.1.4) is not holomorphically convex.*

Exercise **2.6.7***: Let $U \subset \mathbb{C}^n$ be a domain, $f \in \mathcal{O}(U)$, and f is not identically zero. Show that if U is holomorphically convex, then $\widetilde{U} = \{z \in U : f(z) \neq 0\}$ is holomorphically convex. Hint: First see Exercise 1.6.5.*

Exercise **2.6.8***: Suppose $U, V \subset \mathbb{C}^n$ are biholomorphic domains. Prove that U is holomorphically convex if and only if V is holomorphically convex.*

Exercise **2.6.9***: In the definition of holomorphic hull of K, replace U with $\mathbb{C}^n$ and $\mathcal{O}(U)$ with holomorphic polynomials on $\mathbb{C}^n$, to get the* polynomial hull *of K. Prove that the polynomial hull of $K \subset\subset \mathbb{C}^n$ is the same as the holomorphic hull $\widehat{K}_{\mathbb{C}^n}$.*

Exercise **2.6.10***:*
   a) Prove the Hartogs triangle T (see Exercise 2.1.9) is holomorphically convex.
   b) Prove $T \cup B_\epsilon(0)$ (for a small enough $\epsilon > 0$) is not holomorphically convex.

Exercise **2.6.11***: Show that if domains $U_1 \subset \mathbb{C}^n$ and $U_2 \subset \mathbb{C}^n$ are holomorphically convex, then so are all the topological components of $U_1 \cap U_2$.*

Exercise **2.6.12***: Let $n \geq 2$ and $U \subset \mathbb{C}^n$ a domain.*
   a) Let $K \subset\subset U$ be nonempty and compact. Prove $U \setminus K$ is not holomorphically convex.
   b) Prove that if U is bounded and holomorphically convex, then $\mathbb{C}^n \setminus U$ is connected.
   c) Find an unbounded holomorphically convex domain U where $\mathbb{C}^n \setminus U$ is disconnected.

The set $\mathbb{C}^n$ is both holomorphically convex and a domain of holomorphy. These two notions are equivalent also for all other domains in $\mathbb{C}^n$.

Theorem 2.6.3 (Cartan–Thullen). *Let $U \subsetneq \mathbb{C}^n$ be a domain. The following are equivalent:*

(i) *U is a domain of holomorphy.*

(ii) *For all $K \subset\subset U$, $\operatorname{dist}(K, \partial U) = \operatorname{dist}(\widehat{K}_U, \partial U)$.*

(iii) *U is holomorphically convex.*

Proof. We start with (i) $\Rightarrow$ (ii). Consider a $K \subset\subset U$ with $\operatorname{dist}(K, \partial U) > \operatorname{dist}(\widehat{K}_U, \partial U)$. After a possible rotation by a unitary, there is a point $p \in \widehat{K}_U$ and a polydisc $\Delta = \Delta_r(0)$ with polyradius $r = (r_1, \ldots, r_n)$ such that $p + \Delta = \Delta_r(p)$ contains a point of ∂U, but

$$K + \Delta = \bigcup_{q \in K} \Delta_r(q) \subset\subset U.$$

See Figure 2.17.

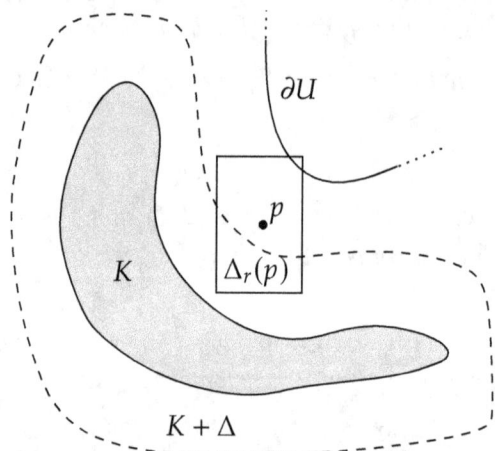

Figure 2.17: Point in the hull closer to the boundary than closest point of K.

If $f \in \mathcal{O}(U)$, then there is an $M > 0$ such that $|f| \leq M$ on $K + \Delta$ as that is a relatively compact set. By the Cauchy estimates for each $q \in K$, we get

$$\left| \frac{\partial^\alpha f}{\partial z^\alpha}(q) \right| \leq \frac{M\alpha!}{r^\alpha}.$$

This inequality therefore holds on $\widehat{K}_U$ and hence at p. The series

$$\sum_\alpha \frac{1}{\alpha!} \frac{\partial^\alpha f}{\partial z^\alpha}(p)(z - p)^\alpha$$

converges in $\Delta_r(p)$. The function f extends to all of $\Delta_r(p)$ and $\Delta_r(p)$ contains points outside of U. In other words, U is not a domain of holomorphy.

The implication (ii) $\Rightarrow$ (iii) is immediate.

Finally, we prove (iii) $\Rightarrow$ (i). Suppose U is holomorphically convex. Let $p \in \partial U$. By convexity, choose nested compact sets $K_{\ell-1} \subsetneq K_\ell \subset\subset U$ such that $\bigcup_\ell K_\ell = U$, and $\widehat{(K_\ell)}_U = K_\ell$. As the sets exhaust U, we can perhaps pass to a subsequence to ensure that there exists a sequence of points $p_\ell \in K_\ell \setminus K_{\ell-1}$ such that $\lim_{\ell\to\infty} p_\ell = p$.

Choose $f_1 \in \mathcal{O}(U)$ so that $f_1(p_1) \geq 1$. Proceed inductively. As p_ℓ is not in the hull of $K_{\ell-1}$, there is a function $f_\ell \in \mathcal{O}(U)$ such that $|f_\ell| \leq 2^{-\ell}$ on $K_{\ell-1}$, but

$$|f_\ell(p_\ell)| \geq \ell + \left| \sum_{k=1}^{\ell-1} f_k(p_\ell) \right|.$$

Finding such a function is left as an exercise below. For every ℓ, the series $\sum_{k=1}^\infty f_k(z)$ converges uniformly on K_ℓ as for all $k > \ell, |f_k| \leq 2^{-k}$ on K_ℓ. As the K_ℓ exhaust U, the series converges uniformly on compact subsets of U. Consequently,

$$f(z) = \sum_{k=1}^\infty f_k(z)$$

is a holomorphic function on U. We bound

$$|f(p_\ell)| \geq |f_\ell(p_\ell)| - \left| \sum_{k=1}^{\ell-1} f_k(p_\ell) \right| - \left| \sum_{k=\ell+1}^\infty f_k(p_\ell) \right| \geq \ell - \sum_{k=\ell+1}^\infty 2^{-k} \geq \ell - 1.$$

So $\lim_{\ell\to\infty} f(p_\ell) = \infty$. Clearly there cannot be any open $W \subset \mathbb{C}^n$ containing p to which f extends (see definition of domain of holomorphy). As every connected open W such that $W \setminus U \neq \emptyset$ and $W \setminus U^c \neq \emptyset$ must contain a point of ∂U, we are done. $\square$

By Exercise 2.6.8, holomorphic convexity is a biholomorphic invariant. Thus, being a domain of holomorphy is also a biholomorphic invariant. This fact is not easy to prove from the definition of a domain of holomorphy, as the biholomorphism is defined only on the interior of our domains.

Holomorphic convexity is an intrinsic notion; it does not require knowing anything about points outside of U. It is a better way to think about domains of holomorphy. Holomorphic convexity generalizes easily to more complicated complex manifolds*, while the notion of a domain of holomorphy only makes sense for domains in $\mathbb{C}^n$.

---

*Manifolds with complex structure, i.e., "manifolds with multiplication by i on the tangent space."

Exercise 2.6.13 (Behnke–Stein again): *Show that the union $\bigcup_\ell U_\ell$ of a nested sequence of holomorphically convex domains $U_{\ell-1} \subset U_\ell \subset \mathbb{C}^n$ is holomorphically convex.*

Exercise 2.6.14: *Prove the existence of the function $f_\ell \in \mathcal{O}(U)$ as indicated in the proof of Cartan–Thullen above.*

Exercise 2.6.15: *Show that if $U \subset \mathbb{C}^n$ is holomorphically convex, then there exists a single function $f \in \mathcal{O}(U)$ that does not extend through any point $p \in \partial U$.*

Exercise 2.6.16: *We know $U = \mathbb{C}^2 \setminus \{z \in \mathbb{C}^2 : z_1 = 0\}$ is a domain of holomorphy. Use part (ii) of the theorem to show that if $W \subset \mathbb{C}^2$ is a domain of holomorphy and $U \subset W$, then either $W = U$ or $W = \mathbb{C}^2$. Hint: Suppose $L \subset W$ is a complex line and K is a circle in L. What is $\widehat{K}_W$?*

In the following series of exercises, which you should do in order, you will solve the Levi problem (and more) for complete Reinhardt domains. Recall that a domain U is a complete Reinhardt domain if whenever $(z_1, \ldots, z_n)$ is in U and $r_k = |z_k|$, then the entire closed polydisc $\overline{\Delta_r(0)} \subset U$. We say a complete Reinhardt domain U is *logarithmically convex* if there exists a (geometrically) convex $C \subset \mathbb{R}^n$ such that $z \in U^*$ if and only if $(\log|z_1|, \ldots, \log|z_n|) \in C$, where $U^* = \{z \in U : z_1 \neq 0, \ldots, z_n \neq 0\}$.

Exercise 2.6.17: *Prove that a logarithmically convex complete Reinhardt domain is the intersection of sets of the form*

$$\{z \in \mathbb{C}^n : \alpha_1 \log|z_1| + \cdots + \alpha_n \log|z_n| < \beta\} = \{z \in \mathbb{C}^n : |z_1|^{\alpha_1} \cdots |z_n|^{\alpha_n} < e^\beta\}$$

for some nonnegative $\alpha_1, \ldots, \alpha_n$, and $\beta \in \mathbb{R}$.

Exercise 2.6.18: *Prove that if a complete Reinhardt domain is Hartogs pseudoconvex, then it is logarithmically convex.*

Exercise 2.6.19: *Let $\alpha_1, \ldots, \alpha_n \geq 0$ and $\beta \in \mathbb{R}$. For each $k \in \mathbb{N}_0$, let $\ell_m^k \in \mathbb{N}_0$ be the smallest nonnegative integer such that $\ell_m^k \geq k\alpha_m$. Prove that the domain of convergence of the power series*

$$\sum_{k=0}^{\infty} e^{-k\beta} z_1^{\ell_1^k} \cdots z_n^{\ell_n^k}$$

is precisely $\{z \in \mathbb{C}^n : |z_1|^{\alpha_1} \cdots |z_n|^{\alpha_n} < e^\beta\}$. Hint: That it diverges outside is easy, what is hard is that it converges inside. It is perhaps useful to notice that $\frac{\ell_m^k}{k} - \alpha_m \leq \frac{1}{k}$, and that for z in the set, there is some $\epsilon > 0$ such that $(1 + \epsilon)|z_1|^{\alpha_1} \cdots |z_n|^{\alpha_n} = e^\beta$.

Exercise 2.6.20: *Prove that if a complete Reinhardt domain is logarithmically convex, then it is holomorphically convex and therefore a domain of holomorphy.*

Exercise 2.6.21: *Prove that a complete Reinhardt domain is a domain of holomorphy if and only if it is the domain of convergence of some power series at the origin. Hint: There is a function that does not extend past any boundary point of a holomorphically convex domain.*

We (you) have proved the following proposition.

Proposition 2.6.4. *Let* $U \subset \mathbb{C}^n$ *be a complete Reinhardt domain. Then the following are equivalent:*

(i) *U is logarithmically convex.*

(ii) *U is a domain of holomorphy.*

(iii) *U is a domain of convergence of some power series at the origin.*

(iv) *U is Hartogs pseudoconvex.*

3 \ CR Functions

3.1 \ Real-analytic functions and complexification

Definition 3.1.1. Let $U \subset \mathbb{R}^n$ be open. A function $f: U \to \mathbb{C}$ is *real-analytic* (or simply *analytic* if clear from context) if at each point $p \in U$, the function f admits a power series that converges (absolutely) to f in some neighborhood of p. A common notation for real-analytic is C^ω.

Before we discuss the connection between real-analytic and holomorphic functions, we prove a simple lemma.

Lemma 3.1.2. *Let $\mathbb{R}^n \subset \mathbb{C}^n$ be the natural inclusion and $V \subset \mathbb{C}^n$ a domain such that $V \cap \mathbb{R}^n \neq \emptyset$. Suppose $f, g: V \to \mathbb{C}$ are holomorphic functions such that $f = g$ on $V \cap \mathbb{R}^n$. Then $f = g$ on V.*

Proof. Considering $f - g$, we may assume that $g = 0$. Let $z = x + iy$ as usual so that $\mathbb{R}^n$ is given by $y = 0$. Our assumption is that $f = 0$ when $y = 0$, so the derivative of f with respect to x_k is zero. When $y = 0$, the Cauchy–Riemann equations say

$$0 = \frac{\partial f}{\partial x_k} = -i \frac{\partial f}{\partial y_k}.$$

Therefore, on $y = 0$,

$$\frac{\partial f}{\partial z_k} = 0.$$

The derivative $\frac{\partial f}{\partial z_k}$ is holomorphic and $\frac{\partial f}{\partial z_k} = 0$ on $y = 0$. By induction, all holomorphic derivatives of f at $p \in \mathbb{R}^n \cap V$ vanish, and f has a zero power series. Hence f is identically zero in a neighborhood of p in $\mathbb{C}^n$. By the identity theorem, f is zero on all of V. $\qquad \square$

We return to $\mathbb{R}^n$ for a moment. We write a power series in $\mathbb{R}^n$ in multi-index notation as usual. Suppose that for some $a \in \mathbb{R}^n$ and some polyradius $r = (r_1, \ldots, r_n)$, the series

$$\sum_\alpha c_\alpha (x - a)^\alpha$$

converges whenever $|x_k - a_k| < r_k$ for all k. Here convergence is absolute convergence. That is,

$$\sum_\alpha |c_\alpha| |x - a|^\alpha$$

converges. If we replace $x_k \in \mathbb{R}$ with $z_k \in \mathbb{C}$ such that $|z_k - a_k| \le |x_k - a_k|$, then the series still converges. Hence the series

$$\sum_\alpha c_\alpha (z - a)^\alpha$$

converges absolutely in $\Delta_r(a) \subset \mathbb{C}^n$.

Proposition 3.1.3 (Complexification part I). *Suppose $U \subset \mathbb{R}^n$ is a domain and $f \colon U \to \mathbb{C}$ is real-analytic. Let $\mathbb{R}^n \subset \mathbb{C}^n$ be the natural inclusion. Then there exists a domain $V \subset \mathbb{C}^n$ such that $U \subset V$ and a unique holomorphic function $F \colon V \to \mathbb{C}$ such that $F|_U = f$.*

One of many consequences of this proposition is that a real-analytic function is C^∞ smooth. Be careful and notice that U is a domain in $\mathbb{R}^n$, but it is not an open set when considered as a subset of $\mathbb{C}^n$. Furthermore, V may be a very "thin" neighborhood around U and there is no way of finding V just from knowing U. You need to also know f. As an example, consider $f(x) = \frac{1}{\epsilon^2 + x^2}$ for $\epsilon > 0$, which is real-analytic on $\mathbb{R}$, but the complexification is not holomorphic at $\pm \epsilon i$.

Proof. We proved the local version already. We must prove that if we extend our f near every point, we always get the same function. That follows from Lemma 3.1.2; any two such functions are equal on $\mathbb{R}^n$, and hence equal. There is a subtle topological technical point in this, so let us elaborate. A key topological fact is that we define V as a union of the polydiscs where the series converges. If a point p is in two different such polydiscs, we need to show that the two definitions of F are the same at p. The intersection of two polydiscs is connected, and in this case it also contains a piece of $\mathbb{R}^n$, and we may apply the lemma. $\qquad\square$

Exercise 3.1.1: Prove the identity theorem for real-analytic functions. That is, if $U \subset \mathbb{R}^n$ is a domain, $f \colon U \to \mathbb{R}$ a real-analytic function, and f is zero on a nonempty open subset of U, then f is identically zero.

Exercise 3.1.2: Suppose $U \subset \mathbb{R}^n$ is a domain and $f \colon U \to \mathbb{R}$ is a real-analytic function such that $f|_W$ is harmonic for some nonempty open $W \subset U$. Prove that f is harmonic.

Exercise 3.1.3: Let $(0, 1) \subset \mathbb{R}$. Construct a real-analytic function on $(0, 1)$ that does not complexify to the rectangle $(0, 1) + i(-\epsilon, \epsilon) \subset \mathbb{C}$ for any $\epsilon > 0$. Why does this not contradict the proposition?

A polynomial $P(x)$ in n real variables $(x_1, \ldots, x_n)$ is homogeneous of degree d if $P(sx) = s^d P(x)$ for all $s \in \mathbb{R}$ and $x \in \mathbb{R}^n$. A homogeneous polynomial of degree d is a polynomial whose every monomial is of total degree d. If f is real-analytic near $a \in \mathbb{R}^n$, then write the power series of f at a as

$$\sum_{m=0}^{\infty} f_m(x - a),$$

where f_m is a homogeneous polynomial of degree m. The f_m is called the *degree m homogeneous part* of f at a.

There is usually a better way to complexify real-analytic functions in $\mathbb{C}^n$. Suppose $U \subset \mathbb{C}^n \cong \mathbb{R}^{2n}$, and $f : U \to \mathbb{C}$ is real-analytic. Assume $a = 0 \in U$ for simplicity. Writing $z = x + iy$, near 0,

$$f(x, y) = \sum_{m=0}^{\infty} f_m(x, y) = \sum_{m=0}^{\infty} f_m \left(\frac{z + \bar{z}}{2}, \frac{z - \bar{z}}{2i} \right).$$

The polynomial f_m becomes a homogeneous polynomial of degree m in the variables z and $\bar{z}$. The series becomes a power series in z and $\bar{z}$. We simply write the function as $f(z, \bar{z})$, and we consider the power series representation in z and $\bar{z}$ rather than in x and y. In multi-index notation, we write a power series at $a \in \mathbb{C}^n$ as

$$\sum_{\alpha, \beta} c_{\alpha, \beta} (z - a)^\alpha (\bar{z} - \bar{a})^\beta.$$

A holomorphic function is real-analytic, but not vice versa. A holomorphic function is a real-analytic function that does not depend on $\bar{z}$.

Before we discuss complexification in terms of z and $\bar{z}$, we need a lemma.

Lemma 3.1.4. *Let $V \subset \mathbb{C}^n \times \mathbb{C}^n$ be a domain, let the coordinates be $(z, \zeta) \in \mathbb{C}^n \times \mathbb{C}^n$, let*

$$D = \left\{ (z, \zeta) \in \mathbb{C}^n \times \mathbb{C}^n : \zeta = \bar{z} \right\},$$

and suppose $D \cap V \neq \emptyset$. Suppose $f, g : V \to \mathbb{C}$ are holomorphic functions such that $f = g$ on $D \cap V$. Then $f = g$ on all of V.

The set D is sometimes called the *diagonal*.

Proof. Without loss of generality, assume that $g = 0$. For $(z, \bar{z}) \in V$, we have $f(z, \bar{z}) = 0$, which is really f composed with the map taking z to $(z, \bar{z})$. This composition is identically zero, so applying Wirtinger operators yields zero. Using the chain rule,

$$0 = \frac{\partial}{\partial \bar{z}_k} \left[f(z, \bar{z}) \right] = \frac{\partial f}{\partial \zeta_k}(z, \bar{z}).$$

Similarly with the z_k,

$$0 = \frac{\partial}{\partial z_k} \left[f(z, \bar{z}) \right] = \frac{\partial f}{\partial z_k}(z, \bar{z}).$$

Either way, we get another holomorphic function in z and ζ that is zero on D. By induction, for all α and β we get

$$0 = \frac{\partial^{|\alpha|+|\beta|}}{\partial z^\alpha \partial \bar{z}^\beta}\left[f(z,\bar{z})\right] = \frac{\partial^{|\alpha|+|\beta|}f}{\partial z^\alpha \partial \zeta^\beta}(z,\bar{z}).$$

All holomorphic derivatives in z and ζ of f are zero on every point $(z,\bar{z})$, so the power series is zero at every point $(z,\bar{z})$, and so f is identically zero in a neighborhood of every point $(z,\bar{z})$. The lemma follows by the identity theorem. □

Let f be a real-analytic function. Suppose the series (in multi-index notation)

$$f(z,\bar{z}) = \sum_{\alpha,\beta} c_{\alpha,\beta}(z-a)^\alpha(\bar{z}-\bar{a})^\beta$$

converges in a polydisc $\Delta_r(a) \subset \mathbb{C}^n$. By convergence we mean absolute convergence,

$$\sum_{\alpha,\beta}|c_{\alpha,\beta}|\,|z-a|^\alpha|\bar{z}-\bar{a}|^\beta$$

converges. The series still converges if we replace $\bar{z}_k$ with ζ_k where $|\zeta_k - \bar{a}| \leq |\bar{z}_k - \bar{a}|$. So the series

$$F(z,\zeta) = \sum_{\alpha,\beta} c_{\alpha,\beta}(z-a)^\alpha(\zeta-\bar{a})^\beta$$

converges (absolutely) for all $(z,\zeta) \in \Delta_r(a) \times \Delta_r(\bar{a})$.

Putting together the discussion above with the lemma we obtain:

Proposition 3.1.5 (Complexification part II). *Suppose $U \subset \mathbb{C}^n$ is a domain and $f \colon U \to \mathbb{C}$ is real-analytic. Then there exists a domain $V \subset \mathbb{C}^n \times \mathbb{C}^n$ such that*

$$\{(z,\zeta) : \zeta = \bar{z} \text{ and } z \in U\} \subset V,$$

and a unique holomorphic function $F \colon V \to \mathbb{C}$ such that $F(z,\bar{z}) = f(z,\bar{z})$ for all $z \in U$.

The function f can be thought of as the restriction of F to the set where $\zeta = \bar{z}$. We will abuse notation and write simply $f(z,\zeta)$ both for f and its extension. The reason for this abuse is evident from the computations above. What we are calling f is a function of $(z,\bar{z})$ if thinking of it as a function on the diagonal where $\zeta = \bar{z}$, or it is a function of z if thinking of it as just the function $z \mapsto f(z,\bar{z})$, or it is the function $(z,\zeta) \mapsto f(z,\zeta)$. We have the following commutative diagram:

$$U \subset \mathbb{C}^n \xrightarrow{\ z \mapsto (z,\bar{z})\ } V \subset \mathbb{C}^n \times \mathbb{C}^n$$

with f (from U) and $f\ (=F)$ (from V) both going to $\mathbb{C}$.

All three ways of going from one place to another in the diagram we are calling f. The arrow from V was called F in the proposition. The notation plays well with differentiation and the Wirtinger operators. Differentiating f (really the F in the proposition) in ζ_k and evaluating at $(z, \bar{z})$ is the same thing as evaluating at $(z, \bar{z})$ and then differentiating in $\bar{z}_k$ using the Wirtinger operator:

$$\frac{\partial F}{\partial \zeta_k}(z, \bar{z}) = \frac{\partial f}{\partial \zeta_k}(z, \bar{z}) = \frac{\partial}{\partial \bar{z}_k}\left[f(z, \bar{z})\right] = \frac{\partial f}{\partial \bar{z}_k}(z, \bar{z}).$$

If we squint our mind's eye, we can't quite see the difference between $\bar{z}$ and ζ. We have already used this idea for smooth functions, but for real-analytic functions we can treat z and $\bar{z}$ as truly independent variables. The abuse of notation is entirely justified, at least once it is understood well.

Remark 3.1.6. The domain V in the proposition is not simply U times the conjugate of U. In general, it is smaller. For example, a real-analytic $f : \mathbb{C}^n \to \mathbb{C}$ does not necessarily complexify to all of $\mathbb{C}^n \times \mathbb{C}^n$. That is because the domain of convergence for a real-analytic function on $\mathbb{C}^n$ is not necessarily all of $\mathbb{C}^n$. In one dimension,

$$f(z, \bar{z}) = \frac{1}{1 + |z|^2}$$

is real-analytic on $\mathbb{C}$, but it is not a restriction to the diagonal of a holomorphic function defined on all of $\mathbb{C}^2$. The problem is that the complexified function

$$f(z, \zeta) = \frac{1}{1 + z\zeta}$$

is undefined on the set where $z\zeta = -1$, which by a fluke never happens when $\zeta = \bar{z}$.

Remark 3.1.7. This form of complexification is sometimes called *polarization* due to its relation to the polarization identities[*]: We can recover a Hermitian matrix A, and therefore the sesquilinear form $\langle Az, w \rangle$ for $z, w \in \mathbb{C}^n$, by simply knowing the value of

$$\langle Az, z \rangle = z^* A z = \sum_{k, \ell = 1}^{n} a_{k\ell} \bar{z}_k z_\ell$$

for all $z \in \mathbb{C}^n$. In fact, under the hood, Proposition 3.1.5 is polarization in an infinite-dimensional Hilbert space, but we digress.

Treating $\bar{z}$ as a separate variable is a very powerful idea, and as we have just seen it is completely natural for real-analytic functions. This is one of the reasons why real-analytic functions play a special role in complex analysis.

Exercise 3.1.4: Let $U \subset \mathbb{C}^n$ be an open set and $\varphi : U \to \mathbb{R}$ a pluriharmonic function. Prove that φ is real-analytic.

---

[*]Such as $4\langle z, w \rangle = \|z + w\|^2 - \|z - w\|^2 + i\left(\|z + iw\|^2 - \|z - iw\|^2\right)$.

Exercise 3.1.5: *Let $U \subset \mathbb{C}^n$ be an open set, $z_0 \in U$. Suppose $\varphi \colon U \to \mathbb{R}$ is a pluriharmonic function. You know that φ is real-analytic. Using complexification, write down a formula for a holomorphic function near z_0 whose real part is φ.*

Exercise 3.1.6: *Let $U \subset \mathbb{C}^n$ be a domain, and suppose $f, g \in \mathcal{O}(U)$. Suppose that $f = \bar{g}$ on U. Use complexification (complexify $f - \bar{g}$) to show that both f and g are constant.*

Example 3.1.8: Not every C^∞ smooth function is real-analytic. For $x \in \mathbb{R}$, define

$$f(x) = \begin{cases} e^{-1/x} & \text{if } x > 0, \\ 0 & \text{if } x \leq 0. \end{cases}$$

The function $f \colon \mathbb{R} \to \mathbb{R}$ is C^∞ and $f^{(k)}(0) = 0$ for all k. The Taylor series of f at the origin does not converge to f in any neighborhood of the origin; it converges to the zero function but not to f. Consequently, there is no neighborhood V of the origin in $\mathbb{C}$ such that f is the restriction to $V \cap \mathbb{R}$ of a holomorphic function in V.

Exercise 3.1.7: *Prove the statements of the example above.*

Definition 3.1.9. A real hypersurface $M \subset \mathbb{R}^n$ is said to be *real-analytic* if locally at every point it is the graph of a real-analytic function. That is, near every point (locally), after perhaps relabeling coordinates, M can be written as a graph

$$y = \varphi(x),$$

where φ is real-analytic, $(x, y) \in \mathbb{R}^{n-1} \times \mathbb{R} = \mathbb{R}^n$.

Compare this definition to Definition 2.2.1. We could define a real-analytic hypersurface as in Definition 2.2.1 and then prove an analogue of Proposition 2.2.9 to show that this new definition is identical to the definition above. We avoid this complication, leaving it to the interested reader.

Exercise 3.1.8: *Show that Definition 3.1.9 is equivalent to an analogue of Definition 2.2.1. That is, state the alternative definition of real-analytic hypersurface and prove the analogue of Proposition 2.2.9.*

A mapping to $\mathbb{R}^m$ is real-analytic if all the components are real-analytic functions. Via complexification we give a simple proof of the following result.

Proposition 3.1.10. *Let $U \subset \mathbb{R}^n$, $V \subset \mathbb{R}^k$ be open and let $f \colon U \to V$ and $g \colon V \to \mathbb{R}^m$ be real-analytic. Then $g \circ f$ is real-analytic.*

Proof. Let $x \in \mathbb{R}^n$ be our coordinates in U and $y \in \mathbb{R}^k$ be our coordinates in V. Complexify $f(x)$ and $g(y)$ by allowing x to be a complex vector in a small neighborhood of U in $\mathbb{C}^n$ and y to be a complex vector in a small neighborhood of V in $\mathbb{C}^k$. So treat f and g as holomorphic functions. On a certain neighborhood of U in $\mathbb{C}^n$, the composition $g \circ f$ makes sense and it is holomorphic, as composition of holomorphic mappings is holomorphic. Restricting the complexified $g \circ f$ back to $\mathbb{R}^n$ we obtain a real-analytic function.   □

The proof demonstrates a simple application of complexification. Many properties of holomorphic functions are easy to prove because holomorphic functions are solutions to certain PDEs (the Cauchy–Riemann equations). There is no PDE that defines real-analytic functions, so complexification provides a useful tool to transfer certain properties of holomorphic functions to real-analytic functions. We must be careful, however. Hypotheses on real-analytic functions only give us those hypotheses at certain points for the complexified holomorphic functions.

Exercise 3.1.9: *Demonstrate the point about complexification we made just above. Find a nonconstant bounded real-analytic $f : \mathbb{R}^n \to \mathbb{R}$ that happens to complexify to $\mathbb{C}^n$.*

Exercise 3.1.10: *Let $U \subset \mathbb{R}^n$ be open. Let $\varphi : (0, 1) \to U$ be a real-analytic function (curve), and let $f : U \to \mathbb{R}$ be real-analytic. Suppose that $(f \circ \varphi)(t) = 0$ for all $t \in (0, \epsilon)$ for some $\epsilon > 0$. Prove that f is zero on the image $\varphi((0, 1))$.*

3.2 \ CR functions

We first need to know what it means for a function $f : X \to \mathbb{C}$ to be smooth if X is not an open set, for example, if X is a hypersurface.

Definition 3.2.1. Let $X \subset \mathbb{R}^n$ be a set. The function $f : X \to \mathbb{C}$ is smooth (resp. real-analytic) if for each point $p \in X$ there is a neighborhood $U \subset \mathbb{R}^n$ of p and a smooth (resp. real-analytic) $F : U \to \mathbb{C}$ such that $F(q) = f(q)$ for $q \in X \cap U$.

For an arbitrary set X, issues surrounding this definition can be rather subtle. The definition is easy to work with, however, if X is nice, such as a hypersurface, or if X is a closure of a domain with smooth boundary.

Proposition 3.2.2. *Suppose $M \subset \mathbb{R}^n$ is a smooth (resp. real-analytic) real hypersurface. A function $f : M \to \mathbb{C}$ is smooth (resp. real-analytic) if and only if whenever near any point we write M in coordinates $(x, y) \in \mathbb{R}^{n-1} \times \mathbb{R}$ as*

$$y = \varphi(x),$$

for a smooth (resp. real-analytic) function φ, then $f(x, \varphi(x))$ is a smooth (resp. real-analytic) function of x.

Exercise **3.2.1:** *Prove the proposition.*

Exercise **3.2.2:** *Prove that if M is a smooth or real-analytic hypersurface, and $f : M \to \mathbb{C}$ is smooth or real-analytic, then the function F from the definition is never unique, even for a fixed neighborhood U.*

Exercise **3.2.3:** *Suppose $M \subset \mathbb{R}^n$ is a smooth hypersurface, $f : M \to \mathbb{C}$ is a smooth function, $p \in M$, and $X_p \in T_p M$. Prove that $X_p f$ is well-defined. That is, suppose U is a neighborhood of p, $F : U \to \mathbb{C}$ and $G : U \to \mathbb{C}$ are smooth functions that both equal f on $U \cap M$. Prove that $X_p F = X_p G$.*

Due to the last exercise, we can apply vectors of $T_p M$ to a smooth function on a hypersurface by simply applying them to any smooth extension. We can similarly apply vectors of $\mathbb{C} T_p M$ to smooth functions on M, as $\mathbb{C} T_p M$ is simply the complex span of vectors in $T_p M$.

Definition 3.2.3. Let $M \subset \mathbb{C}^n$ be a smooth real hypersurface. A smooth $f : M \to \mathbb{C}$ is a *smooth CR function* if

$$X_p f = 0$$

for all $p \in M$ and all vectors $X_p \in T_p^{(0,1)} M$.

Remark 3.2.4. One only needs one derivative (rather than C^∞) in the definition above. One can even define a continuous CR function if the derivative is taken in the distribution sense, but we digress.

Remark 3.2.5. When $n = 1$, a real hypersurface $M \subset \mathbb{C}$ is a curve and $T_p^{(0,1)} M$ is trivial. Therefore, all smooth functions $f : M \to \mathbb{C}$ are CR functions.

Proposition 3.2.6. *Let $M \subset U$ be a smooth (resp. real-analytic) real hypersurface in an open $U \subset \mathbb{C}^n$. Suppose $F : U \to \mathbb{C}$ is a holomorphic function, then the restriction $f = F|_M$ is a smooth (resp. real-analytic) CR function.*

Proof. First let us prove that f is smooth. The function F is smooth and defined on a neighborhood of every point of M, and so it can be used in the definition. Similarly for real-analytic.

Let us show f is CR at some $p \in M$. Differentiating f with vectors in $\mathbb{C} T_p M$ is the same as differentiating F. As $T_p^{(0,1)} M \subset T_p^{(0,1)} \mathbb{C}^n$, we have

$$X_p f = X_p F = 0 \qquad \text{for all} \quad X_p \in T_p^{(0,1)} M. \qquad \square$$

On the other hand, not every smooth CR function is a restriction of a holomorphic function.

Example 3.2.7: Take the smooth function $f\colon \mathbb{R} \to \mathbb{R}$ we defined before that is not real-analytic at the origin. Let $M \subset \mathbb{C}^2$ be the real-analytic hypersurface defined by $\operatorname{Im} z_2 = 0$. Clearly, $T_p^{(0,1)}M$ is one-complex-dimensional, and at each $p \in M$, $\frac{\partial}{\partial \bar{z}_1}\big|_p$ is tangent and spans $T_p^{(0,1)}M$. Define $g\colon M \to \mathbb{C}$ by

$$g(z_1, z_2, \bar{z}_1, \bar{z}_2) = f(\operatorname{Re} z_2).$$

Then g is CR as it is independent of $\bar{z}_1$. If $G\colon U \subset \mathbb{C}^2 \to \mathbb{C}$ is a holomorphic function where U is some open set containing the origin, then G restricted to M must be real-analytic (a power series in $\operatorname{Re} z_1$, $\operatorname{Im} z_1$, and $\operatorname{Re} z_2$) and therefore G cannot equal to g on M.

Exercise 3.2.4: Suppose $M \subset \mathbb{C}^n$ is a smooth real hypersurface and $f\colon M \to \mathbb{C}$ is a CR function that is a restriction of a holomorphic function $F\colon U \to \mathbb{C}$ defined in some neighborhood $U \subset \mathbb{C}^n$ of M. Show that F is unique, that is, if $G\colon U \to \mathbb{C}$ is another holomorphic function such that $G|_M = f = F|_M$, then $G = F$.

Exercise 3.2.5: Show that there is no maximum principle of CR functions. In fact, find a smooth real hypersurface $M \subset \mathbb{C}^n$, $n \geq 2$, and a smooth CR function f on M such that $|f|$ attains a strict maximum at a point.

Exercise 3.2.6: Suppose $M \subset \mathbb{C}^n$, $n \geq 2$, is the real hypersurface given by $\operatorname{Im} z_n = 0$. Show that every smooth CR function on M is holomorphic in the variables $z_1, \ldots, z_{n-1}$. Use this to show that for no smooth CR function f on M can $|f|$ attain a strict maximum on M. Show that there does exist a smooth nonconstant CR function f such that $|f|$ attains a (nonstrict) maximum on M.

Real-analytic CR functions on a real-analytic hypersurface M always extend to holomorphic functions of a neighborhood of M. To prove this fact, we complexify everything, that is, we treat the zs and $\bar{z}$s as separate variables. The standard way of writing a hypersurface as a graph is not as convenient for this setting, so we prove that a real-analytic hypersurface is a graph of a holomorphic function in the complexified variables restricted to the diagonal. That is, using variables (z, w), we write M as a graph of $\bar{w}$ over z, $\bar{z}$, and w. We can then eliminate $\bar{w}$ in any real-analytic expression.

Proposition 3.2.8. *Suppose $M \subset \mathbb{C}^n$ is a real-analytic hypersurface and $p \in M$. Then after a translation and rotation by a unitary matrix, $p = 0$, and near the origin in coordinates $(z, w) \in \mathbb{C}^{n-1} \times \mathbb{C}$, the hypersurface M is given by*

$$\bar{w} = \Phi(z, \bar{z}, w),$$

where $\Phi(z, \zeta, w)$ is a holomorphic function defined on a neighborhood of the origin in $\mathbb{C}^{n-1} \times \mathbb{C}^{n-1} \times \mathbb{C}$, such that Φ, $\frac{\partial \Phi}{\partial z_k}$, $\frac{\partial \Phi}{\partial \zeta_k}$ vanish at the origin for all k, and $w = \bar{\Phi}(\zeta, z, \Phi(z, \zeta, w))$ for all z, ζ, and w.

A local basis for $T^{(0,1)}M$ vector fields is given by

$$\frac{\partial}{\partial \bar{z}_k} + \frac{\partial \Phi}{\partial \bar{z}_k}\frac{\partial}{\partial \bar{w}} \quad \left(= \frac{\partial}{\partial \bar{z}_k} + \frac{\partial \Phi}{\partial \zeta_k}\frac{\partial}{\partial \bar{w}} \right), \qquad k = 1, \ldots, n-1.$$

Finally, let $\mathcal{M}$ be the set in $(z, \zeta, w, \omega) \in \mathbb{C}^{n-1} \times \mathbb{C}^{n-1} \times \mathbb{C} \times \mathbb{C}$ coordinates given near the origin by $\omega = \Phi(z, \zeta, w)$. Then $\mathcal{M}$ is the unique complexification of M near the origin in the sense that if $f(z, \bar{z}, w, \bar{w})$ is a real-analytic function vanishing on M near the origin, then $f(z, \zeta, w, \omega)$ vanishes on $\mathcal{M}$ near the origin.

Again as a slight abuse of notation, Φ refers to both the function $\Phi(z, \zeta, w)$ and $\Phi(z, \bar{z}, w)$.

Proof. Translate and rotate so that M is given by

$$\operatorname{Im} w = \varphi(z, \bar{z}, \operatorname{Re} w),$$

where φ is $O(2)$. Write the defining function as $r(z, \bar{z}, w, \bar{w}) = -\frac{w - \bar{w}}{2i} + \varphi\left(z, \bar{z}, \frac{w + \bar{w}}{2}\right)$. Complexifying, consider $r(z, \zeta, w, \omega)$ as a holomorphic function of $2n$ variables, and let $\mathcal{M}$ be the set defined by $r(z, \zeta, w, \omega) = 0$. The derivative of r in ω (that is $\bar{w}$) does not vanish near the origin. Use the implicit function theorem for holomorphic functions to write $\mathcal{M}$ near the origin as

$$\omega = \Phi(z, \zeta, w).$$

Restrict to the diagonal, $\bar{w} = \omega$ and $\bar{z} = \zeta$, to get $\bar{w} = \Phi(z, \bar{z}, w)$. This is order 2 in the z and the $\bar{z}$ since φ is $O(2)$.

Because r is real-valued, then $r(z, \bar{z}, w, \bar{w}) = \overline{r(z, \bar{z}, w, \bar{w})} = \bar{r}(\bar{z}, z, \bar{w}, w)$. Complexify to obtain $r(z, \zeta, w, \omega) = \bar{r}(\zeta, z, \omega, w)$ for all (z, ζ, w, ω) near the origin. If $r(z, \zeta, w, \omega) = 0$, then

$$0 = \overline{r(z, \zeta, w, \omega)} = \overline{\bar{r}(\zeta, z, \omega, w)} = r(\bar{\zeta}, \bar{z}, \bar{\omega}, \bar{w}) = 0.$$

So, $(z, \zeta, w, \omega) \in \mathcal{M}$ if and only if $(\bar{\zeta}, \bar{z}, \bar{\omega}, \bar{w}) \in \mathcal{M}$. Near the origin, $(z, \zeta, w, \omega) \in \mathcal{M}$ if and only if $\omega = \Phi(z, \zeta, w)$, and hence if and only if $\bar{w} = \Phi(\bar{\zeta}, \bar{z}, \bar{\omega})$. Conjugating, we get that $\mathcal{M}$ is also given by

$$w = \bar{\Phi}(\zeta, z, \omega).$$

As $\big(z, \zeta, w, \Phi(z, \zeta, w)\big) \in \mathcal{M}$, for all z, ζ, and w,

$$w = \bar{\Phi}\big(\zeta, z, \Phi(z, \zeta, w)\big).$$

The vector field $X_k = \frac{\partial}{\partial \bar{z}_k} + \frac{\partial \Phi}{\partial \bar{z}_k}\frac{\partial}{\partial \bar{w}}$ annihilates the function $\Phi(z, \bar{z}, w) - \bar{w}$, but that is not enough. The vector field must annihilate a real defining function such as the real or imaginary part of $\Phi(z, \bar{z}, w) - \bar{w}$. So X_k must also annihilate the conjugate $\bar{\Phi}(\bar{z}, z, \bar{w}) - w$, at least on M. Compute, for $(z, w) \in M$,

$$
\begin{aligned}
X_k\big[\bar{\Phi}(\bar{z}, z, \bar{w}) - w\big] &= \frac{\partial\bar{\Phi}}{\partial\bar{z}_k}(\bar{z}, z, \bar{w}) + \frac{\partial\Phi}{\partial\bar{z}_k}(z, \bar{z}, w)\frac{\partial\bar{\Phi}}{\partial\bar{w}}(\bar{z}, z, \bar{w}) \\
&= \frac{\partial\bar{\Phi}}{\partial\bar{z}_k}(\bar{z}, z, \Phi(z, \bar{z}, \bar{w})) + \frac{\partial\Phi}{\partial\bar{z}_k}(z, \bar{z}, w)\frac{\partial\bar{\Phi}}{\partial\bar{w}}(\bar{z}, z, \Phi(z, \bar{z}, \bar{w})) \\
&= \frac{\partial}{\partial\bar{z}_k}\big[\bar{\Phi}(\bar{z}, z, \Phi(z, \bar{z}, w))\big] = \frac{\partial}{\partial\bar{z}_k}\big[w\big] = 0.
\end{aligned}
$$

The last claim of the proposition is left as an exercise.   □

Why do we say the last claim in the proposition proves the "uniqueness" of the complexification? Suppose we defined a complexification $\mathcal{M}'$ by another holomorphic equation $f = 0$. By the claim, $\mathcal{M} \subset \mathcal{M}'$, at least near the origin. If the derivative df is nonzero at the origin, then $f(z, \zeta, w, \Phi(z, \zeta, w)) = 0$ implies that $\frac{\partial f}{\partial\omega}$ is nonzero at the origin. Using the holomorphic implicit function theorem we can uniquely solve $f = 0$ for ω near the origin, that unique solution is Φ, and hence $\mathcal{M}' = \mathcal{M}$ near the origin.

As an example, recall that the sphere (minus a point) in $\mathbb{C}^2$ is biholomorphic to the hypersurface given by $\operatorname{Im} w = |z|^2$. That is, $\frac{w - \bar{w}}{2i} = z\bar{z}$. Solving for $\bar{w}$ and using ζ and ω obtains the equation for the complexification $\omega = -2iz\zeta + w$. Then $\Phi(z, \zeta, w) = -2iz\zeta + w$, and $\bar{\Phi}(\zeta, z, \omega) = 2i\zeta z + \omega$. Let us check that Φ is the right sort of function: $\bar{\Phi}(\zeta, z, \Phi(z, \zeta, w)) = 2i\zeta z + (-2iz\zeta + w) = w$. The CR vector field is given by $\frac{\partial}{\partial\bar{z}} + 2iz\frac{\partial}{\partial\bar{w}}$.

Exercise 3.2.7: *Finish the proof of the proposition: Let $M \subset \mathbb{C}^n$ be a real-analytic hypersurface given by $\bar{w} = \Phi(z, \bar{z}, w)$ near the origin, as in the proposition. Let $f(z, \bar{z}, w, \bar{w})$ be a real-analytic function such that $f = 0$ on M. Prove that the complexified $f(z, \zeta, w, \omega)$ vanishes on $\mathcal{M}$.*

Exercise 3.2.8: *In the proposition we only rotated and translated. Sometimes the following change of coordinates is also done. Prove that one can change coordinates (no longer linear) so that the Φ in the proposition is such that $\Phi(z, 0, w) = \Phi(0, \zeta, w) = w$ for all z, ζ, and w. These coordinates are called* normal coordinates.

Exercise 3.2.9: *Suppose Φ is a holomorphic function defined on a neighborhood of the origin in $\mathbb{C}^{n-1} \times \mathbb{C}^{n-1} \times \mathbb{C}$.*
- *a) Show that $\bar{w} = \Phi(z, \bar{z}, w)$ defines a real-analytic hypersurface near the origin if and only if $w = \bar{\Phi}(\zeta, z, \Phi(z, \zeta, w))$ for all z, ζ, and w. Hint: One direction was proved already.*
- *b) As an example, show that $\bar{w} = z\bar{z}$ does not satisfy the condition above, nor does it define a real hypersurface.*

Let us prove that real-analytic CR functions on real-analytic hypersurfaces are restrictions of holomorphic functions. To motivate the proof, consider a real-analytic function f on the circle $|z|^2 = z\bar{z} = 1$ (f is vacuously CR). This f is a restriction of a real-analytic function on a neighborhood of the circle, that we write $f(z, \bar{z})$. On the circle $\bar{z} = 1/z$. Thus, $F(z) = f(z, 1/z)$ is a holomorphic function defined on a neighborhood of the circle and equal to f on the circle. Our strategy then is to solve for one of the barred variables via Proposition 3.2.8 and hope the CR conditions take care of the rest of the barred variables in more than one dimension.

Theorem 3.2.9 (Severi). *Suppose $M \subset \mathbb{C}^n$ is a real-analytic hypersurface and $p \in M$. For every real-analytic CR function $f \colon M \to \mathbb{C}$, there exists a holomorphic function $F \in \mathcal{O}(U)$ for a neighborhood U of p such that $F(q) = f(q)$ for all $q \in M \cap U$.*

Proof. Write M near p as $\bar{w} = \Phi(z, \bar{z}, w)$. Let $\mathcal{M}$ be the set in the $2n$ variables (z, w, ζ, ω) given by $\omega = \Phi(z, \zeta, w)$. Take f and consider any real-analytic extension of f to a neighborhood of p and write it $f(z, w, \bar{z}, \bar{w})$. Complexify* as before to $f(z, w, \zeta, \omega)$. On $\mathcal{M}$ we have $f(z, w, \zeta, \omega) = f\big(z, w, \zeta, \Phi(z, \zeta, w)\big)$. Let

$$F(z, w, \zeta) = f\big(z, w, \zeta, \Phi(z, \zeta, w)\big).$$

Clearly $F(z, w, \bar{z})$ equals f on M. As f is a CR function, it is annihilated by $\frac{\partial}{\partial \bar{z}_k} + \frac{\partial \Phi}{\partial \bar{z}_k}\frac{\partial}{\partial \bar{w}}$ on M. So

$$\frac{\partial F}{\partial \zeta_k} + \frac{\partial \Phi}{\partial \zeta_k}\frac{\partial F}{\partial \omega} = \frac{\partial F}{\partial \zeta_k} = 0$$

on $M \subset \mathcal{M}$. We have a real-analytic function $\frac{\partial F}{\partial \zeta_k}(z, w, \bar{z})$ that is zero on M, so $\frac{\partial F}{\partial \zeta_k}(z, w, \zeta) = 0$ on $\mathcal{M}$ (Proposition 3.2.8 again). As $\frac{\partial F}{\partial \zeta_k}$ is a function only of z, w, and ζ (and not of ω), $\frac{\partial F}{\partial \zeta_k} = 0$ for all (z, w, ζ) in a neighborhood of the origin. Consequently, F does not depend on ζ, and F is actually a holomorphic function of z and w only and $F = f$ on M. $\qquad\square$

The most important place where we find CR functions that aren't necessarily real-analytic is as boundary values of holomorphic functions.

Proposition 3.2.10. *Suppose $U \subset \mathbb{C}^n$ is an open set with smooth boundary. Suppose $f \colon \overline{U} \to \mathbb{C}$ is a smooth function, holomorphic on U. Then $f|_{\partial U}$ is a smooth CR function.*

Proof. The function $f|_{\partial U}$ is clearly smooth.

Suppose $p \in \partial U$. If $X_p \in T_p^{(0,1)}\partial U$ is such that

$$X_p = \sum_{k=1}^{n} a_k \frac{\partial}{\partial \bar{z}_k}\Big|_p,$$

---

*At this point f stands for three distinct objects: the function on M, its real-analytic extension to a neighborhood in $\mathbb{C}^n$, and its complexification to a neighborhood of $(p, \bar{p})$ in $\mathbb{C}^n \times \mathbb{C}^n$.

then take a sequence $\{q_\ell\}$ in U that approaches p. Consider

$$X_{q_\ell} = \sum_{k=1}^{n} a_k \frac{\partial}{\partial \bar{z}_k}\bigg|_{q_\ell}.$$

Then $X_{q_\ell} f = 0$ for all ℓ and by continuity $X_p f = 0$.   □

The boundary values of a holomorphic function define the function uniquely. That is, if two holomorphic functions continuous up to the (smooth) boundary are equal on an open set of the boundary, then they are equal in the domain:

Proposition 3.2.11. *Suppose $U \subset \mathbb{C}^n$ is a domain with smooth boundary and $f \colon \overline{U} \to \mathbb{C}$ is a continuous function, holomorphic on U. If $f = 0$ on a nonempty open subset of ∂U, then $f = 0$ on all of U.*

Proof. Take $p \in \partial U$ such that $f = 0$ on a neighborhood of p in ∂U. Consider a small neighborhood Δ of p such that f is zero on $\partial U \cap \Delta$. Define $g \colon \Delta \to \mathbb{C}$ by setting $g(z) = f(z)$ if $z \in U$ and $g(z) = 0$ otherwise. See Figure 3.1. It is not hard to see that g is continuous, and it is clearly holomorphic where it is not zero. Radó's theorem (Theorem 2.4.12) says that g is holomorphic, and as it is zero on a nonempty open subset of Δ, it is identically zero on Δ, meaning f is zero on a nonempty open subset of U, and we are done by identity.

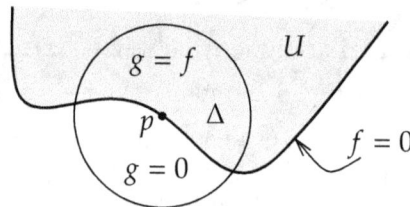

Figure 3.1: Extending a function zero on the boundary.

□

Exercise 3.2.10: Find a domain $U \subset \mathbb{C}^n$, $n \geq 2$, with smooth boundary and a smooth CR function $f \colon \partial U \to \mathbb{C}$ such that there is no holomorphic function on U or $\mathbb{C}^n \setminus U$ continuous up to the boundary and whose boundary values are f.

Exercise 3.2.11:
  a) Suppose $U \subset \mathbb{C}^n$ is a bounded open set with smooth boundary, $f \colon \overline{U} \to \mathbb{C}$ is a continuous function, holomorphic in U, and $f|_{\partial U}$ is real-valued. Show that f is constant.
  b) Find a counterexample to the statement if you allow U to be unbounded.

> *Exercise 3.2.12:* *Find a smooth CR function on the sphere $S^{2n-1} \subset \mathbb{C}^n$ that is not a restriction of a holomorphic function of a neighborhood of S^{2n-1}.*
>
> *Exercise 3.2.13:* *Show a global version of Severi. Given a real-analytic hypersurface $M \subset \mathbb{C}^n$ and a real-analytic CR function $f : M \to \mathbb{C}$, show that there exists a neighborhood U of M, and an $F \in \mathcal{O}(U)$ such that $F|_U = f$.*

A problem we tackle next is to try to extend a smooth CR function from the boundary of a domain to a holomorphic function inside. This is a PDE problem where the PDEs are the Cauchy–Riemann equations, and the function on the boundary is the boundary condition. Cauchy–Riemann equations are *overdetermined*, that is, there are too many equations. Not every data on the boundary gives a solution. Proposition 3.2.10 says that the data being CR is a necessary condition for a solution (it is not sufficient in general). Proposition 3.2.11 says the solution is unique if it exists.

3.3 \ Approximation of CR functions

The following theorem (proved circa 1980) holds in much more generality, but we state its simplest version. One of the simplifications we make is that we consider only smooth CR functions here, although the theorem holds even for continuous CR functions where the CR conditions are interpreted in the sense of distributions.

Theorem 3.3.1 (Baouendi–Trèves). *Suppose $M \subset \mathbb{C}^n$ is a smooth real hypersurface, $p \in M$ is a point, and $z = (z_1, \dots, z_n)$ are the holomorphic coordinates of $\mathbb{C}^n$. Then there exists a compact neighborhood $K \subset M$ of p, such that for every smooth CR function $f : M \to \mathbb{C}$, there exists a sequence $\{p_\ell\}$ of polynomials in z such that*

$$p_\ell(z) \to f(z) \qquad \text{uniformly in } K.$$

A key point is that K cannot be chosen arbitrarily—it depends on p and M. On the other hand, K does not depend on f. Given M and $p \in M$ there is a K such that *every* CR function on M is approximated uniformly on K by holomorphic polynomials. The theorem applies in one dimension, although in that case the theorem of Mergelyan (see Theorem B.31) is much more general.

Example 3.3.2: Let us show that K cannot possibly be arbitrary. For simplicity $n = 1$. Let $S^1 \subset \mathbb{C}$ be the unit circle (boundary of the disc), then every smooth function on S^1 is a smooth CR function. Let f be a nonconstant real function such as $\operatorname{Re} z$. Suppose for contradiction that we could take $K = S^1$ in the theorem. Then $f(z) = \operatorname{Re} z$ could be uniformly approximated on S^1 by holomorphic polynomials. By the maximum principle, the polynomials would converge on $\mathbb{D}$ to a holomorphic function on $\mathbb{D}$ continuous on $\overline{\mathbb{D}}$. This function would have nonconstant real boundary values, which is impossible. Clearly K cannot be the entire circle.

The example is easily extended to $\mathbb{C}^n$ by considering $M = S^1 \times \mathbb{C}^{n-1}$. Then $\operatorname{Re} z_1$ is a smooth CR function on M that cannot be approximated uniformly on $S^1 \times \{0\}$ by holomorphic polynomials.

The technique of the example above will be used later in a more general situation, to extend CR functions using Baouendi–Trèves.

Remark 3.3.3. It is important to note the difference between Baouendi–Trèves (and similar theorems in complex analysis) and the Weierstrass approximation theorem. In Baouendi–Trèves we obtain an approximation by holomorphic polynomials, while Weierstrass gives us polynomials in the real variables, or in z and $\bar{z}$. For example, via Weierstrass, every continuous function is uniformly approximable on S^1 via polynomials in $\operatorname{Re} z$ and $\operatorname{Im} z$, and therefore by polynomials in z and $\bar{z}$. These polynomials do not in general converge anywhere but on S^1.

> *Exercise 3.3.1:* Let $z = x + iy$ as usual in $\mathbb{C}$. Find a sequence of polynomials in x and y that converge uniformly to e^{x-y} on S^1, but diverge everywhere else.

The proof is an ingenious use of the standard technique used to prove the Weierstrass approximation theorem. Also, as we have seen mollifiers before, the technique will not be completely foreign even to the reader who does not know the Weierstrass approximation theorem. Basically what we do is use the standard convolution argument, this time against a holomorphic function. Letting $z = x + iy$ we only do the convolution in the x variables keeping the "free components" of y zero. Then we use the fact that the function is CR to show that we get an approximation even for other y.

In the formulas below, given a vector $v = (v_1, \ldots, v_n)$, it will be useful to write

$$[v]^2 \overset{\text{def}}{=} v_1^2 + \cdots + v_n^2.$$

The following lemma is a neat application of ideas from several complex variables to solve a problem that does not at first seem to involve holomorphic functions.

Lemma 3.3.4. *Let W be the set of $n \times n$ complex matrices A such that*

$$\|(\operatorname{Im} A)x\| < \|(\operatorname{Re} A)x\|$$

for all nonzero $x \in \mathbb{R}^n$ and $\operatorname{Re} A$ is positive definite. Then for all $A \in W$,

$$\int_{\mathbb{R}^n} e^{-[Ax]^2} \det A \, dx = \pi^{n/2}.$$

Proof. Suppose A has real entries and A is positive definite (so A is also invertible). By a change of coordinates

$$\int_{\mathbb{R}^n} e^{-[Ax]^2} \det A \, dx = \int_{\mathbb{R}^n} e^{-[x]^2} \, dx = \left(\int_{\mathbb{R}} e^{-x_1^2} \, dx_1 \right) \cdots \left(\int_{\mathbb{R}} e^{-x_n^2} \, dx_n \right) = (\sqrt{\pi})^n.$$

Next suppose A is any matrix in W. There is some $\epsilon > 0$ such that $\|(\operatorname{Im} A)x\|^2 \leq (1 - \epsilon^2)\|(\operatorname{Re} A)x\|^2$ for all $x \in \mathbb{R}^n$. That is because we only need to check this for x in the unit sphere, which is compact (exercise). By reality of $\operatorname{Re} A$, $\operatorname{Im} A$, and x we get $[(\operatorname{Re} A)x]^2 = \|(\operatorname{Re} A)x\|^2$ and $[(\operatorname{Im} A)x]^2 = \|(\operatorname{Im} A)x\|^2$. So

$$\left| e^{-[Ax]^2} \right| = e^{-\operatorname{Re}[Ax]^2} \leq e^{-[(\operatorname{Re} A)x]^2 + [(\operatorname{Im} A)x]^2} \leq e^{-\epsilon^2 [(\operatorname{Re} A)x]^2}.$$

Therefore, the integral exists for all A in W by a similar computation as above.

The expression

$$\int_{\mathbb{R}^n} e^{-[Ax]^2} \det A \, dx$$

is a well-defined holomorphic function in the entries of A, thinking of W as a domain (see exercises below) in $\mathbb{C}^{n^2}$. We have a holomorphic function that is constantly equal to $\pi^{n/2}$ on $W \cap \mathbb{R}^{n^2}$ and hence it is equal to $\pi^{n/2}$ everywhere on W. $\qquad \square$

Exercise 3.3.2: Prove the existence of $\epsilon > 0$ in the proof above.

Exercise 3.3.3: Show that $W \subset \mathbb{C}^{n^2}$ in the proof above is a domain (open and connected).

Exercise 3.3.4: Prove that we can really differentiate under the integral to show that the integral is holomorphic in the entries of A.

Exercise 3.3.5: Show that some hypotheses are needed for the lemma. In particular, take $n = 1$ and find the exact set of A (now a complex number) for which the conclusion of the lemma is true.

Given an $n \times n$ matrix A, let $\|A\|$ denote the operator norm,

$$\|A\| = \sup_{\|v\|=1} \|Av\| = \sup_{v \in \mathbb{C}^n, v \neq 0} \frac{\|Av\|}{\|v\|}.$$

Exercise 3.3.6: Let W be as in Lemma 3.3.4. Let B be an $n \times n$ real matrix such that $\|B\| < 1$. Show that $I + iB \in W$.

We will be using differential forms, and the following lemma says that as far as the exterior derivative is concerned, all CR functions behave as restrictions of holomorphic functions.

Lemma 3.3.5. *Let $M \subset \mathbb{C}^n$ be a smooth real hypersurface, $f \colon M \to \mathbb{C}$ be a smooth CR function, and $(z_1, \ldots, z_n)$ be the holomorphic coordinates of $\mathbb{C}^n$. Then at each point $p \in M$, the exterior derivative df is a linear combination of $dz_1, \ldots, dz_n$, thinking of $z_1, \ldots, z_n$ as functions on M. Namely,*

$$d(f \, dz) = df \wedge dz = 0.$$

Recall the notation $dz = dz_1 \wedge dz_2 \wedge \cdots \wedge dz_n$.

Proof. After a complex affine change of coordinates, we simply need to prove the result at the origin. Let $\xi_1, \ldots, \xi_n$ be the new holomorphic coordinates and suppose the $T_0^{(1,0)}M$ tangent space is spanned by $\frac{\partial}{\partial \xi_1}\big|_0, \ldots, \frac{\partial}{\partial \xi_{n-1}}\big|_0$, and such that $\frac{\partial}{\partial \operatorname{Re} \xi_n}\big|_0$ is tangent and $\frac{\partial}{\partial \operatorname{Im} \xi_n}\big|_0$ is normal. At the origin, the CR conditions are $\frac{\partial f}{\partial \bar{\xi}_k}(0) = 0$ for all k, so

$$df(0) = \frac{\partial f}{\partial \xi_1}(0)\, d\xi_1(0) + \cdots + \frac{\partial f}{\partial \xi_{n-1}}(0)\, d\xi_{n-1}(0) + \frac{\partial f}{\partial \operatorname{Re} \xi_n}(0)\, d(\operatorname{Re} \xi_n)(0).$$

Also, at the origin $d\xi_n(0) = d(\operatorname{Re} \xi_n)(0) + i\, d(\operatorname{Im} \xi_n)(0) = d(\operatorname{Re} \xi_n)(0)$. So $df(0)$ is a linear combination of $d\xi_1(0), \ldots, d\xi_n(0)$. As ξ is a complex affine function of z, each $d\xi_k$ is a linear combination of dz_1 through dz_n, and the claim follows. So if f is a CR function, then $d(f\, dz) = df \wedge dz = 0$ since $dz_k \wedge dz_k = 0$. $\qquad\square$

Proof of the theorem of Baouendi–Trèves. Suppose $M \subset \mathbb{C}^n$ is a smooth real hypersurface, and without loss of generality suppose $p = 0 \in M$. Let $z = (z_1, \ldots, z_n)$ be the holomorphic coordinates, write $z = x + iy$, $y = (y', y_n)$, and suppose M is given by

$$y_n = \psi(x, y'),$$

where ψ is $O(2)$. The variables (x, y') parametrize M near 0:

$$z_k = x_k + iy_k, \quad \text{for } k = 1, \ldots, n-1, \quad \text{and} \quad z_n = x_n + i\psi(x, y').$$

Define

$$\varphi(x, y') = \big(y_1, \ldots, y_{n-1}, \psi(x, y')\big).$$

Write $(x, y') \mapsto z = x + i\varphi(x, y')$ as the parametrization. That is, think of z as a function of (x, y').

Let $r > 0$ and $d > 0$ be small numbers to be determined later. Assume they are small enough so that f and φ are defined and smooth on some neighborhood of the set where $\|x\| \le r$ and $\|y'\| \le d$. There exists a smooth $g : \mathbb{R}^n \to [0, 1]$ such that $g \equiv 1$ on $B_{r/2}(0)$ and $g \equiv 0$ outside of $B_r(0)$. See Figure 3.2. An explicit formula can be given. Alternatively, g can be obtained by using mollifiers on a function that is identically one on $B_{3r/4}(0)$ and zero elsewhere. Such a g is called a *cutoff function*.

Exercise 3.3.7: Find an explicit formula for g without using mollifiers.

Let

$$K' = \{(x, y') : \|x\| \le r/4, \|y'\| \le d\}.$$

Let $K = z(K')$, that is the image of K' under the mapping $z(x, y')$.

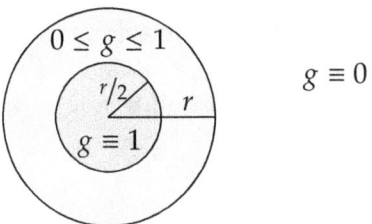

Figure 3.2: Cutoff function.

Consider the CR function f a function of (x, y') and write $f(x, y')$. For $\ell \in \mathbb{N}$, let α_ℓ be a differential n-form defined (thinking of $w \in \mathbb{C}^n$ as a constant parameter) by

$$\alpha_\ell(x, y') = \left(\frac{\ell}{\pi}\right)^{n/2} e^{-\ell[w-z]^2} g(x) f(x, y') \, dz$$

$$= \left(\frac{\ell}{\pi}\right)^{n/2} e^{-\ell[w-x-i\varphi(x,y')]^2} g(x) f(x, y')$$

$$(dx_1 + idy_1) \wedge \cdots \wedge (dx_{n-1} + idy_{n-1}) \wedge \left(dx_n + id\psi(x, y')\right).$$

The key is the exponential, which looks like the bump function mollifier, except that now we have w and z possibly complex. The exponential is also holomorphic in w, and that will give us entire holomorphic approximating functions.

Fix y' with $0 < \|y'\| < d$ and let D be defined by

$$D = \left\{(x, s) \in \mathbb{R}^n \times \mathbb{R}^{n-1} : \|x\| < r \text{ and } s = ty' \text{ for } t \in (0, 1)\right\}.$$

D is an $(n + 1)$-dimensional "cylinder." We take a ball in the x directions, then take a single fixed point y' in the s variables and make a cylinder. See Figure 3.3.

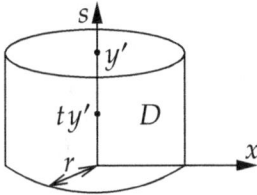

Figure 3.3: Cylinder D.

Orient D in the standard way as if it sat in the (x, t) variables in $\mathbb{R}^n \times \mathbb{R}$. Stokes' theorem says

$$\int_D d\alpha_\ell(x, s) = \int_{\partial D} \alpha_\ell(x, s).$$

Since $g(x) = 0$ if $\|x\| \geq r$, α_ℓ is zero on the sides of the cylinder D, so the integral over ∂D only needs to consider the top and bottom of the cylinder. And because of g, the integral over the top and bottom can be taken over $\mathbb{R}^n$. As is usual in these sorts of arguments, we do the slight abuse of notation ignoring that f and φ are undefined where g is identically zero:

$$
\int_{\partial D} \alpha_\ell(x, s)
$$
$$
= \left(\frac{\ell}{\pi}\right)^{n/2} \int_{x \in \mathbb{R}^n} e^{-\ell[w-x-i\varphi(x,y')]^2} g(x) f(x, y') \, dx_1 \wedge \cdots \wedge dx_{n-1} \wedge \left(dx_n + i d_x \psi(x, y')\right)
$$
$$
- \left(\frac{\ell}{\pi}\right)^{n/2} \int_{x \in \mathbb{R}^n} e^{-\ell[w-x-i\varphi(x,0)]^2} g(x) f(x, 0) \, dx_1 \wedge \cdots \wedge dx_{n-1} \wedge \left(dx_n + i d_x \psi(x, 0)\right),
$$

$$(3.1)$$

where d_x means the derivative in the x directions only. I.e., $d_x \psi = \frac{\partial \psi}{\partial x_1} dx_1 + \cdots + \frac{\partial \psi}{\partial x_n} dx_n$.

We will show that as $\ell \to \infty$, the left-hand side of (3.1) goes to zero uniformly for $w \in K$ and the first term on the right-hand side goes to $f(\tilde{x}, y')$ if $w = z(\tilde{x}, y')$ is in M. Hence, we define entire functions that we will show approximate f:

$$
f_\ell(w) = \left(\frac{\ell}{\pi}\right)^{n/2} \int_{x \in \mathbb{R}^n} e^{-\ell[w-x-i\varphi(x,0)]^2} g(x) f(x, 0) \, dx_1 \wedge \cdots \wedge dx_{n-1} \wedge \left(dx_n + i d_x \psi(x, 0)\right).
$$

Clearly each f_ℓ is holomorphic and defined for all $w \in \mathbb{C}^n$.

In the next claim it is important that f is a CR function.

Claim 3.3.6. *We have*

$$
d\alpha_\ell(x, s) = \left(\frac{\ell}{\pi}\right)^{n/2} e^{-\ell[w-z(x,s)]^2} f(x, s) \, dg(x) \wedge dz(x, s),
$$

and for sufficiently small $r > 0$ and $d > 0$,

$$
\lim_{\ell \to \infty} \left(\frac{\ell}{\pi}\right)^{n/2} \int_{(x,s) \in D} e^{-\ell[w-z(x,s)]^2} f(x, s) \, dg(x) \wedge dz(x, s) = 0
$$

uniformly as a function of $w \in K$ and $y' \in B_d(0)$ (recall that D depends on y').

Proof. The function $(x, s) \mapsto e^{-\ell[w-z(x,s)]^2}$ is CR (as a function on M), and so is $f(x, s)$. Therefore, using Lemma 3.3.5,

$$
d\alpha_\ell(x, s) = \left(\frac{\ell}{\pi}\right)^{n/2} e^{-\ell[w-z(x,s)]^2} f(x, s) \, dg(x) \wedge dz(x, s).
$$

Since dg is zero for $\|x\| \leq r/2$, the integral

$$
\int_D d\alpha_\ell(x, s) = \left(\frac{\ell}{\pi}\right)^{n/2} \int_D e^{-\ell[w-z(x,s)]^2} f(x, s) \, dg(x) \wedge dz(x, s)
$$

is only evaluated for the subset of D where $\|x\| > r/2$.

Suppose $w \in K$ and $(x, s) \in D$ with $\|x\| > r/2$. Let $w = z(\tilde{x}, \tilde{s})$. We need to estimate

$$\left| e^{-\ell[w - z(x,s)]^2} \right| = e^{-\ell \operatorname{Re}[w - z(x,s)]^2}.$$

Then

$$-\operatorname{Re}[w - z]^2 = -\|\tilde{x} - x\|^2 + \|\varphi(\tilde{x}, \tilde{s}) - \varphi(x, s)\|^2.$$

By the mean value theorem

$$\|\varphi(\tilde{x}, \tilde{s}) - \varphi(x, s)\| \leq \|\varphi(\tilde{x}, \tilde{s}) - \varphi(x, \tilde{s})\| + \|\varphi(x, \tilde{s}) - \varphi(x, s)\| \leq a\|\tilde{x} - x\| + A\|\tilde{s} - s\|,$$

where a and A are

$$a = \sup_{\|\hat{x}\| \leq r, \|\hat{y}'\| \leq d} \left\| \left[\frac{\partial \varphi}{\partial x}(\hat{x}, \hat{y}') \right] \right\|, \qquad A = \sup_{\|\hat{x}\| \leq r, \|\hat{y}'\| \leq d} \left\| \left[\frac{\partial \varphi}{\partial y'}(\hat{x}, \hat{y}') \right] \right\|.$$

Here $\left[\frac{\partial \varphi}{\partial x} \right]$ and $\left[\frac{\partial \varphi}{\partial y'} \right]$ are the derivatives (matrices) of φ with respect to x and y' respectively, and the norm we are taking is the operator norm. Because $\left[\frac{\partial \varphi}{\partial x} \right]$ is zero at the origin, we pick r and d small enough (and hence K small enough) so that $a \leq 1/4$. We furthermore pick d possibly even smaller to ensure that $d \leq \frac{r}{32A}$. We have that $r/2 \leq \|x\| \leq r$, but $\|\tilde{x}\| \leq r/4$ (recall $w \in K$), so

$$\frac{r}{4} \leq \|\tilde{x} - x\| \leq \frac{5r}{4}.$$

Also, $\|\tilde{s} - s\| \leq 2d$ by triangle inequality.

Therefore,

$$\begin{aligned} -\operatorname{Re}[w - z(x, s)]^2 &\leq -\|\tilde{x} - x\|^2 + a^2\|\tilde{x} - x\|^2 + A^2\|\tilde{s} - s\|^2 + 2aA\|\tilde{x} - x\|\|\tilde{s} - s\| \\ &\leq \frac{-15}{16}\|\tilde{x} - x\|^2 + A^2\|\tilde{s} - s\|^2 + \frac{A}{2}\|\tilde{x} - x\|\|\tilde{s} - s\| \\ &\leq \frac{-r^2}{64}. \end{aligned}$$

In other words,

$$\left| e^{-\ell[w - z(x,s)]^2} \right| \leq e^{-\ell r^2/64},$$

or

$$\left| \left(\frac{\ell}{\pi} \right)^{n/2} \int_{(x,s) \in D} e^{-\ell[w - z(x,s)]^2} f(x, s) \, dg(x) \wedge dz(x, s) \right| \leq C\ell^{n/2} e^{-\ell r^2/64},$$

for some constant C. Note that D depends on y'. The set of all y' with $\|y'\| \leq d$ is a compact set, so we can make C large enough to not depend on the y' that was chosen. The claim follows. $\qquad\square$

Claim 3.3.7. *For the given $r > 0$ and $d > 0$,*

$$\lim_{\ell \to \infty} \left(\frac{\ell}{\pi}\right)^{n/2} \int_{x \in \mathbb{R}^n} e^{-\ell[\tilde{x}+i\varphi(\tilde{x},y')-x-i\varphi(x,y')]^2}$$

$$g(x)f(x,y')dx_1 \wedge \cdots \wedge dx_{n-1} \wedge \left(dx_n + id_x\psi(x,y')\right) = f(\tilde{x},y')$$

uniformly in $(\tilde{x}, y') \in K'$.

That is, we look at (3.1) and we plug in $w = z(\tilde{x}, y') \in K$. The g (as usual) makes sure we never evaluate f, ψ, or φ at points where they are not defined.

Proof. The change of variables formula implies

$$dx_1 \wedge \cdots \wedge dx_{n-1} \wedge \left(dx_n + id_x\psi(x,y')\right) = d_x z(x,y') = \det\left[\frac{\partial z}{\partial x}(x,y')\right]dx,$$

where $\left[\frac{\partial z}{\partial x}(x,y')\right]$ is the matrix corresponding to the derivative of the mapping z with respect to the x variables evaluated at (x,y').

Let us change variables of integration via $\xi = \sqrt{\ell}(x - \tilde{x})$:

$$\left(\frac{\ell}{\pi}\right)^{n/2} \int_{x \in \mathbb{R}^n} e^{-\ell[\tilde{x}+i\varphi(\tilde{x},y')-x-i\varphi(x,y')]^2} g(x)f(x,y') \det\left[\frac{\partial z}{\partial x}(x,y')\right]dx =$$

$$\left(\frac{1}{\pi}\right)^{n/2} \int_{\xi \in \mathbb{R}^n} e^{-\left[\xi + i\sqrt{\ell}\left(\varphi\left(\tilde{x}+\frac{\xi}{\sqrt{\ell}},y'\right)-\varphi(\tilde{x},y')\right)\right]^2}$$

$$g\left(\tilde{x}+\frac{\xi}{\sqrt{\ell}}\right)f\left(\tilde{x}+\frac{\xi}{\sqrt{\ell}},y'\right) \det\left[\frac{\partial z}{\partial x}\left(\tilde{x}+\frac{\xi}{\sqrt{\ell}},y'\right)\right]d\xi.$$

We now wish to take a limit as $\ell \to \infty$ and for this we apply the dominated convergence theorem. So we need to dominate the integrand. The second half of the integrand is uniformly bounded independent of ℓ as

$$x \mapsto g(x)f(x,y') \det\left[\frac{\partial z}{\partial x}(x,y')\right]$$

is a continuous function with compact support (because of g). Hence it is enough to worry about the exponential term. We also only consider those ξ where the integrand is not zero. Recall that r and d are small enough that

$$\sup_{\|\hat{x}\| \le r, \|\hat{y}'\| \le d} \left\|\left[\frac{\partial \varphi}{\partial x}(\hat{x},\hat{y}')\right]\right\| \le \frac{1}{4},$$

and as $\|\tilde{x}\| \le r/4$ (as $(\tilde{x}, y') \in K$) and $\left\|\tilde{x}+\frac{\xi}{\sqrt{\ell}}\right\| \le r$ (because g is zero otherwise), then

$$\left\|\varphi\left(\tilde{x}+\frac{\xi}{\sqrt{\ell}},y'\right) - \varphi(\tilde{x},y')\right\| \le \frac{1}{4}\left\|\tilde{x}+\frac{\xi}{\sqrt{\ell}}-\tilde{x}\right\| = \frac{\|\xi\|}{4\sqrt{\ell}}.$$

So under the same conditions,

$$\left| e^{-\left[\xi+i\sqrt{\ell}\left(\varphi\left(\tilde{x}+\frac{\xi}{\sqrt{\ell}},y'\right)-\varphi(\tilde{x},y')\right)\right]^2} \right| = e^{-\operatorname{Re}\left[\xi+i\sqrt{\ell}\left(\varphi\left(\tilde{x}+\frac{\xi}{\sqrt{\ell}},y'\right)-\varphi(\tilde{x},y')\right)\right]^2}$$

$$= e^{-\|\xi\|^2+\ell\left\|\varphi\left(\tilde{x}+\frac{\xi}{\sqrt{\ell}},y'\right)-\varphi(\tilde{x},y')\right\|^2}$$

$$\leq e^{-(15/16)\|\xi\|^2}.$$

And that is integrable. Thus, take the pointwise limit under the integral to obtain

$$\left(\frac{1}{\pi}\right)^{n/2} \int_{\xi\in\mathbb{R}^n} e^{-\left[\xi+i\left[\frac{\partial\varphi}{\partial x}(\tilde{x},y')\right]\xi\right]^2} g(\tilde{x})f(\tilde{x},y')\det\left[\frac{\partial z}{\partial x}(\tilde{x},y')\right]d\xi.$$

In the exponent, we have an expression for the derivative in the ξ direction with y' fixed. If $(\tilde{x},y')\in K'$, then $g(\tilde{x})=1$, and so we can ignore g.

Let $A = I + i\left[\frac{\partial\varphi}{\partial x}(\tilde{x},y')\right]$. Lemma 3.3.4 says

$$\left(\frac{1}{\pi}\right)^{n/2} \int_{\xi\in\mathbb{R}^n} e^{-\left[\xi+i\left[\frac{\partial\varphi}{\partial x}(\tilde{x},y')\right]\xi\right]^2} f(\tilde{x},y')\det\left[\frac{\partial z}{\partial x}(\tilde{x},y')\right]d\xi = f(\tilde{x},y').$$

That the convergence is uniform in $(\tilde{x},y')\in K'$ is left as an exercise. □

Exercise 3.3.8: In the claim above, finish the proof that the convergence is uniform in $(\tilde{x},y')\in K'$. Hint: It may be easier to use the form of the integral before the change of variables and prove that the sequence is uniformly Cauchy.

We are essentially done with the proof of the theorem. The two claims together with (3.1) show that f_ℓ are entire holomorphic functions that approximate f uniformly on K. Entire holomorphic functions can be approximated by polynomials uniformly on compact subsets; simply take the partial sums of Taylor series at the origin. □

Exercise 3.3.9: Explain why being approximable on K by (holomorphic) polynomials does not necessarily mean that f is real-analytic.

Exercise 3.3.10: Suppose $M\subset\mathbb{C}^n$ is given by $\operatorname{Im} z_n = 0$. Use the standard Weierstrass approximation theorem to show that given a $K\subset\subset M$, and a smooth CR function $f: M \to \mathbb{C}$, then f can be uniformly approximated by holomorphic polynomials on K.

3.4 \ Extension of CR functions

We will apply the so-called "technique of analytic discs" together with Baouendi–Trèves to prove the Lewy extension theorem. Lewy's original proof was different and predates Baouendi–Trèves. A local extension theorem of this type was first proved by Helmut Knesser in 1936.

Theorem 3.4.1 (Lewy). *Suppose $M \subset \mathbb{C}^n$ is a smooth real hypersurface and $p \in M$. There exists a neighborhood U of p with the following property. Suppose $r: U \to \mathbb{R}$ is a smooth defining function for $M \cap U$, denote by $U_- \subset U$ the set where r is negative and $U_+ \subset U$ the set where r is positive. Let $f: M \to \mathbb{C}$ be a smooth CR function. Then:*

(i) *If the Levi form with respect to r has a positive eigenvalue at p, then f extends to a holomorphic function on U_- continuous up to M (continuous on $\{z \in U : r(z) \leq 0\}$).*

(ii) *If the Levi form with respect to r has a negative eigenvalue at p, then f extends to a holomorphic function on U_+ continuous up to M (continuous on $\{z \in U : r(z) \geq 0\}$).*

(iii) *If the Levi form with respect to r has eigenvalues of both signs at p, then f extends to a function holomorphic on U.*

So if the Levi form has eigenvalues of both signs, then near p all CR functions are restrictions of holomorphic functions. The function r can be any defining function for M. Either we can extend it to all of U or we can take a smaller U such that r is defined on U. As we noticed before, once we pick sides (where r is positive and where it is negative), then the number of positive eigenvalues and the number of negative eigenvalues of the Levi form is fixed. A different r may flip U_- and U_+, but the conclusion of the theorem is exactly the same.

Proof. We prove the first item, and the second item follows by considering $-r$. Suppose $p = 0$ and M is given in some neighborhood Ω of the origin as

$$\operatorname{Im} w = |z_1|^2 + \sum_{k=2}^{n-1} \epsilon_k |z_k|^2 + E(z_1, z', \bar{z}_1, \bar{z}', \operatorname{Re} w),$$

where $z' = (z_2, \ldots, z_{n-1})$, $\epsilon_k = -1, 0, 1$, and E is $O(3)$. Let Ω_- be given by

$$0 > r = |z_1|^2 + \sum_{k=2}^{n-1} \epsilon_k |z_k|^2 + E(z_1, z', \bar{z}_1, \bar{z}', \operatorname{Re} w) - \operatorname{Im} w.$$

The (real) Hessian of the function (of one variable)

$$z_1 \mapsto |z_1|^2 + E(z_1, 0, \bar{z}_1, 0, 0)$$

is positive definite in a neighborhood of the origin and the function has a strict minimum at 0. There is some small disc $D \subset \mathbb{C}$ such that this function is strictly positive on ∂D.

Therefore, for $(z', w) \in W$ in some small neighborhood $W \subset \mathbb{C}^{n-1}$ of the origin, the function

$$z_1 \mapsto |z_1|^2 + \sum_{k=2}^{n-1} \epsilon_k |z_k|^2 + E(z_1, z', \bar{z}_1, \bar{z}', \mathrm{Re}\, w) - \mathrm{Im}\, w$$

is still strictly positive on ∂D.

We wish to apply Baouendi–Trèves and so let K be the compact neighborhood of the origin from the theorem. Take D and W small enough such that $(D \times W) \cap M \subset K$. Find the polynomials p_ℓ that approximate f uniformly on K. Consider $z_1 \in D$ and $(z', w) \in W$ such that $(z_1, z', w) \in \Omega_-$. Let $\Delta = \big(D \times \{(z', w)\}\big) \cap \Omega_-$. Denote by $\partial \Delta$ the boundary of Δ in the subspace topology of $\mathbb{C} \times \{(z', w)\}$.

The set Ω_+ where $r > 0$ is open and it contains $(\partial D) \times \{(z', w)\}$. Therefore, $\partial \Delta$ contains no points of $(\partial D) \times \{(z', w)\}$. Consequently, $\partial \Delta$ contains only points where $r = 0$, that is, $\partial \Delta \subset M$. Also $\partial \Delta \subset D \times W$. As $(D \times W) \cap M \subset K$, we have $\partial \Delta \subset K$. See Figure 3.4.

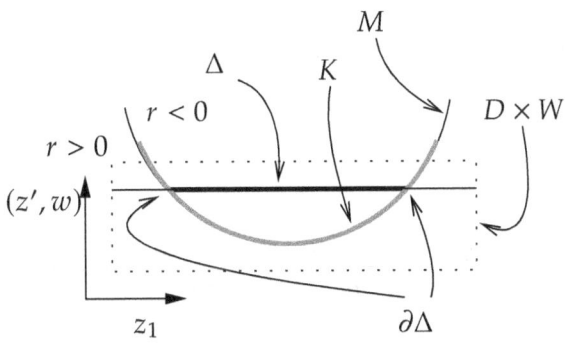

Figure 3.4: Proof of Lewy extension.

As $p_\ell \to f$ uniformly on K, we have that $p_\ell \to f$ uniformly on $\partial \Delta$. The p_ℓ are holomorphic, so by the maximum principle, p_ℓ converge uniformly on all of Δ. In fact, as (z_1, z', w) was an arbitrary point in $(D \times W) \cap \Omega_-$, the polynomials p_ℓ converge uniformly on $(D \times W) \cap \overline{\Omega_-}$. Let $U = D \times W$, then $U_- = (D \times W) \cap \Omega_-$. Notice U depends on K, but not on f. So p_ℓ converge to a continuous function F on $\overline{U_-} \cap U$ and F is holomorphic on U_-. Clearly F equals f on $M \cap U$.

To prove the last item, pick a side, and then use one of the first two items to extend the function to that side. Via the tomato can principle (Theorem 2.3.11) the function also extends across M and therefore to a whole neighborhood of p. □

If you were wondering what happened to the analytic discs we promised, the Δ in the above is an analytic disc (simply connected) for a small enough U, but it was not necessary to prove that fact.

We state the next corollary for a strongly convex domain, even though it holds with far more generality. It is a special case of the *Hartogs–Bochner** theorem. Later, in Exercise 4.3.4, you will prove it for strongly pseudoconvex domains. However, the theorem is true for every bounded domain with connected smooth boundary with no assumptions on the Levi form, but a different approach would have to be taken.

Corollary 3.4.2. *Suppose $U \subset \mathbb{C}^n$, $n \geq 2$, is a bounded domain with smooth boundary that is strongly convex and $f : \partial U \to \mathbb{C}$ is a smooth CR function. Then there exists a continuous function $F : \overline{U} \to \mathbb{C}$ holomorphic in U such that $F|_{\partial U} = f$.*

Proof. A strongly convex domain is strongly pseudoconvex, so f must extend to the inside locally near every point. The extension is locally unique as any two extensions have the same boundary values. Therefore, there exists a set $K \subset\subset U$ such that f extends to $U \setminus K$. Via an exercise below we can assume that K is strongly convex and therefore we can apply the special case of Hartogs phenomenon that you proved in Exercise 2.1.8 to find an extension holomorphic in U. $\square$

Exercise 3.4.1: Prove the existence of the strongly convex K in the proof of Corollary 3.4.2.

Exercise 3.4.2: Show by example that the corollary is not true when $n = 1$. Explain where in the proof have we used that $n \geq 2$.

Exercise 3.4.3: Suppose $f : \partial \mathbb{B}_2 \to \mathbb{C}$ is a smooth CR function. Write down an explicit formula for the extension F.

Exercise 3.4.4: A smooth real hypersurface $M \subset \mathbb{C}^3$ is defined by $\operatorname{Im} w = |z_1|^2 - |z_2|^2 + O(3)$ and f is a nonconstant smooth CR function on M. Show that $|f|$ does not attain a maximum at the origin.

Exercise 3.4.5: A real-analytic hypersurface $M \subset \mathbb{C}^n$, $n \geq 3$, is such that the Levi form at $p \in M$ has eigenvalues of both signs. Show that every smooth CR function f on M is, in fact, real-analytic in a neighborhood of p.

Exercise 3.4.6: Let $M \subset \mathbb{C}^3$ be defined by $\operatorname{Im} w = |z_1|^2 - |z_2|^2$.
   a) Show that for this M, the conclusion of Baouendi–Trèves holds with an arbitrary compact subset $K \subset\subset M$.
   b) Use this to show that every smooth CR function $f : M \to \mathbb{C}$ is a restriction of an entire holomorphic function $F : \mathbb{C}^3 \to \mathbb{C}$.

Exercise 3.4.7: Find an $M \subset \mathbb{C}^n$, $n \geq 2$, such that near some $p \in M$, for every neighborhood W of p in M, there is a CR function $f : W \to \mathbb{C}$ that does not extend holomorphically to either side of M at p.

---

*What is called Hartogs–Bochner is the C^1 version of this theorem where the domain is only assumed to be bounded and the boundary connected, and it was proved by neither Hartogs nor Bochner, but by Martinelli in 1961.

Exercise 3.4.8: *Suppose $f \colon \partial \mathbb{B}_n \to \mathbb{C}$ is a smooth function and $n \geq 2$. Prove that f is a CR function if and only if*

$$\int_0^{2\pi} f(e^{i\theta} v) e^{ik\theta} \, d\theta = 0 \qquad \text{for all } v \in \partial \mathbb{B}_n \text{ and all } k \in \mathbb{N}.$$

Exercise 3.4.9: *Prove the third item in the Lewy extension theorem without the use of the tomato can principle. That is, prove in a more elementary way that if $M \subset U \subset \mathbb{C}^n$ is a smooth real hypersurface in an open set U and $f \colon U \to \mathbb{C}$ is continuous and holomorphic in $U \setminus M$, then f is holomorphic.*

Remark 3.4.3. Studying solutions to nonhomogeneous CR equations of the form $Xf = \psi$ for a CR vector field X, and the fact that such conditions can guarantee that a function must be real-analytic, led Lewy to a famous, very surprising, and rather simple example of a linear partial differential equation with smooth coefficients that has no solution on any open set[*]. The example is surprising because when a linear PDE has real-analytic coefficients, a solution always exists by the theorem of Cauchy–Kowalevski. Furthermore, if X is a real vector field (X is in TM, not simply in $\mathbb{C}TM$), then a solution to $Xf = \psi$ exists by the method of characteristics, even if X and ψ are only smooth.

---

[*]Lewy, Hans, *An example of a smooth linear partial differential equation without solution*, Annals of Mathematics, **66** (1957), 155–158.

4 | The $\bar{\partial}$-problem

4.1 | The generalized Cauchy integral formula

Before we get into the $\bar{\partial}$-problem, let us prove a more general version of Cauchy's formula using Stokes' theorem (really Green's theorem). This version is called the *Cauchy–Pompeiu integral formula*. We only need the theorem for smooth functions, but as it is often applied in less regular contexts and it is just an application of Stokes' theorem, we state it more generally. In applications, the boundary is often only piecewise smooth, and again that is all we need for Stokes.

Theorem 4.1.1 (Cauchy–Pompeiu). *Let $U \subset \mathbb{C}$ be a bounded open set with piecewise-C^1 boundary ∂U oriented positively (see appendix B), and let $f \colon \overline{U} \to \mathbb{C}$ be continuous with bounded continuous partial derivatives in U. Then for $z \in U$:*

$$f(z) = \frac{1}{2\pi i} \int_{\partial U} \frac{f(\zeta)}{\zeta - z}\, d\zeta + \frac{1}{2\pi i} \int_{U} \frac{\frac{\partial f}{\partial \bar{\zeta}}(\zeta)}{\zeta - z}\, d\zeta \wedge d\bar{\zeta}.$$

If f is holomorphic, then the second term is zero, and we have the standard Cauchy formula. If $\zeta = x + iy$, then the standard orientation on $\mathbb{C}$ is the one corresponding to the area form $dA = dx \wedge dy$. The form $d\zeta \wedge d\bar{\zeta}$ is the area form up to a scalar. That is,

$$d\zeta \wedge d\bar{\zeta} = (dx + i\,dy) \wedge (dx - i\,dy) = (-2i)dx \wedge dy = (-2i)dA.$$

As we want to use Stokes, we need to write the standard exterior derivative in terms of z and $\bar{z}$. For $z = x + iy$, we compute:

$$d\psi = \frac{\partial \psi}{\partial x}dx + \frac{\partial \psi}{\partial y}dy = \frac{\partial \psi}{\partial z}dz + \frac{\partial \psi}{\partial \bar{z}}d\bar{z}.$$

Exercise 4.1.1: Observe the singularity in the second term of the Cauchy–Pompeiu formula. Prove that the integral still makes sense (the function is integrable). Hint: polar coordinates.

Exercise 4.1.2: Why can we not differentiate in $\bar{z}$ under the integral in the second term of the Cauchy–Pompeiu formula? Notice that it would lead to an impossible result.

Proof. Fix $z \in U$. We wish to apply Stokes' theorem*, but the integrand is not smooth at z. Let $\Delta_r(z)$ be a small disc such that $\Delta_r(z) \subset\subset U$. See Figure 4.1.

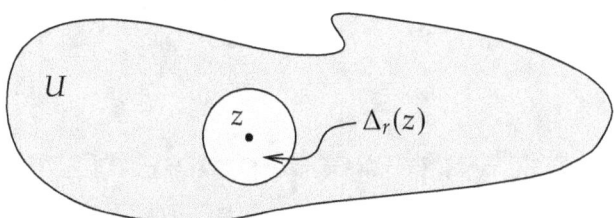

Figure 4.1: Proof of Cauchy–Pompeiu.

Stokes applies on $U \setminus \Delta_r(z)$, writing the exterior derivative with respect to ζ and $\bar{\zeta}$,

$$\int_{\partial U} \frac{f(\zeta)}{\zeta - z} \, d\zeta - \int_{\partial \Delta_r(z)} \frac{f(\zeta)}{\zeta - z} \, d\zeta = \int_{U \setminus \Delta_r(z)} d\left(\frac{f(\zeta)}{\zeta - z} \, d\zeta \right) = \int_{U \setminus \Delta_r(z)} \frac{\frac{\partial f}{\partial \bar{\zeta}}(\zeta)}{\zeta - z} \, d\bar{\zeta} \wedge d\zeta.$$

The second equality follows because the derivative in ζ has a $d\zeta$ and when we wedge it with $d\zeta$ we get zero. We now wish to let the radius r go to zero. Via the exercise above, $\frac{\frac{\partial f}{\partial \bar{\zeta}}(\zeta)}{\zeta - z} \, d\bar{\zeta} \wedge d\zeta$ is integrable over all of U. Therefore,

$$\lim_{r \to 0} \int_{U \setminus \Delta_r(z)} \frac{\frac{\partial f}{\partial \bar{\zeta}}(\zeta)}{\zeta - z} \, d\bar{\zeta} \wedge d\zeta = \int_U \frac{\frac{\partial f}{\partial \bar{\zeta}}(\zeta)}{\zeta - z} \, d\bar{\zeta} \wedge d\zeta = - \int_U \frac{\frac{\partial f}{\partial \bar{\zeta}}(\zeta)}{\zeta - z} \, d\zeta \wedge d\bar{\zeta}.$$

The second equality is simply swapping the order of $d\zeta$ and $d\bar{\zeta}$. By continuity of f, we finish the proof,

$$\lim_{r \to 0} \frac{1}{2\pi i} \int_{\partial \Delta_r(z)} \frac{f(\zeta)}{\zeta - z} \, d\zeta = \lim_{r \to 0} \frac{1}{2\pi} \int_0^{2\pi} f(z + re^{i\theta}) \, d\theta = f(z). \qquad \square$$

Exercise 4.1.3:

a) *Let $U \subset \mathbb{C}$ be a bounded open set with piecewise-C^1 boundary and suppose $f : \overline{U} \to \mathbb{C}$ is a C^1 function such that*

$$\int_U \frac{\frac{\partial f}{\partial \bar{\zeta}}(\zeta)}{\zeta - z} \, dA(\zeta) = 0$$

for every $z \in \partial U$. Prove that $f|_{\partial U}$ gives the boundary values of a function continuous on $\overline{U}$ and holomorphic in U.

b) *Given arbitrary $\epsilon > 0$, find a C^1 function f on the closed unit disc $\overline{\mathbb{D}}$, such that $\frac{\partial f}{\partial \bar{z}}$ is identically zero outside an ϵ-neighborhood of the origin, yet $f|_{\partial \mathbb{D}}$ does not represent the boundary values of a holomorphic function.*

---

*We are really using Green's theorem, which is the generalized Stokes' theorem in 2 dimensions, see Theorem B.2.

Exercise 4.1.4: *Let $U \subset \mathbb{C}$ and f be as in the theorem, but let $z \notin \overline{U}$. Show that*

$$\frac{1}{2\pi i} \int_{\partial U} \frac{f(\zeta)}{\zeta - z} \, d\zeta + \frac{1}{2\pi i} \int_U \frac{\frac{\partial f}{\partial \bar{\zeta}}(\zeta)}{\zeta - z} \, d\zeta \wedge d\bar{\zeta} = 0.$$

4.2 \ Compactly supported $\bar{\partial}$-problem

For a smooth function ψ, consider the exterior derivative in terms of z and $\bar{z}$,

$$d\psi = \frac{\partial \psi}{\partial z_1} dz_1 + \cdots + \frac{\partial \psi}{\partial z_n} dz_n + \frac{\partial \psi}{\partial \bar{z}_1} d\bar{z}_1 + \cdots + \frac{\partial \psi}{\partial \bar{z}_n} d\bar{z}_n.$$

Let us give a name to the two parts of the derivative:

$$\partial \psi \stackrel{\text{def}}{=} \frac{\partial \psi}{\partial z_1} dz_1 + \cdots + \frac{\partial \psi}{\partial z_n} dz_n, \qquad \bar{\partial} \psi \stackrel{\text{def}}{=} \frac{\partial \psi}{\partial \bar{z}_1} d\bar{z}_1 + \cdots + \frac{\partial \psi}{\partial \bar{z}_n} d\bar{z}_n.$$

Then $d\psi = \partial \psi + \bar{\partial} \psi$. Notice ψ is holomorphic if and only if $\bar{\partial} \psi = 0$.

The so-called *inhomogeneous $\bar{\partial}$-problem* ($\bar{\partial}$ is pronounced "dee bar") is to solve the equation

$$\bar{\partial} \psi = g,$$

for ψ, given a one-form

$$g = g_1 d\bar{z}_1 + \cdots + g_n d\bar{z}_n.$$

Such a g is called a $(0,1)$-*form*. The fact that the partial derivatives of ψ commute forces certain compatibility conditions on g for us to have any hope of getting a solution (see below).

Exercise 4.2.1: *Find an explicit example of a g in $\mathbb{C}^2$ such that no corresponding ψ exists.*

On any open set where $g = 0$, ψ is holomorphic. So for a general g, what we are doing is finding a function that is not holomorphic in a specific way.

Theorem 4.2.1. *Suppose g is a $(0,1)$-form on $\mathbb{C}^n$, $n \geq 2$, given by*

$$g = g_1 d\bar{z}_1 + \cdots + g_n d\bar{z}_n,$$

where $g_j \colon \mathbb{C}^n \to \mathbb{C}$ are compactly supported smooth functions satisfying the compatibility conditions

$$\frac{\partial g_k}{\partial \bar{z}_\ell} = \frac{\partial g_\ell}{\partial \bar{z}_k} \qquad \text{for all } k, \ell = 1, 2, \ldots, n. \tag{4.1}$$

Then there exists a unique compactly supported smooth function $\psi \colon \mathbb{C}^n \to \mathbb{C}$ such that

$$\bar{\partial} \psi = g.$$

The compatibility conditions on g are necessary, but the compactness is not. Without compactness, the domain where the equation lives would come into play. Let us not worry about this, and prove that this simple compactly supported version always has a solution. The compactly supported solution is unique: If ψ_1 and ψ_2 are solutions, then $\bar{\partial}(\psi_1 - \psi_2) = g - g = 0$, and so $\psi_1 - \psi_2$ is holomorphic. The only holomorphic compactly supported function is 0, and hence the compactly supported solution ψ is unique.

Proof. Really there are n smooth functions, $g_1, \ldots, g_n$, so the equation $\bar{\partial}\psi = g$ consists of the n equations

$$\frac{\partial \psi}{\partial \bar{z}_k} = g_k,$$

where the functions g_k satisfy the compatibility conditions (4.1).

We claim that the following is an explicit solution:

$$\psi(z) = \frac{1}{2\pi i} \int_{\mathbb{C}} \frac{g_1(\zeta, z_2, \ldots, z_n)}{\zeta - z_1} d\zeta \wedge d\bar{\zeta}$$
$$= \frac{1}{2\pi i} \int_{\mathbb{C}} \frac{g_1(\zeta + z_1, z_2, \ldots, z_n)}{\zeta} d\zeta \wedge d\bar{\zeta}.$$

To show that the singularity does not matter for integrability is the same idea as for the generalized Cauchy formula.

Let us check that ψ is the solution. We use the generalized Cauchy formula on the z_1 variable. Take R large enough so that $g_k(\zeta, z_2, \ldots, z_n)$ is zero when $|\zeta| \geq R$ for all k. For every k,

$$g_k(z) = \frac{1}{2\pi i} \int_{|\zeta|=R} \frac{g_k(\zeta, z_2, \ldots, z_n)}{\zeta - z_1} d\zeta + \frac{1}{2\pi i} \int_{|\zeta| \leq R} \frac{\frac{\partial g_k}{\partial \bar{z}_1}(\zeta, z_2, \ldots, z_n)}{\zeta - z_1} d\zeta \wedge d\bar{\zeta}$$
$$= \frac{1}{2\pi i} \int_{\mathbb{C}} \frac{\frac{\partial g_k}{\partial \bar{z}_1}(\zeta, z_2, \ldots, z_n)}{\zeta - z_1} d\zeta \wedge d\bar{\zeta}.$$

Using the second form of the definition of ψ, the compatibility conditions (4.1), and the computation above we get

$$\frac{\partial \psi}{\partial \bar{z}_k}(z) = \frac{1}{2\pi i} \int_{\mathbb{C}} \frac{\frac{\partial g_1}{\partial \bar{z}_k}(\zeta + z_1, z_2, \ldots, z_n)}{\zeta} d\zeta \wedge d\bar{\zeta}$$
$$= \frac{1}{2\pi i} \int_{\mathbb{C}} \frac{\frac{\partial g_k}{\partial \bar{z}_1}(\zeta + z_1, z_2, \ldots, z_n)}{\zeta} d\zeta \wedge d\bar{\zeta}$$
$$= \frac{1}{2\pi i} \int_{\mathbb{C}} \frac{\frac{\partial g_k}{\partial \bar{z}_1}(\zeta, z_2, \ldots, z_n)}{\zeta - z_1} d\zeta \wedge d\bar{\zeta} = g_k(z).$$

Exercise 4.2.2: *Show that we were allowed to differentiate ψ under the integral in the computation above, but only in the second form of the integral.*

That ψ has compact support follows because g_1 has compact support together with the identity theorem. In particular, ψ is holomorphic for large z since $\bar{\partial}\psi = g = 0$ when z is large. When at least one of $z_2, \ldots, z_n$ is large, then ψ is identically zero simply from its definition. See Figure 4.2.

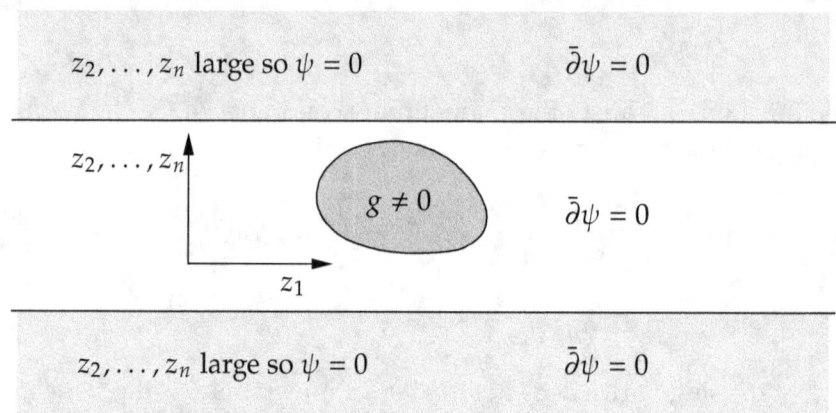

Figure 4.2: Far enough, $\bar{\partial}\psi = 0$.

As $\bar{\partial}\psi = 0$ on the light gray and white areas in the diagram, ψ is holomorphic there. As ψ is zero on the light gray region, it is zero also on the white region by the identity theorem. That is, ψ is zero on the unbounded component of the set where $g = 0$, and so ψ has compact support.   □

The first part of the proof still works when $n = 1$; we get a solution ψ. However, the last bit of the proof does not work in one dimension, so ψ need not have compact support.

Exercise 4.2.3:
   a) *Show that if g is supported in $K \subset\subset \mathbb{C}^n$, $n \geq 2$, then ψ is supported in the complement of the unbounded component of $\mathbb{C}^n \setminus K$. In particular, show that if K is the support of g and $\mathbb{C}^n \setminus K$ is connected, then the support of ψ is K.*
   b) *Find an explicit example where the support of ψ is strictly larger than the support of g.*

Exercise 4.2.4: *Find an example of a smooth function $g \colon \mathbb{C} \to \mathbb{C}$ with compact support, such that no solution $\psi \colon \mathbb{C} \to \mathbb{C}$ to $\frac{\partial \psi}{\partial \bar{z}} = g$ (at least one of which always exists) is of compact support.*

4.3 \ The general Hartogs phenomenon

We can now prove the general Hartogs phenomenon as an application of the solution of the compactly supported inhomogeneous $\bar{\partial}$-problem. We proved special versions of this phenomenon using Hartogs figures before. The proof of the theorem has a complicated history as Hartogs' original proof from 1906 contained gaps. A fully working proof was finally supplied by Fueter in 1939 for $n = 2$ and independently by Bochner and Martinelli for higher n in the early 40s. The proof we give is the standard one given nowadays due to Leon Ehrenpreis from 1961.

Theorem 4.3.1 (Hartogs phenomenon). *Let $U \subset \mathbb{C}^n$ be a domain, $n \geq 2$, and let $K \subset\subset U$ be a compact set such that $U \setminus K$ is connected. Every holomorphic $f : U \setminus K \to \mathbb{C}$ extends uniquely to a holomorphic function on U. See Figure 4.3.*

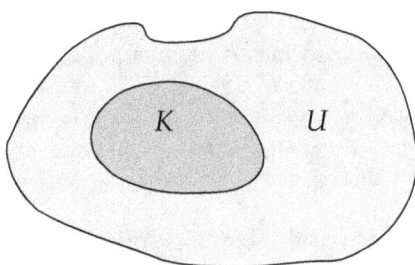

Figure 4.3: Hartogs phenomenon.

The idea of the proof is extending in any way whatsoever and then using the solution to the $\bar{\partial}$-problem to correct the result to make it holomorphic.

Proof. First find a smooth function φ that is 1 in a neighborhood of K and is compactly supported in U (exercise below). Let $f_0 = (1 - \varphi)f$ on $U \setminus K$ and $f_0 = 0$ on K. The function f_0 is smooth on U and it is holomorphic and equal to f near the boundary of U, where φ is 0. We let $g = \bar{\partial}f_0$ on U, that is $g_k = \frac{\partial f_0}{\partial \bar{z}_k}$, and we let $g = 0$ outside U. As g_k are identically zero near ∂U, we find that each g_k is smooth on $\mathbb{C}^n$. The compatibility conditions (4.1) are satisfied because partial derivatives commute. Let us see why g is compactly supported. The only place to check is on $U \setminus K$ as elsewhere we have $g = 0$ automatically. Note that f is holomorphic on $U \setminus K$ and compute

$$\frac{\partial f_0}{\partial \bar{z}_k} = \frac{\partial}{\partial \bar{z}_k}\big((1 - \varphi)f\big) = \frac{\partial f}{\partial \bar{z}_k} - \varphi\frac{\partial f}{\partial \bar{z}_k} - \frac{\partial \varphi}{\partial \bar{z}_k}f = -\frac{\partial \varphi}{\partial \bar{z}_k}f.$$

The function $\frac{\partial \varphi}{\partial \bar{z}_k}$ is compactly supported in $U \setminus K$ by construction. Apply the solution of the compactly supported $\bar{\partial}$-problem to find a compactly supported function ψ

such that $\bar{\partial}\psi = g$. Set $F = f_0 - \psi$. We check that F is the desired extension. First, it is holomorphic:

$$\frac{\partial F}{\partial \bar{z}_k} = \frac{\partial f_0}{\partial \bar{z}_k} - \frac{\partial \psi}{\partial \bar{z}_k} = g_k - g_k = 0.$$

Next, Exercise 4.2.3 and the fact that $U \setminus K$ is connected reveal that ψ must be compactly supported in U. This means that F agrees with f near the boundary (in particular on an open set) and thus everywhere in $U \setminus K$ since $U \setminus K$ is connected. □

The hypotheses on dimension and on connectedness of $U \setminus K$ are necessary. No such theorem is true in one dimension. If $U \setminus K$ is disconnected, a simple counterexample can be constructed. See the exercise below.

Exercise 4.3.1: Show that φ exists. Hint: Use mollifiers.

Exercise 4.3.2: Suppose $U \subset \mathbb{C}^n$ is a domain and $K \subset U$ is a compact set (perhaps $U \setminus K$ is disconnected). Prove that given $f \in \mathcal{O}(U \setminus K)$ there exists an $F \in \mathcal{O}(U)$ that equals f on the intersection of U and the unbounded component of $\mathbb{C}^n \setminus K$.

Exercise 4.3.3: Suppose $U \subset \mathbb{C}^n$ is a domain and $K \subset U$ is a compact set such that $U \setminus K$ is disconnected. Find a counterexample to the conclusion to Hartogs.

One of many consequences of the Hartogs phenomenon is that the zero set of a holomorphic function f is never compact in dimension 2 or higher, although there exist easier proofs of that fact, see Exercise 1.6.6. If it were compact, $\frac{1}{f}$ would provide a contradiction, see also Exercise 1.6.5.

Corollary 4.3.2. *Suppose $U \subset \mathbb{C}^n$, $n \geq 2$, is a domain and $f : U \to \mathbb{C}$ is holomorphic. If the zero set $f^{-1}(0)$ is not empty, then it is not compact.*

Replacing $U \setminus K$ with a hypersurface is usually called the Hartogs–Bochner theorem (when the hypersurface is C^1 or smooth). The real-analytic case was stated first by Severi in 1931.

Corollary 4.3.3 (Severi). *Suppose $U \subset \mathbb{C}^n$, $n \geq 2$, is a bounded domain with connected real-analytic boundary and $f : \partial U \to \mathbb{C}$ is a real-analytic CR function. Then there exists some neighborhood $U' \subset \mathbb{C}^n$ of $\overline{U}$ and a holomorphic function $F : U' \to \mathbb{C}$ for which $F|_{\partial U} = f$.*

Proof. By Severi's result (Theorem 3.2.9), for every $p \in \partial U$, there is a small ball B_p centered at p, such that f extends to B_p. Cover ∂U by finitely many such balls so that if B_p intersects B_q, then the (connected) intersection $B_p \cap B_q$ contains points of ∂U. The extensions in B_p and in B_q then agree on a piece of a hypersurface ∂U, and hence agree. Taking a union of the B_p, we find a unique extension in single neighborhood of ∂U. We write this neighborhood as $U' \setminus K$ for some compact K and a connected U' such that $\overline{U} \subset U'$. Consider the topological components of $\mathbb{C}^n \setminus K$. As ∂U is connected and U is bounded, the unbounded component of $\mathbb{C}^n \setminus K$ must contain all of ∂U. By boundedness of U, all the other components are relatively compact in U. If we add them to K, then K is still compact and $U' \setminus K$ is connected. We apply the Hartogs phenomenon.   □

Exercise 4.3.4 (Hartogs–Bochner again)**:** *Let $U \subset \mathbb{C}^n$, $n \geq 2$, be a bounded domain with connected strongly pseudoconvex smooth boundary and let $f : \partial U \to \mathbb{C}$ be a smooth CR function. Prove that there exists a continuous function $F : \overline{U} \to \mathbb{C}$ holomorphic in U such that $F|_{\partial U} = f$. Note: Strong pseudoconvexity is not needed ("bounded with smooth boundary" will do), but that is much more difficult to prove.*

Exercise 4.3.5: *Suppose $U \subset \mathbb{C}^n$, $n \geq 2$, is a bounded domain of holomorphy. Show that $\mathbb{C}^n \setminus U$ is connected using the Hartogs phenomenon.*

Exercise 4.3.6: *Suppose $W \subset U \subset \mathbb{C}^n$, $n \geq 3$, are domains such that for each fixed $z_3^0, z_4^0, \ldots, z_n^0$,*

$$\left\{ (z_1, z_2) \in \mathbb{C}^2 : (z_1, z_2, z_3^0, \ldots, z_n^0) \in U \setminus W \right\}$$
$$\subset\subset \left\{ (z_1, z_2) \in \mathbb{C}^2 : (z_1, z_2, z_3^0, \ldots, z_n^0) \in U \right\}.$$

Prove that every $f \in \mathcal{O}(W)$ extends to a holomorphic function on U. Note: The fact that W is connected is important.

Exercise 4.3.7:
    a) *Prove that if $n \geq 2$, no domain of the form $U = \mathbb{C}^n \setminus K$ for a compact K is biholomorphic to a bounded domain.*
    b) *Prove that every domain of the form $U = \mathbb{C} \setminus K$ for a compact K with nonempty interior is biholomorphic to a bounded domain.*

Exercise 4.3.8: *Suppose $U \subset \mathbb{C}^n$, $n \geq 2$, is a domain such that for some affine $A : \mathbb{C}^2 \to \mathbb{C}^n$ the set $A^{-1}(\mathbb{C}^n \setminus U)$ has a bounded topological component. Prove that U is not a domain of holomorphy.*

4.4   Solvability of the $\bar{\partial}$-problem in the polydisc

Let us tackle the solvability of the $\bar{\partial}$-problem for differential forms. In general, the problem is equivalent to holomorphic convexity, although it is rather involved, and thus we content ourselves with polydiscs and other simple examples. To work with differential forms, we, as before, split the derivatives into the holomorphic and antiholomorphic parts. For higher order forms, we work with multi-indices for simplicity, although the way that multi-indices are applied is slightly different.

Definition 4.4.1. Suppose $U \subset \mathbb{C}^n$ is open and let p and q be integers, $0 \leq p, q \leq n$. Let α and β be ordered p- and q-tuples of distinct integers such that $1 \leq \alpha_1 < \alpha_2 < \cdots < \alpha_p \leq n$ and $1 \leq \beta_1 < \beta_2 < \cdots < \beta_q \leq n$. Write

$$dz_\alpha = dz_{\alpha_1} \wedge \cdots \wedge dz_{\alpha_p} \quad \text{and} \quad d\bar{z}_\beta = d\bar{z}_{\beta_1} \wedge \cdots \wedge d\bar{z}_{\beta_q}.$$

A differential form of *bidegree* (p, q), or a (p, q)-*form* for short, is

$$\eta = \sum_{\alpha, \beta} \eta_{\alpha\beta} \, dz_\alpha \wedge d\bar{z}_\beta,$$

where the α and β run over all p- and q-tuples as above and $\eta_{\alpha\beta}$ are smooth functions on U. Note that $(0,0)$-forms are simply smooth functions on U. A general k-form can be written as a sum of (p, q)-forms for different p and q where $p + q = k$. Define

$$\partial \eta \overset{\text{def}}{=} \sum_{\alpha, \beta} \sum_{k=1}^n \frac{\partial \eta_{\alpha\beta}}{\partial z_k} dz_k \wedge dz_\alpha \wedge d\bar{z}_\beta \quad \text{and} \quad \bar{\partial} \eta \overset{\text{def}}{=} \sum_{\alpha, \beta} \sum_{k=1}^n \frac{\partial \eta_{\alpha\beta}}{\partial \bar{z}_k} d\bar{z}_k \wedge dz_\alpha \wedge d\bar{z}_\beta.$$

If η is of bidegree (p, q), then $\partial \eta$ is of bidegree $(p + 1, q)$ and $\bar{\partial} \eta$ is of bidegree $(p, q + 1)$. We get the total exterior derivative $d\eta = \partial \eta + \bar{\partial} \eta$ as before.

Exercise 4.4.1: Prove $d\eta = \partial \eta + \bar{\partial} \eta$ and prove the Leibniz rule: If η is a (p, q)-form, then

$$\partial(\eta \wedge \omega) = \partial \eta \wedge \omega + (-1)^{p+q} \eta \wedge \partial \omega \quad \text{and} \quad \bar{\partial}(\eta \wedge \omega) = \bar{\partial} \eta \wedge \omega + (-1)^{p+q} \eta \wedge \bar{\partial} \omega.$$

Exercise 4.4.2: Show that $\bar{\partial}^2 = 0$, that is, prove that $\bar{\partial}^2 \eta = \bar{\partial}\bar{\partial}\eta = 0$ for every form η. Similarly, show $\partial^2 = 0$.

Exercise 4.4.3: Given a hypersurface $M \subset \mathbb{C}^n$ with a defining function r, compute $\partial\bar{\partial}r$ and show that it gives the Levi form. That is, for a $T^{(1,0)}M$ vector field Z, the Levi form is given by $\langle \partial\bar{\partial}r, Z \wedge \bar{Z} \rangle$. Hint: See appendix C on how to evaluate differential forms as multilinear forms.

A form η is a $\bar{\partial}$-*exact form* if there exists a form ω such that $\bar{\partial}\omega = \eta$. A form η is a $\bar{\partial}$-*closed form* if $\bar{\partial}\eta = 0$. For an open set $U \subset \mathbb{C}^n$, we define the *Dolbeault cohomology groups* (quotient of complex vector spaces)

$$H^{(p,q)}(U) = \frac{\{\bar{\partial}\text{-closed forms of bidegree } (p, q) \text{ on } U\}}{\{\bar{\partial}\text{-exact forms of bidegree } (p, q) \text{ on } U\}}.$$

By convention, the only $(0,0)$-form that is exact is the identically zero form.

Exercise 4.4.4: Show that the $\bar{\partial}$-closed forms of degree (p, q) are a subspace of the vector space of all (p, q)-forms and similarly that the $\bar{\partial}$-exact forms of degree (p, q) are a subspace.

Exercise 4.4.5: Prove that if two domains $U, V \subset \mathbb{C}^n$ are biholomorphic, then for all (p, q), then $H^{(p,q)}(U)$ and $H^{(p,q)}(V)$ are isomorphic as vector spaces.

The $\bar{\partial}$-problem for (p, q)-forms is then the solvability of the equation $\bar{\partial}\omega = \eta$ for every (p, q)-form η given the necessary compatibility conditions $\bar{\partial}\eta = 0$. This problem can then be stated as the cohomology condition $H^{(p,q)}(U) = 0$, where by 0, we mean the trivial vector space. The Dolbeault cohomology is, in a sense, a refinement of the so-called de Rham cohomology, which is the usual smooth cohomology measuring the normal topology of U, where the Dolbeault cohomology also takes into account the complex structure. Note that $H^{(0,0)}(U)$ is just the set of $\bar{\partial}$-closed $(0, 0)$-forms, that is, it is the set $\mathcal{O}(U)$ of holomorphic functions on U. More generally, $H^{(p,0)}(U)$ is the set of $\bar{\partial}$-closed $(p, 0)$-forms, that is, forms

$$\eta = \sum_{\alpha} \eta_\alpha \, dz_\alpha \quad \text{such that} \quad \bar{\partial}\eta = \sum_{\alpha} \sum_{k=1}^{n} \frac{\partial \eta_\alpha}{\partial \bar{z}_k} d\bar{z}_k \wedge dz_\alpha = 0.$$

That is, all the functions η_α are holomorphic.

Solvability of the equation $\bar{\partial}\omega = \eta$ for every (p, q)-form η such that $\bar{\partial}\eta = 0$ whenever $q \geq 1$ is equivalent to holomorphic convexity, although the complete proof is beyond the scope of this book. You will prove one direction of this theorem in $\mathbb{C}^2$ in the exercises in this section, and we will prove the general version of this direction in the next section; see Theorem 4.5.6. That is, we will prove that the vanishing of the cohomology groups implies domain of holomorphy. Let us state this theorem without proof.

Theorem 4.4.2. *A domain $U \subset \mathbb{C}^n$ is a domain of holomorphy (and hence holomorphically convex) if and only if $H^{(0,q)}(U) = 0$ whenever $1 \leq q \leq n - 1$.*

If U is a domain of holomorphy, it is in fact true that $H^{(p,q)}(U) = 0$ whenever $q \geq 1$. We will not prove this fact, but we will prove it for a polydisc, and we saw above that $H^{(p,0)}(U)$ is never trivial.

Exercise 4.4.6: Prove that it is sufficient to consider $p = 0$, that is, $H^{(0,q)}(U) = 0$ if and only if $H^{(p,q)}(U) = 0$ for all p.

Exercise 4.4.7: Suppose $U \subset \mathbb{C}^n$ is open. Show that $H^{(p,q)}(U) = 0$ if and only if $H^{(p,q)}(W) = 0$ for every connected component W of U.

Example 4.4.3: As we mentioned, if U is not a domain of holomorphy, the $\bar{\partial}$ problem is not always solvable, and hence the Dolbeault cohomology groups may be nonzero. Let us show that $H^{(0,1)}(\mathbb{C}^2 \setminus \{0\})$ contains a nonzero element.

Let $r = |z|^2 + |w|^2$. Write

$$\frac{1}{zw} = \frac{\bar{w}}{zr} - \frac{-\bar{z}}{wr}.$$

That is, the two functions on the right-hand side differ by a holomorphic function wherever z and w are both not zero and hence their $\bar{\partial}$'s are equal (where they are both defined). The left-hand term is defined when $z \neq 0$ and the right-hand term is defined when $w \neq 0$. So the following form η is well-defined on $\mathbb{C}^2 \setminus \{0\}$

$$\eta = \bar{\partial}\left(\frac{\bar{w}}{zr}\right) \quad \text{if } z \neq 0, \quad \text{and} \quad \eta = \bar{\partial}\left(\frac{-\bar{z}}{wr}\right) \quad \text{if } w \neq 0.$$

That η is $\bar{\partial}$-closed follows by $\bar{\partial}^2 = 0$. Suppose for contradiction that there existed a smooth $f\colon \mathbb{C}^2 \setminus \{0\} \to \mathbb{C}$ such that $\bar{\partial}f = \eta$. Define $g = zf - \bar{w}/r$. When $z \neq 0$, then $g/z = f - \bar{w}/zr$, and so $\bar{\partial}(g/z) = 0$. Therefore, g is holomorphic where $z \neq 0$, but g is, and this is really where the contradiction comes in, smooth on $\mathbb{C}^2 \setminus \{0\}$ and hence satisfies the Cauchy–Riemann equations on $\mathbb{C}^2 \setminus \{0\}$ and so g is holomorphic on $\mathbb{C}^2 \setminus \{0\}$. By Hartogs phenomenon (any version), g extends to be holomorphic in $\mathbb{C}^2$. In particular, g is holomorphic near 0, which contradicts the fact that $g(0, w) = -1/w$.

Example 4.4.4: It is not simply the topology (as we know already) that determines the $H^{(p,q)}$ groups. The previous example still works in $\mathbb{C}^2 \setminus \mathbb{R}^2 = \{(z, w) \in \mathbb{C}^2 : \text{Im } z \neq 0 \text{ or Im } w \neq 0\}$. That is, the function g from the previous example can be defined in $\mathbb{C}^2 \setminus \mathbb{R}^2$ and hence extends to $\mathbb{C}^2$ (see Exercise 2.1.10), leading again to a contradiction. Notice that the domain $\mathbb{C}^2 \setminus \{z = 0\}$ has the same exact topology as $\mathbb{C}^2 \setminus \mathbb{R}^2$. However, $\mathbb{C}^2 \setminus \{z = 0\}$ is a domain of holomorphy, and via an exercise below, $H^{(0,1)}(\mathbb{C}^2 \setminus \{z = 0\}) = 0$. The thing is that the f actually exists by construction: $f = \bar{w}/zr$. Then g is identically zero, so we do not get any contradiction.

Exercise 4.4.8: Prove that if $U \subset \mathbb{C}^2$ is a domain and $K \subset U$ is compact, then $H^{(0,1)}(U \setminus K)$ is nontrivial.

Exercise 4.4.9: Give another example for why topology is not enough. Consider the Hartogs figure $H = \{(z, w) \in \mathbb{D}^2 : |z| > 1/2 \text{ or } |w| < 1/2\}$. Show that while H is homeomorphic to the polydisc (has trivial topology), $H^{(0,1)}(H)$ is nontrivial.

Exercise 4.4.10: Suppose $U \subset \mathbb{C}^2$ is a domain such that $H^{(0,1)}(U) = 0$, then U is a domain of holomorphy. Hint: Prove the contrapositive. Suppose that U is not a domain of holomorphy and show that $H^{(0,1)}(U) \neq 0$ using the reasoning from the examples.

Exercise 4.4.11: Suppose $U \subset \mathbb{C}^n$ is a domain with smooth boundary, $0 \in \partial U$, $\{z_1 = z_2 = 0\} \cap U = \emptyset$, and the Levi form at the origin has a negative eigenvalue. Show that $H^{(0,1)}(U) \neq 0$.

Exercise 4.4.12: Prove that $H^{(0,1)}(\mathbb{C}^2 \setminus \{0\})$ is not just nontrivial, it is an infinite-dimensional vector space.

We will prove the solvability of the $\bar{\partial}$-problem for the polydisc. We will allow some of the factors in the polydisc to be $\mathbb{C}$, so we define *possibly unbounded polydisc* $\Delta \subset \mathbb{C}^n$ to mean $\Delta = D_1 \times \cdots \times D_n$ where each D_k is either a disc or $\mathbb{C}$. In this way, we also achieve a solution on $\mathbb{C}^n$ itself. This is the theorem we will actually prove:

Theorem 4.4.5. *Let $\Delta \subset \mathbb{C}^n$ be a possibly unbounded polydisc, let $p \geq 0$ and $q \geq 1$ be integers, and let η be a smooth (p,q)-form on Δ such that $\bar{\partial}\eta = 0$, then there exists a $(p, q-1)$-form ω such that $\bar{\partial}\omega = \eta$. In other words, $H^{(p,q)}(\Delta) = 0$ for $q \geq 1$.*

Before tackling the proof of the theorem, let us solve a simple $\bar{\partial}$-problem in one dimension using the Cauchy–Pompeiu formula.

Lemma 4.4.6. *Let $U \subset \mathbb{C}$ be a bounded open set with piecewise-C^1 boundary, and let $g \colon \overline{U} \to \mathbb{C}$ be a smooth function (restriction of a smooth function on a neighborhood of $\overline{U}$). Then $\psi \colon U \to \mathbb{C}$ given by*

$$\psi(z) = \frac{1}{2\pi i} \int_U \frac{g(\zeta)}{\zeta - z} d\zeta \wedge d\bar{\zeta}$$

is a smooth function such that $\frac{\partial \psi}{\partial \bar{z}} = g$.

By taking conjugates, we can similarly solve the $\frac{\partial \psi}{\partial z} = g$ problem. Moreover, since the solution is given as an integral, we can solve the problem with parameters. That is, if g depends on some other variables in a smooth or holomorphic way, then the solution ψ also depends on those variables smoothly or holomorphically. Compare the expression for ψ to the one used in the proof of Theorem 4.2.1. The proof is based on the fact that $\psi(z) = \log|z|^2$ solves the problem for $g(z) = 1/\bar{z}$. This fact may be surprising, as doing calculus blindly, one would arrive at the multivalued function $\log \bar{z} +$ holomorphic function, but $\log \bar{z} + \log z = \log|z|^2$ is single valued as needed. The lemma, via the exercises, leads to showing that $H^{(0,1)}(U) = 0$ for every domain $U \subset \mathbb{C}$. Every $(0,1)$-form $g \, d\bar{z}$ in U is $\bar{\partial}$-closed, so $H^{(0,1)}(U) = 0$ is equivalent to showing that $\frac{\partial \psi}{\partial \bar{z}} = g$ is solvable for any smooth function g on U.

Exercise 4.4.13: The fact that the functions are complex-valued is important. Show that for $g(x + iy) = \frac{y}{x^2+y^2}$ for $\mathbb{C} \setminus \{0\}$, there is no real-valued $\psi \colon \mathbb{C} \setminus \{0\} \to \mathbb{R}$ such that $\frac{\partial \psi}{\partial z} = g$ or $\frac{\partial \psi}{\partial \bar{z}} = g$.

Proof of the lemma. Fix $z \in U$ for a moment and take a small disc $\Delta_r(z)$ such that $\overline{\Delta_r(z)} \subset U$, see Figure 4.4. Then via Stokes,

$$\int_{\partial U} g(\zeta) \log|\zeta - z|^2 d\bar{\zeta} - \int_{\partial \Delta_r(z)} g(\zeta) \log|\zeta - z|^2 d\bar{\zeta} = \int_{U \setminus \Delta_r(z)} d\big(g(\zeta) \log|\zeta - z|^2 d\bar{\zeta}\big)$$

$$= \int_{U \setminus \Delta_r(z)} \frac{\partial g}{\partial \zeta}(\zeta) \log|\zeta - z|^2 d\zeta \wedge d\bar{\zeta} + \int_{U \setminus \Delta_r(z)} \frac{g(\zeta)}{\zeta - z} d\zeta \wedge d\bar{\zeta}.$$

Note that both $\log|\zeta - z|^2$ and $\frac{1}{\zeta - z}$ are integrable over U with respect to the area measure. As $r \to 0$, the last term above goes to $\int_U \frac{g(\zeta)}{\zeta - z} d\zeta \wedge d\bar{\zeta}$, which is $2\pi i \psi(z)$. Next, the function g is bounded on $\overline{U}$, by say M, so

$$\left| \int_{\partial \Delta_r(z)} g(\zeta) \log|\zeta - z|^2 d\bar{\zeta} \right| \leq \int_{\partial \Delta_r(z)} M \left| \log(r^2) \right| |d\bar{\zeta}| = 2\pi r M \left| \log(r^2) \right| \underset{\text{as } r \to 0}{\to} 0.$$

Thus, taking the limit we get

$$\psi(z) = \frac{1}{2\pi i} \int_{\partial U} g(\zeta) \log|\zeta - z|^2 d\bar{\zeta} - \frac{1}{2\pi i} \int_U \frac{\partial g}{\partial \zeta}(\zeta) \log|\zeta - z|^2 d\zeta \wedge d\bar{\zeta}.$$

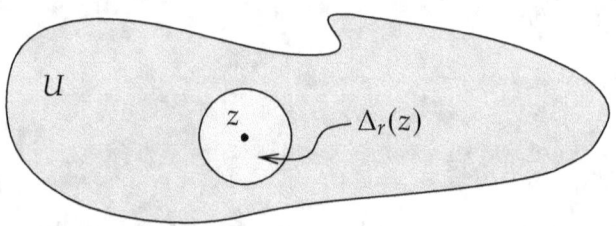

Figure 4.4: Using Stokes.

Taking partial derivatives (in Re z and Im z) still leaves the integrands integrable, and hence we can pass them under the integral sign. In particular, the function ψ is C^1 and we can take the $\bar{z}$ derivative. We then apply the Cauchy–Pompeiu formula (actually its conjugate applied to $\bar{g}$)

$$\frac{\partial \psi}{\partial \bar{z}}(z) = \frac{-1}{2\pi i} \int_{\partial U} \frac{g(\zeta)}{\bar{\zeta} - \bar{z}} d\bar{\zeta} + \frac{1}{2\pi i} \int_U \frac{\frac{\partial g}{\partial \bar{\zeta}}(\zeta)}{\bar{\zeta} - \bar{z}} d\zeta \wedge d\bar{\zeta} = g(z).$$

So we are done with the existence of this solution, we need to show that it is also smooth. As g is smooth, $\frac{\partial \psi}{\partial \bar{z}}$ is also smooth. If we prove that $\frac{\partial \psi}{\partial z}$ is also smooth, then ψ must be smooth. We take the z derivative instead of the $\bar{z}$ derivative to find

$$\frac{\partial \psi}{\partial z}(z) = \frac{-1}{2\pi i} \int_{\partial U} \frac{g(\zeta)}{\zeta - z} d\bar{\zeta} + \frac{1}{2\pi i} \int_U \frac{\frac{\partial g}{\partial \zeta}(\zeta)}{\zeta - z} d\zeta \wedge d\bar{\zeta}.$$

The first integral is clearly smooth. The second integral is precisely the sort of integral we have just shown is C^1 (with g replaced by $\frac{\partial g}{\partial \zeta}$), so ψ is C^2. By induction, $\frac{\partial \psi}{\partial z}$ is smooth, and hence ψ is smooth. $\qquad \square$

Moving to several variables, we prove that we can solve the problem on a subpolydisc of any polydisc, which is usually called the *Dolbeault lemma* or *Dolbeault–Grothendieck lemma*. As it is the analogue of the Poincaré lemma, it is sometimes called the $\bar{\partial}$-Poincaré lemma.

Lemma 4.4.7 (Dolbeault–Grothendieck). *Let $\Delta_s(w) \subset \Delta_r(w) \subset \mathbb{C}^n$ be polydiscs where $0 < s_\ell < r_\ell < \infty$ for each ℓ. Let $p \geq 0$ and $q \geq 1$ be integers, and let η be a smooth (p,q)-form on $\Delta_r(w)$ such that $\bar\partial\eta = 0$, then there exists a smooth $(p, q-1)$-form ω on $\Delta_s(w)$ such that $\bar\partial\omega = \eta$.*

Proof. Let k be an integer such that η only involves $d\bar z_1, \ldots, d\bar z_k$ from the barred differentials. If $k = 0$, then the lemma is true trivially as $q \geq 1$, so η would just have to be zero. We will proceed by induction.

Write
$$\eta = d\bar z_k \wedge \tau + \theta,$$
where τ and θ only involve the barred differentials $d\bar z_1, \ldots, d\bar z_{k-1}$. Now
$$0 = \bar\partial\eta = \bar\partial(d\bar z_k \wedge \tau + \theta) = -d\bar z_k \wedge \bar\partial\tau + \bar\partial\theta.$$

Hence the coefficients of τ and θ must be holomorphic in $z_{k+1}, \ldots, z_n$ as their derivatives in the bars of those variables are zero. In particular, if $\tau_{\alpha\beta}$ is one of the coefficients of τ, then it is a smooth function of the larger polydisc, but also holomorphic in the variables $z_{k+1}, \ldots, z_n$.

> *Exercise 4.4.14:* Check the assertion that the coefficients of τ and θ are holomorphic in the variables $z_{k+1}, \ldots, z_n$.

By Lemma 4.4.6, there is a smooth function $\psi_{\alpha\beta}$ in the z_k variable such that $\frac{\partial\psi_{\alpha\beta}}{\partial\bar z_k} = \tau_{\alpha\beta}$. Moreover, each $\psi_{\alpha\beta}$ is also a smooth function in
$$\Delta_{r_1}(w_1) \times \cdots \times \Delta_{r_{k-1}}(w_{k-1}) \times \Delta_t(w_k) \times \Delta_{r_{k+1}}(w_{k+1}) \times \cdots \times \Delta_{r_n}(w_n)$$
for some t such that $s_k < t < r_k$. It is also holomorphic in the variables $z_{k+1}, \ldots, z_n$. From the $\psi_{\alpha\beta}$ functions we construct a $(p, q-1)$-form ψ. We compute $\bar\partial\psi$:

$$\bar\partial\left(\sum_{\alpha\beta} \psi_{\alpha\beta}\, dz_\alpha \wedge d\bar z_\beta\right) = \sum_{\alpha\beta} \frac{\partial\psi_{\alpha\beta}}{\partial\bar z_k} d\bar z_k \wedge dz_\alpha \wedge d\bar z_\beta + \sum_{\alpha\beta}\sum_{\ell=1}^{k-1} \frac{\partial\psi_{\alpha\beta}}{\partial\bar z_\ell} d\bar z_\ell \wedge dz_\alpha \wedge d\bar z_\beta$$

$$= d\bar z_k \wedge \left(\sum_{\alpha\beta} \tau_{\alpha\beta}\, dz_\alpha \wedge d\bar z_\beta\right) + \sum_{\alpha\beta}\sum_{\ell=1}^{k-1} \frac{\partial\psi_{\alpha\beta}}{\partial\bar z_\ell} d\bar z_\ell \wedge dz_\alpha \wedge d\bar z_\beta$$

$$= d\bar z_k \wedge \tau + \delta,$$

where δ also does not contain the barred differentials other than $d\bar z_1, \ldots, d\bar z_{k-1}$. Now note that
$$\bar\partial(\theta - \delta) = \bar\partial(\eta - \bar\partial\psi) = \bar\partial\eta - \bar\partial^2\psi = 0.$$

Since θ and δ both only contain the barred differentials $d\bar z_1, \ldots, d\bar z_{k-1}$, we can apply the induction hypothesis to find a $(p, q-1)$-form φ such that $\bar\partial\varphi = \theta - \delta$. Then we let $\omega = \psi + \varphi$ and note that we are done:
$$\bar\partial\omega = \bar\partial(\psi + \varphi) = d\bar z_k \wedge \tau + \delta + \theta - \delta = \eta. \qquad \square$$

Exercise **4.4.15:** *Prove that the construction from the proof of the lemma reproduces the compactly supported solution for compactly supported $(0, 1)$-forms from section 4.2 provided we start with a sufficiently large polydisc, of course.*

Exercise **4.4.16:** *Prove the lemma if you replace the polydiscs $\Delta_s(w)$ and $\Delta_r(w)$ with two nested balls with center w.*

We can now prove the theorem itself.

Proof of Theorem 4.4.5. The proof splits in two cases. First, suppose that $q > 1$. Pick a sequence of polydiscs Δ_k all centered at the origin, such that $\bigcup_k \Delta_k = \Delta$ and such that $\overline{\Delta_k} \subset \Delta_{k+1}$ for all k. Using the Dolbeault–Grothendieck lemma with Δ_3 and Δ_4, we find a smooth form ω_1 defined on Δ_3 such that $\bar{\partial}\omega_1 = \eta$. We will construct a sequence of forms $\{\omega_k\}$ each ω_k defined in Δ_{k+2} such that $\omega_{k+1}|_{\Delta_k} = \omega_k|_{\Delta_k}$. Suppose we have defined $\omega_1, \ldots, \omega_k$. Using Δ_{k+3} and Δ_{k+4} with the lemma again, define a new form σ on Δ_{k+3} such that $\bar{\partial}\sigma = \eta$. What we need to do is to correct σ so that it equals ω_k on Δ_k. On Δ_{k+2}, we have $\bar{\partial}(\omega_k - \sigma) = \eta - \eta = 0$, so $\omega_k - \sigma$ is closed. The lemma gives a new form θ on Δ_{k+1} such that $\bar{\partial}\theta = \omega_k - \sigma$. Define a smooth bump function φ on $\mathbb{C}^n$ such that $\varphi = 1$ on $\overline{\Delta_k}$ and φ is compactly supported in Δ_{k+1}. Define $\omega_{k+1} = \sigma + \bar{\partial}(\varphi\theta)$, which can be defined as a smooth form on Δ_{k+3} as φ is identically zero where θ is undefined. The fact that $\bar{\partial}^2 = 0$ ensures that $\bar{\partial}\omega_{k+1} = \bar{\partial}\sigma = \eta$ on Δ_{k+3}. On Δ_k, we have $\omega_{k+1} = \sigma + \bar{\partial}\theta = \sigma + \omega_k - \sigma = \omega_k$. See Figure 4.5. The sequence $\{\omega_k\}$ is defined. We define ω on Δ by simply letting $\omega = \omega_k$ on Δ_k.

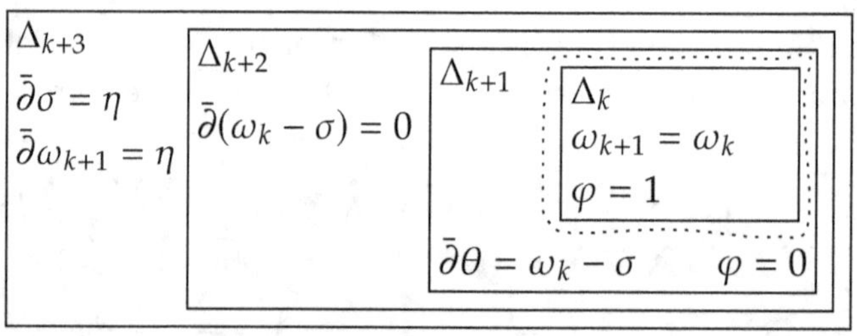

Figure 4.5: Diagram for defining $\omega_{k+1} = \sigma + \bar{\partial}(\varphi\theta)$. The dotted line gives supp φ.

Now assume $q = 1$. By Exercise 4.4.6, it is enough to consider $p = 0$ to simplify notation. That is, we are now looking for a smooth function ω so that $\bar{\partial}\omega = \eta$. We take the polydiscs Δ_k as before and define ω_1 on Δ_2 using η on Δ_3. But instead of ensuring that ω_{k+1} and ω_k are equal on Δ_k, we will ask that

$$|\omega_{k+1}(z) - \omega_k(z)| < 2^{-k} \quad \text{for } z \in \overline{\Delta_k}.$$

Suppose that such $\omega_1, \ldots, \omega_k$ have been defined. Use the lemma with Δ_{k+3} and Δ_{k+2} to obtain a smooth function σ on Δ_{k+2} such that $\bar{\partial}\sigma = \eta$. The function $\sigma - \omega_k$ is holomorphic (on Δ_{k+1} where ω_k is defined) as $\bar{\partial}(\sigma - \omega_k) = \eta - \eta = 0$, and hence it has a power series representation converging uniformly on $\overline{\Delta_k}$. Thus there exists a holomorphic polynomial P such that

$$|\sigma(z) - \omega_k(z) - P(z)| < 2^{-k} \quad \text{for } z \in \overline{\Delta_k}.$$

So let $\omega_{k+1} = \sigma - P$. As P is holomorphic, we have $\bar{\partial}\omega_{k+1} = \eta$, and we satisfied the required properties.

On any particular $\overline{\Delta_\ell}$, the sequence $\{\omega_k\}$ is uniformly Cauchy as if $m > k \geq \ell$, then $|\omega_m(z) - \omega_k(z)| < 2^{-k} + 2^{-k-1} + \cdots + 2^{-m+1} < 2^{-k+1}$. So the sequence converges uniformly on compact subsets of Δ to a function $\omega \colon \Delta \to \mathbb{C}$. We need to show that ω is smooth and satisfies $\bar{\partial}\omega = \eta$. The functions $\omega_m - \omega_\ell$ converge uniformly as $m \to \infty$ on Δ_ℓ. As $\bar{\partial}(\omega_m - \omega_\ell) = \eta - \eta = 0$, these functions are holomorphic, and so the limit $\omega - \omega_\ell$ is holomorphic. In particular, ω is smooth as ω_ℓ is smooth and $\bar{\partial}\omega = \bar{\partial}\omega_\ell = \eta$. $\quad\square$

Exercise 4.4.17: Prove the theorem also holds for a ball, say the unit ball $\mathbb{B}_n \subset \mathbb{C}^n$. Hint: Use Exercise 4.4.16.

Exercise 4.4.18: For any disc $D \subset \mathbb{C}$, let D^* denote the punctured disc $D^* = D \setminus \{a\}$ where a is the center. Prove that the theorem also holds for $U = D_1 \times \cdots \times D_k \times D^*_{k+1} \times \cdots \times D^*_n$ for some (possibly unbounded) discs $D_1, \ldots, D_n$. Hint: You may have to prove a slightly more general version of the Dolbeault–Grothendieck lemma. Also, see Exercise 1.2.4.

Exercise 4.4.19: Prove that in one variable, for any domain $U \subset \mathbb{C}$, we have $H^{(0,1)}(U) = 0$.

Exercise 4.4.20: Prove the theorem for the Hartogs triangle $T = \{(z, w) \in \mathbb{D}^2 : |z| > |w|\}$.

Exercise 4.4.21: Suppose $U \subset \mathbb{C}^n$ is a domain and $p \in \partial U$. Suppose there is some polydisc Δ centered at p such that $H^{(0,q)}(U \cap \Delta) \neq 0$, then $H^{(0,q)}(U) \neq 0$. Hint: Find a smooth function φ on $U \cap \Delta$ such that φ is identically zero near the boundary of U and identically one near the boundary of Δ. And use the theorem.

Exercise 4.4.22: Use the Exercise 4.4.21 and Exercise 4.4.11 to prove that if $U \subset \mathbb{C}^n$, $n \geq 2$, is a domain with smooth boundary and $p \in \partial U$ such that the Levi form at p has a negative eigenvalue and $n - 2$ positive eigenvalues, then $H^{(0,1)}(U) \neq 0$.

Exercise 4.4.23: Being a subset gives us no relation between the Dolbeault cohomology groups: Find domains $U \subset V \subset W \subset \mathbb{C}^2$ such that $H^{(0,1)}(U) = H^{(0,1)}(W) = 0$, but $H^{(0,1)}(V)$ is nontrivial.

4.5 \ Extension from an affine subspace

Let $L \subset \mathbb{C}^n$ be a (complex) *affine subspace* of dimension k, that is, the set defined by $Mz = c$ for an $(n - k) \times n$ matrix M of rank $n - k$ and $c \in \mathbb{C}^{n-k}$. After a complex linear transformation and translation, we assume that $L = \{z \in \mathbb{C}^n : z_{k+1} = \cdots = z_n = 0\}$. We say a function f defined on an open subset of L is holomorphic if after this change of coordinates it is holomorphic in the $z_1, \ldots, z_k$ variables. If $k = n - 1$, we call L a (complex) *affine hyperplane*, and if $k = 1$, we call L a *complex line*.

It is easy to see that if $U \subset \mathbb{C}^n$ is open, L is an affine subspace, and $F \in \mathcal{O}(U)$, then $F|_{U \cap L}$ is holomorphic. It is not hard to see that being able to do the inverse may be quite useful in proofs by induction on dimension, that is, starting with a holomorphic function f on $U \cap L$ and finding a holomorphic function on U whose restriction is f. If f is just smooth, then finding such a smooth extension is not difficult no matter what U looks like, however, in the holomorphic category it is not that easy (or even always possible). A holomorphic extension from an affine subspace is possible for domains of holomorphy; in fact, it is a defining characteristic of such domains.

Example 4.5.1: Suppose $U = \mathbb{C}^2 \setminus \{0\}$ and $L = \{(z, w) \in \mathbb{C}^2 : w = 0\}$. Then $U \cap L$ is the punctured plane. Let $f(z) = 1/z$ be the function on $U \cap L$. Suppose that there is an $F \in \mathcal{O}(U)$ whose restriction to $U \cap L$ was f. Such an F extends to all of $\mathbb{C}^2$, and hence f extends to all of L (which is just the z-plane), which is impossible. Note that we proved that $H^{(0,1)}(U)$ is nontrivial.

Example 4.5.2: Suppose $U = B_2\big((0,1)\big) \subset \mathbb{C}^2$ and $L = \{(z, w) \in \mathbb{C}^2 : w = 0\}$. Then $U \cap L$ is a disc of radius $\sqrt{3}$ in the z-plane. Take a holomorphic $f : U \cap L \to \mathbb{C}$ (as a function of z) that does not extend past any point in the boundary of the disc. The obvious way to extend f to an F would be to simply take a function $F(z, w) = f(z)$, but that is not defined on all of U. Extension to all of U is somewhat harder to prove. See Figure 4.6. It is possible to do this explicitly in this specific case (exercise below). Note that $H^{(0,1)}(U) = 0$, which you proved as an exercise in the previous section.

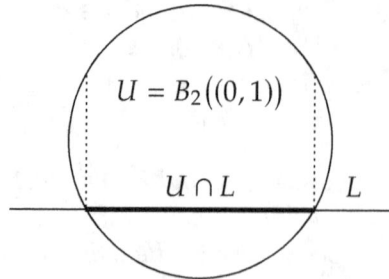

Figure 4.6: Extending f from a hyperplane to the ball. It is easy to extend to within the dotted lines.

Exercise **4.5.1:** *Explicitly define the extension of the holomorphic function f to F in the setup of Example 4.5.2.*

Exercise **4.5.2:** *Come up with an example domain U and hyperplane L (in $\mathbb{C}^2$) where $H^{(0,1)}(U)$ is nontrivial, but where every $f \in \mathcal{O}(U \cap L)$ extends holomorphically to U.*

Exercise **4.5.3:** *Multiple hyperplanes are possible. Let L be the set $\mathbb{C}^2$ where $z = 0$, M be the set where $w = 0$, and N be the set where $z = w$. Let f, g, h be holomorphic functions on L, M, and N. Some extra hypothesis is necessary to tie the three functions together. For simplicity, suppose that the functions and all their first derivatives vanish at the origin. Prove that there is a holomorphic F on $\mathbb{C}^2$ such that $F|_L = f$, $F|_M = g$, $F|_N = h$.*

Exercise **4.5.4:** *Show that you cannot replace the hyperplane with something that has no complex structure. Find an example of a domain of holomorphy $U \subset \mathbb{C}^2$, and a real-analytic $f : U \cap \mathbb{R}^2 \to \mathbb{C}$ that is not a restriction of a holomorphic function on U. Note that f does extend holomorphically to some neighborhood of $U \cap \mathbb{R}^2$ in U, just perhaps not all of U.*

The way to find the extension in general is to start with a simple extension to some neighborhood of L, such as $F(z, w) = f(z)$, then extend in a smooth way via a cutoff function to all of U. To make the extension to U holomorphic, we then correct via the appropriate $\bar{\partial}$-problem. We discussed the extension of holomorphic functions, but it is no harder to prove a more general version of the problem about $\bar{\partial}$-closed (p, q)-forms—holomorphic functions are the $\bar{\partial}$-closed $(0, 0)$-forms.

To be able to prove this more general theorem, we need to know what it means to restrict (p, q)-forms to L, not just functions. Instead of making this too complicated, let us do the same simplification as we did above for functions. Informally, we will restrict the values of the form to L and only take the parts of the form that "point along L." Let us, also for simplicity, suppose that L is a hyperplane; the more general case then follows. As before, after a linear map and a translation, assume $L = \{z \in \mathbb{C}^n : z_n = 0\}$. Write a (p, q)-form as

$$\eta = \sum_{\alpha\beta} \eta_{\alpha\beta}\, dz_\alpha \wedge d\bar{z}_\beta + \omega_2 \wedge dz_n + \omega_3 \wedge d\bar{z}_n,$$

where α and β do not include n (so no dz_n nor $d\bar{z}_n$ in the first term). Then

$$\eta|_L = \sum_{\alpha\beta} \eta_{\alpha\beta}|_L\, dz_\alpha \wedge d\bar{z}_\beta.$$

Basically, we restrict the components to L, and throw out the differentials that would take vectors that do not point along L. If η is a function (a $(0, 0)$-form), then, of course, this is just the restriction of the function.

Theorem 4.5.3. *Suppose that $U \subset \mathbb{C}^n$ is open with $H^{(p,q+1)}(U) = 0$ and L is an affine hyperplane. If ψ is a smooth $\bar{\partial}$-closed (p, q)-form on $U \cap L$, then there exists a smooth $\bar{\partial}$-closed (p, q)-form Ψ on U such that $\Psi|_{U \cap L} = \psi$.*

So to extend holomorphic functions from a hyperplane, we need $H^{(0,1)}(U) = 0$.

Proof. After a translation and a linear map, assume $L = \{z \in \mathbb{C}^n : z_n = 0\}$. Write $z = (z', z_n)$ as usual. Let $\chi \colon U \to \mathbb{R}$ be the function that is identically 1 in some neighborhood of $U \cap L$ and such that it is identically 0 in some neighborhood of $\{z \in U : (z', 0) \notin U \cap L\}$. See Figure 4.7.

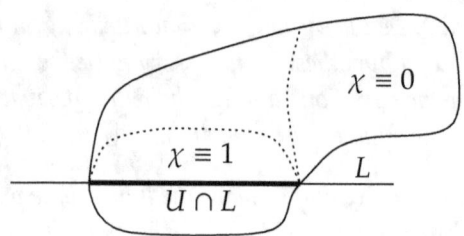

Figure 4.7: The cutoff function χ. Note that $\chi \equiv 0$ on a neighborhood of where it is not trivial to extend ψ.

We will define
$$\Psi(z) = \chi(z)\psi(z') + z_n \eta(z),$$
where η will be chosen appropriately to make Ψ $\bar{\partial}$-closed. This form Ψ is defined in U as χ is identically zero where ψ is undefined. Clearly, if such an η can be found, then $\Psi|_{U \cap L} = \psi$. Let us compute $\bar{\partial}\Psi$ to see what is required of η. Where ψ is defined we get
$$\bar{\partial}\Psi = \bar{\partial}\chi \wedge \psi + \chi\overbrace{\bar{\partial}\psi}^{0} + \overbrace{\bar{\partial}z_n}^{0} \wedge \eta + z_n\bar{\partial}\eta = \bar{\partial}\chi \wedge \psi + z_n\bar{\partial}\eta.$$
As $\bar{\partial}\chi$ is 0 in a neighborhood of $U \cap L$, the form $\frac{-\bar{\partial}\chi \wedge \psi}{z_n}$ extends smoothly through $U \cap L$. Moreover, it is a smooth $(p, q+1)$-form on U as $\bar{\partial}\chi$ is identically zero in a neighborhood of where ψ is undefined. The form is $\bar{\partial}$-closed as $\bar{\partial}^2 = 0$, $\bar{\partial}\psi = 0$, and $1/z_n$ is holomorphic. By hypothesis, we find an η such that $\bar{\partial}\eta = \frac{-\bar{\partial}\chi \wedge \psi}{z_n}$, and we are done. $\qquad\square$

We remark that an extension works for zero sets of holomorphic functions, that is, subvarieties (see chapter 6), not just hyperplanes, a result which is called the Cartan extension theorem, but we will not prove this fact. However, as an exercise prove the extension for two hyperplanes.

Exercise 4.5.5: *Suppose that $U \subset \mathbb{C}^n$ is open with $H^{(p,q+1)}(U) = 0$ and L_1 and L_2 are two affine hyperplanes such that $U \cap L_1 \cap L_2 = \emptyset$. Let ψ_1 be a smooth $\bar{\partial}$-closed (p, q)-form on $U \cap L_1$ and ψ_2 be a smooth $\bar{\partial}$-closed (p, q)-form on $U \cap L_2$. Show that there is a smooth $\bar{\partial}$-closed (p, q)-form Ψ on U such that $\Psi|_{U \cap L_1} = \psi_1$ and $\Psi|_{U \cap L_2} = \psi_2$.*

In what follows, when we talk about $U \cap L$ as an open set for a hyperplane L, we think of it as an open set in $\mathbb{C}^{n-1}$. More generally, if L is a k-dimensional affine subspace, then we will treat $U \cap L$ as an open set in $\mathbb{C}^{n-k}$.

Corollary 4.5.4. *Suppose $U \subset \mathbb{C}^n$ is open and L is an affine hyperplane. If $H^{(p,q)}(U) = 0$ and $H^{(p,q+1)}(U) = 0$ then $H^{(p,q)}(U \cap L) = 0$.*

Proof. Let ψ be a $\bar{\partial}$-closed (p,q)-form on $U \cap L$. As $H^{(p,q+1)}(U) = 0$, via the theorem, there is a $\bar{\partial}$-closed (p,q)-form Ψ on U such that $\Psi|_{U \cap L} = \psi$. As $H^{(p,q)}(U) = 0$, there is a $(p, q-1)$-form Φ on U such that $\bar{\partial}\Phi = \Psi$. It is not difficult to see that $(\bar{\partial}\Phi)|_{U \cap L} = \bar{\partial}(\Phi|_{U \cap L})$ (true for any form), so $\varphi = \Phi|_{U \cap L}$ is the solution to $\bar{\partial}\varphi = \psi$. $\square$

Exercise 4.5.6: Prove the claim that for any (p,q)-form, $(\bar{\partial}\Phi)|_{U \cap L} = \bar{\partial}(\Phi|_{U \cap L})$.

We have the following immediate corollary.

Corollary 4.5.5. *Suppose that $U \subset \mathbb{C}^n$ is open such that $H^{(0,q)}(U) = 0$ whenever $1 \leq q \leq n-1$ and L is an affine hyperplane. Then $H^{(0,q)}(U \cap L) = 0$ whenever $1 \leq q \leq n-2$.*

Exercise 4.5.7: Prove a more general corollary. Suppose that $U \subset \mathbb{C}^n$ is a domain and suppose that L is a k-dimensional affine subspace $1 \leq k \leq n-1$. Suppose that $H^{(0,q)}(U) = 0$ whenever $1 \leq q \leq n-1$. Then $H^{(0,q)}(U \cap L) = 0$ whenever $1 \leq q \leq k-1$, and every $f \in \mathcal{O}(U \cap L)$ has a holomorphic extension to U.

We now prove the general version of one direction of the theorem mentioned in the previous section. As an exercise, you proved by direct construction that if $H^{(0,1)}(U) = 0$ in $\mathbb{C}^2$, then U is a domain of holomorphy. We extend this result to $\mathbb{C}^n$.

Theorem 4.5.6. *Suppose $U \subset \mathbb{C}^n$ is a domain such that $H^{(0,q)}(U) = 0$ whenever $1 \leq q \leq n-1$. Then U is a domain of holomorphy.*

Proof. We induct on dimension. When $n = 1$, the hypothesis just says that U is a domain, and any domain in $\mathbb{C}$ is a domain of holomorphy. Assume the theorem holds for any domain in $\mathbb{C}^{n-1}$.

Let $U \subset \mathbb{C}^n$ be a domain and suppose that V and W are open sets in $\mathbb{C}^n$ such that $\emptyset \neq V \subset U \cap W$, W is connected, and W contains points outside of U. That is, V and W are like in the definition of the domain of holomorphy. We want to show that there must exist at least one $F \in \mathcal{O}(U)$ for which there does not exist any $G \in \mathcal{O}(W)$ such that $F = G$ on V. Consider a component of $W \cap U$ that contains some component of V. Without loss of generality we could take V to be any small ball in this component of $W \cap U$. It is not difficult to check (exercise) that there exists such a V (move it around if you must) so that there is a point $z_0 \in \partial U \cap \partial V \cap W$.

Exercise **4.5.8:** *Prove that such a V exists.*

There is some affine hyperplane L through z_0 such that the boundary of $V \cap L$ (in the topology of L) includes z_0, and hence the boundary of $U \cap L$ (in the topology of L) includes z_0. Here we use the fact that V is a ball; pick an L that includes the normal direction to the boundary of V at z_0. See Figure 4.8. By Corollary 4.5.5, $H^{(0,q)}(U \cap L) = 0$ for $1 \leq q \leq n - 2$. By the induction hypothesis (and Exercise 4.4.7), every component of $U \cap L$ is a domain of holomorphy. Since z_0 is in the boundary of $U \cap L$, there exists a holomorphic function $f \in \mathcal{O}(U \cap L)$ that does not extend (along L) through z_0. As $H^{(0,1)}(U) = 0$, there exists an $F \in \mathcal{O}(U)$ such that $F|_{U \cap L} = f$. If there existed a $G \in \mathcal{O}(W)$ that agreed with F on V, then $G|_{W \cap L}$ would be an extension of f through z_0, which we know is impossible, so no such G exists. $\square$

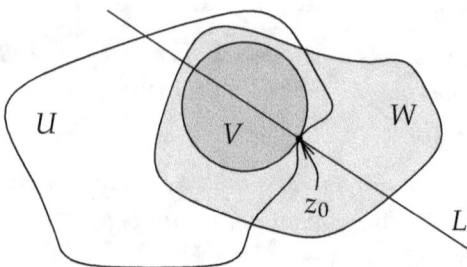

Figure 4.8: Location of z_0 and the placement of V and L with respect to U and W.

If you think about what we really needed in the proof, it was not the cohomology vanishing, we needed the extension. It is sufficient to extend from complex lines, that is, affine subspaces of dimension 1, since the intersection of $U \cap L$ in that case is always a domain of holomorphy. On the other hand, it is not sufficient to have extension from hyperplanes, see Exercise 4.5.12.

Theorem 4.5.7. *Suppose $U \subset \mathbb{C}^n$ is a domain such that for every complex line L and every $f \in \mathcal{O}(U \cap L)$ there exists an $F \in \mathcal{O}(U)$ where $F|_{U \cap L} = f$. Then U is a domain of holomorphy.*

Exercise **4.5.9:** *Prove Theorem 4.5.7.*

Exercise **4.5.10:** *Suppose $U \subset \mathbb{C}^n$ is a domain of holomorphy and L is an affine subspace of dimension k. Show that every component of $U \cap L$ is a domain of holomorphy.*

Exercise **4.5.11:** *Suppose $U \subset \mathbb{C}^n$ is a domain and $k \in \mathbb{N}$ is such that $H^{(0,q)}(U) = 0$ for all $1 \leq q \leq n - k$, and every component of $U \cap L$ is a domain of holomorphy for every affine subspace L of dimension k. Prove that U is a domain of holomorphy.*

Exercise 4.5.12: *Extension from hyperplanes is not enough: Find a domain $U \subset \mathbb{C}^3$ that is not a domain of holomorphy, such that for every affine hyperplane L and every $f \in \mathcal{O}(U \cap L)$ there is an $F \in \mathcal{O}(U)$ such that $F|_{U \cap L} = f$. Hint: Modify Example 4.5.1.*

Remark 4.5.8. We have by now stated several equivalent conditions for a domain to be a domain of holomorphy, although we have not proved all the implications in this book. In particular, for a domain $U \subset \mathbb{C}^n$, the following are equivalent:

(i) U is a domain of holomorphy.

(ii) U is Levi pseudoconvex (if U has smooth boundary).

(iii) U is Hartogs pseudoconvex (continuous plurisubharmonic exhaustion function).

(iv) $-\log \rho(z)$ is plurisubharmonic (ρ is distance to ∂U).

(v) U is convex with respect to plurisubharmonic functions.

(vi) U is holomorphically convex.

(vii) $\mathrm{dist}(K, \partial U) = \mathrm{dist}(\hat{K}_U, \partial U)$ for every $K \subset\subset U$.

(viii) $H^{(0,q)}(U) = 0$ for all $1 \leq q \leq n - 1$.

(ix) Every $f \in \mathcal{O}(U \cap L)$ extends holomorphically to U for every complex line L.

4.6 \ The Cousin problems

A chapter in many a book on one complex variable is devoted to the Mittag-Leffler theorem on finding a meromorphic function with prescribed poles, and another one on a theorem of Weierstrass for finding a holomorphic function with a prescribed zero set. The analogues of these results in several complex variables are the so-called Cousin I and Cousin II problems* respectively. The Cousin I problem is an additive version of the problem and obtains an analogue to Mittag-Leffler. The Cousin II problem is the multiplicative version to obtain an analogue of Weierstrass.

Definition 4.6.1 (Cousin I). Suppose $U \subset \mathbb{C}^n$ is open. Let $\{U_\iota\}_{\iota \in I}$ be an open covering of U, and when $U_\iota \cap U_\kappa \neq \emptyset$, let $h_{\iota \kappa} \in \mathcal{O}(U_\iota \cap U_\kappa)$ be such that

$$h_{\iota \kappa} + h_{\kappa \iota} = 0 \qquad \text{in } U_\iota \cap U_\kappa,$$
$$h_{\iota \kappa} + h_{\kappa \lambda} + h_{\lambda \iota} = 0 \quad \text{in } U_\iota \cap U_\kappa \cap U_\lambda.$$

The covering and the functions $h_{\iota \kappa}$ are called *Cousin I data*. The *solution of the Cousin I problem* is a set of holomorphic functions $f_\iota \in \mathcal{O}(U_\iota)$ such that

$$h_{\iota \kappa} = f_\iota - f_\kappa.$$

*Named for Pierre Cousin, the French mathematician, and not for some family scuffle.

Clearly if the functions f_ι exist, then their differences must satisfy the two conditions. The point is going the other way: finding f_ι given $h_{\iota\kappa}$. We remark that we could always take ι, κ, and λ to be distinct, as the first condition means that $h_{\iota\iota} = 0$ for all ι otherwise, and the second condition then just reduces to the first if two of the indices are equal.

Exercise 4.6.1: *The triple sum is enough to force a similar condition on 4 or more summands, but the double sum is not enough. That is,*
  a) *Find an example where the first condition $h_{\iota\kappa} + h_{\kappa\iota} = 0$ is satisfied but the second condition $h_{\iota\kappa} + h_{\kappa\lambda} + h_{\lambda\iota} = 0$ is not.*
  b) *Show that if both conditions are satisfied (we have valid Cousin I data), then $h_{\iota\kappa} + h_{\kappa\lambda} + h_{\lambda\mu} + h_{\mu\iota} = 0$, and similarly for any number of terms.*

To see how Cousin I relates to Mittag-Leffler, note that Mittag-Leffler could be stated by giving meromorphic functions locally and then trying to piece them together. Recall that a meromorphic function is locally a ratio of holomorphic functions. So suppose that we have a covering $\{U_\iota\}$ and in each U_ι we have a meromorphic function g_ι such that $g_\iota - g_\kappa$ is holomorphic in $U_\iota \cap U_\kappa$ (or more precisely, extends to be holomorphic on that set). That is, if we were in one dimension, the two functions would have the same principal part. The solution is to find a global meromorphic function with the same singular behavior (principal part).

On $U_\iota \cap U_\kappa$, let

$$h_{\iota\kappa} = g_\iota - g_\kappa.$$

It is easy to see that we obtain Cousin I data. Suppose the Cousin I problem is solvable in U. Then we would find holomorphic f_ι as above. We define a meromorphic function f on U by defining it in each U_ι as

$$f = g_\iota - f_\iota.$$

The function is well-defined. Indeed, on $U_\iota \cap U_\kappa$ the possible definitions are $g_\iota - f_\iota$ and $g_\kappa - f_\kappa$, and their difference is zero:

$$(g_\iota - f_\iota) - (g_\kappa - f_\kappa) = (g_\iota - g_\kappa) + (f_\kappa - f_\iota) = h_{\iota\kappa} + h_{\kappa\iota} = 0.$$

The function f has the same singularity as g_ι in U_ι since $f - g_\iota$ is holomorphic.

So the data gives some local solution to some problem and the solution of the Cousin problem gives a way of gluing the local data together into a global solution. We state this as a proposition (the converse also holds but we do not prove it here):

Proposition 4.6.2. *Suppose $U \subset \mathbb{C}^n$ is open, $\{U_\iota\}_{\iota \in I}$ an open covering of U, and g_ι are meromorphic functions on each U_ι such that $g_\iota - g_\kappa$ is holomorphic on $U_\iota \cap U_\kappa$. Suppose that the Cousin I problem is solvable on U. Then there exists a meromorphic function f on U such that on each U_ι, $f - g_\iota$ is holomorphic.*

In one dimension, Cousin I (that is, Mittag-Leffler) is solvable for any domain in $\mathbb{C}$. In several variables, Cousin I is solvable on domains of holomorphy in $\mathbb{C}^n$, in particular, it is solvable when we can solve the $\bar{\partial}$-problem for $(0,1)$-forms.

Example 4.6.3: Let us see an example where the Cousin I problem is not solvable. Let $U = \mathbb{C}^2 \setminus \{0\}$, where we know that $H^{(0,1)}(U)$ is nontrivial. Write $U = U_1 \cup U_2$ where $U_1 = \{(z,w) \in \mathbb{C}^2 : z \neq 0\}$ and $U_2 = \{(z,w) \in \mathbb{C}^2 : w \neq 0\}$. On $U_1 \cap U_2$, define

$$h_{12}(z,w) = \frac{1}{zw}, \quad h_{21}(z,w) = \frac{-1}{zw}.$$

These are holomorphic functions giving Cousin I data. Suppose the problem were solvable and we find $f_1 \in \mathcal{O}(U_1)$ and $f_2 \in \mathcal{O}(U_2)$ such that on $U_1 \cap U_2$, we have

$$f_1 - f_2 = h_{12} = \frac{1}{zw}.$$

In other words, $f_1 = \frac{1}{zw} + f_2$ on $U_1 \cap U_2$. But that means that $z f_1$ is holomorphic in U_2 and therefore in U, and thus extends to $\mathbb{C}^2$. Similarly, $w f_2$ extends to a holomorphic function on $\mathbb{C}^2$. Thus $zw f_1 - zw f_2$ is a holomorphic function on $\mathbb{C}^2$ that vanishes at the origin, but $zw f_1 - zw f_2 = zw(h_{12}) \equiv 1$ on $U_1 \cap U_2$, which leads to a contradiction.

The Cousin I problem with smooth data is smoothly solvable in any domain.

Lemma 4.6.4. *Suppose $U \subset \mathbb{C}^n$ is open, $\{U_\iota\}_{\iota \in I}$ be an open covering of U, and let $h_{\iota\kappa}$ be smooth (not necessarily holomorphic) Cousin I data. Then there exist smooth (not necessarily holomorphic) solution functions f_ι.*

Proof. Find a smooth partition of unity $\{\varphi_\gamma\}_{\gamma \in \Gamma}$ subordinate to the cover $\{U_\iota\}_{\iota \in I}$. That is, φ_γ are smooth functions of U valued in $[0,1]$ that add up to 1 at every point, in a neighborhood of any point only finitely many φ_γ are nonzero, and each φ_γ is supported in some U_κ, so denote such $\kappa \in I$ as κ_γ. For $z \in U_\iota$, let

$$f_\iota(z) = \sum_{\gamma \in \Gamma} \varphi_\gamma(z) h_{\iota\kappa_\gamma}(z).$$

Note why this is well-defined and smooth: For $z \in U_\iota$, given a γ and hence κ_γ, either $\varphi_\gamma = 0$ in some neighborhood of z, or $z \in U_{\kappa_\gamma} \cap U_\iota$. So each term can be interpreted as a smooth function on U_ι, and in a neighborhood of z, we are adding up at most finitely many smooth functions, so f_ι is smooth in U_ι. For $z \in U_\iota \cap U_\lambda$,

$$f_\iota(z) - f_\lambda(z) = \sum_{\gamma \in \Gamma} \varphi_\gamma(z)\big(h_{\iota\kappa_\gamma}(z) - h_{\lambda\kappa_\gamma}(z)\big) = \sum_{\gamma \in \Gamma} \varphi_\gamma(z) h_{\iota\lambda}(z) = h_{\iota\lambda}(z). \qquad \square$$

The (holomorphic) Cousin I problem is solvable on any domain of holomorphy and in general on any domain with a trivial first Dolbeault cohomology group.

Theorem 4.6.5. *Suppose $U \subset \mathbb{C}^n$ is a domain with $H^{(0,1)}(U) = 0$. Then the Cousin I problem is solvable.*

Proof. Let $\{U_\iota\}_{\iota\in I}$ be an open covering of U, and let $h_{\iota\kappa}$ be (holomorphic) Cousin I data. Using the lemma, find the smooth solutions f_ι. The functions f_ι need not be holomorphic, but $f_\iota - f_\kappa = h_{\iota\kappa}$ are holomorphic, and so $\bar{\partial}f_\iota - \bar{\partial}f_\kappa = \bar{\partial}(f_\iota - f_\kappa) = 0$. The $(0, 1)$-form η given by

$$\eta = \bar{\partial}f_\iota$$

is therefore well-defined on U. Moreover, η is $\bar{\partial}$-closed, so by assumption on U, there exists a smooth function ψ such that $\bar{\partial}\psi = \eta$. On each U_ι,

$$F_\iota = f_\iota - \psi.$$

On $U_\iota \cap U_\kappa$, we have

$$F_\iota - F_\kappa = f_\iota - f_\kappa = h_{\iota\kappa},$$

and on U_ι, we have

$$\bar{\partial}F_\iota = \bar{\partial}f_\iota - \bar{\partial}\psi = \eta - \eta = 0.$$

Therefore, the F_ι give a solution to the Cousin I problem. $\square$

Corollary 4.6.6. *Cousin I problem is solvable on any possibly unbounded polydisc in $\mathbb{C}^n$.*

Exercise 4.6.2: Use the solution of the $\bar{\partial}$-problem in the disc to show that if the Cousin I problem is solvable in a domain $U \subset \mathbb{C}^n$, then $H^{(0,1)}(U) = 0$. Hint: Solve locally and then follow the same idea as trying to piece together the meromorphic function above.

Exercise 4.6.3: Formulate a version of Cousin I problem for integer-valued continuous functions on domains in $\mathbb{R}^2$. Prove that the problem is not always solvable in $\mathbb{R}^2 \setminus \{(0,0)\}$.

Let us briefly mention the second Cousin problem and its relation to the Cousin I problem and to the theorem of Weierstrass.

Definition 4.6.7 (Cousin II). Suppose $U \subset \mathbb{C}^n$ is open. Let $\{U_\iota\}_{\iota\in I}$ be an open covering of U, and when $U_\iota \cap U_\kappa \neq \emptyset$, let $h_{\iota\kappa} \in \mathcal{O}(U_\iota \cap U_\kappa)$ be nonvanishing functions such that

$$h_{\iota\kappa}h_{\kappa\iota} = 1 \qquad \text{in } U_\iota \cap U_\kappa,$$
$$h_{\iota\kappa}h_{\kappa\lambda}h_{\lambda\iota} = 1 \quad \text{in } U_\iota \cap U_\kappa \cap U_\lambda.$$

The covering and the functions $h_{\iota\kappa}$ are called *Cousin II data*. The *solution of the Cousin II problem* is a set of nonvanishing holomorphic functions $f_\iota \in \mathcal{O}(U_\iota)$ such that

$$h_{\iota\kappa} = \frac{f_\iota}{f_\kappa}.$$

Cousin II is the analogue of the Weierstrass product theorem, that is, finding a function with a prescribed zero set. Suppose that $M \subset U$ is locally given by the vanishing of a single holomorphic function with a nonvanishing derivative (a complex submanifold of codimension 1)*, that is, for every $p \in U$, there is a neighborhood U_ι and a holomorphic g_ι with $dg_\iota \neq 0$ such that $g_\iota^{-1}(0) = M \cap U_\iota$. On $U_\iota \cap U_\kappa$, let

$$h_{\iota\kappa} = \frac{g_\iota}{g_\kappa}.$$

Exercise 4.6.4: Prove that $h_{\iota\kappa}$ is holomorphic and nonvanishing.

Thus we have Cousin II data. If the Cousin II problem is solvable, we have f_ι as above. Define a holomorphic function f on U by defining it on each U_ι via

$$f = \frac{g_\iota}{f_\iota}.$$

Similarly as before, this gives a well-defined function, and clearly it vanishes precisely on M. Moreover, the derivative is nonzero on M. We state this result as a proposition.

Proposition 4.6.8. *Suppose $U \subset \mathbb{C}^n$ is a domain on which the Cousin II problem is solvable and $M \subset U$ is a complex hypersurface (locally the zero set of a holomorphic function with nonvanishing derivative). Then there exists an $f \in \mathcal{O}(U)$ such that $f^{-1}(0) = M$ and $df \neq 0$ on M.*

The second Cousin problem is not always solvable on every domain of holomorphy like the Cousin I problem. An extra condition on the topology of the domain is necessary. Interestingly, on a domain of holomorphy, if the Cousin II problem is solvable just continuously, then it is solvable. We will skip the proof of this fact, but let us describe how the topological obstruction arises. Suppose we have Cousin II data $h_{\iota\kappa}$. We refine the covering to make the sets U_ι and their intersections $U_\iota \cap U_\kappa$ simply connected. Then we take logarithms $g_{\iota\kappa} = \log h_{\iota\kappa}$, where we pick the correct branch to also get $g_{\kappa\iota} = -g_{\iota\kappa}$, so $g_{\iota\kappa} + g_{\kappa\iota} = 0$. For the triple sum we get

$$g_{\iota\kappa} + g_{\kappa\lambda} + g_{\lambda\iota} = 2\pi i m_{\iota\kappa\lambda}$$

for some integer $m_{\iota\kappa\lambda}$. It is not always possible to pick the branches in such a way as to make $m_{\iota\kappa\lambda} = 0$ for all indices. Were it possible, we could try to solve this corresponding Cousin I problem. This is a question of plain old singular cohomology, that is, topology.[†]

Exercise 4.6.5: Suppose $U \subset \mathbb{C}^n$ is a domain with $H^{(0,1)}(U) = 0$ and $M \subset U$ is a complex hypersurface. Suppose there is a continuous function g on U such that locally near every $p \in M$, if r is a defining function for M (holomorphic with nonvanishing derivative), then r/g extends to be continuous and nonvanishing in a neighborhood of p. Prove that there exists an $f \in \mathcal{O}(U)$ such that $f^{-1}(0) = M$ and $df \neq 0$ on M. Hint: Cover with balls U_ι and in each ball define $\log(f_\iota/g)$, then obtain a Cousin I problem.

---

*The nonvanishing condition on the derivative is not necessary. We use it for simplicity.
[†]For the interested reader, the needed extra cohomology condition is $H^2(U, \mathbb{Z}) = 0$.

Exercise **4.6.6:** *Let* $U \subset \mathbb{C}$ *be a domain and assume the Cousin II problem is solvable (it always is in* $\mathbb{C}$*). Prove the classical theorem of Weierstrass using Cousin II. That is, given a countable set of points in* U *and multiplicities, and assuming the set has no limit points in* U*, find a function* $f \in \mathcal{O}(U)$ *that has zeros precisely at the given points of precisely the given multiplicities.*

5 | Integral Kernels

5.1 | The Bochner–Martinelli kernel

A generalization of Cauchy's formula to several variables is called the Bochner–Martinelli integral formula, which reduces to Cauchy's (Cauchy–Pompeiu) formula when $n = 1$. As for Cauchy's formula, we will prove the formula for all smooth functions via Stokes' theorem. First, let us define the *Bochner–Martinelli kernel*:

$$\omega(\zeta, z) \stackrel{\text{def}}{=} \frac{(n-1)!}{(2\pi i)^n} \sum_{k=1}^{n} \frac{\bar{\zeta}_k - \bar{z}_k}{\|\zeta - z\|^{2n}} \, d\bar{\zeta}_1 \wedge d\zeta_1 \wedge \cdots \wedge \widehat{d\bar{\zeta}_k} \wedge d\zeta_k \wedge \cdots \wedge d\bar{\zeta}_n \wedge d\zeta_n.$$

The notation $\widehat{d\bar{\zeta}_k}$ means that this term is simply left out.

Theorem 5.1.1 (Bochner–Martinelli). *Let $U \subset \mathbb{C}^n$ be a bounded open set with smooth boundary and let $f \colon \overline{U} \to \mathbb{C}$ be a smooth function. Then for $z \in U$,*

$$f(z) = \int_{\partial U} f(\zeta)\omega(\zeta, z) - \int_U \bar{\partial} f(\zeta) \wedge \omega(\zeta, z).$$

In particular, if $f \in \mathcal{O}(U)$, then

$$f(z) = \int_{\partial U} f(\zeta)\omega(\zeta, z).$$

Recall that if $\zeta = x + iy$ are the coordinates in $\mathbb{C}^n$, the orientation that we assigned to $\mathbb{C}^n$ in this book* is the one corresponding to the volume form

$$dV = dx_1 \wedge dy_1 \wedge dx_2 \wedge dy_2 \wedge \cdots \wedge dx_n \wedge dy_n.$$

With this orientation,

$$d\zeta_1 \wedge d\bar{\zeta}_1 \wedge d\zeta_2 \wedge d\bar{\zeta}_2 \wedge \cdots \wedge d\zeta_n \wedge d\bar{\zeta}_n = (-2i)^n dV,$$

and hence

$$d\bar{\zeta}_1 \wedge d\zeta_1 \wedge d\bar{\zeta}_2 \wedge d\zeta_2 \wedge \cdots \wedge d\bar{\zeta}_n \wedge d\zeta_n = (2i)^n dV.$$

---

*Again, there is no canonical orientation of $\mathbb{C}^n$, and not all authors follow this (perhaps more prevalent) convention.

Exercise **5.1.1:** *Similarly to the Cauchy–Pompeiu formula, note the singularity in the second term of the Bochner–Martinelli formula. Prove that the integral still makes sense (the function is integrable).*

Exercise **5.1.2:** *Check that for $n = 1$, the Bochner–Martinelli formula reduces to the standard Cauchy–Pompeiu formula.*

Recall the definition of ∂ and $\bar{\partial}$ from Definition 4.4.1, and recall that $d\eta = \partial\eta + \bar{\partial}\eta$.

Proof of Bochner–Martinelli. The structure of the proof is essentially the same as that of the Cauchy–Pompeiu theorem for $n = 1$, although some of the formulas are more involved.

Let $z \in U$ be fixed. Suppose $r > 0$ is small enough so that $\overline{B_r(z)} \subset U$. Orient both ∂U and $\partial B_r(z)$ positively. As $f(\zeta)\omega(\zeta, z)$ contains all the holomorphic $d\zeta_k$,

$$
\begin{aligned}
d\big(f(\zeta)\omega(\zeta, z)\big) &= \bar{\partial}\big(f(\zeta)\omega(\zeta, z)\big) \\
&= \bar{\partial}f(\zeta) \wedge \omega(\zeta, z) \\
&\quad + f(\zeta)\frac{(n-1)!}{(2\pi i)^n} \sum_{k=1}^{n} \frac{\partial}{\partial\bar{\zeta}_k}\left[\frac{\bar{\zeta}_k - \bar{z}_k}{\|\zeta - z\|^{2n}}\right] d\bar{\zeta}_1 \wedge d\zeta_1 \wedge \cdots \wedge d\bar{\zeta}_n \wedge d\zeta_n.
\end{aligned}
$$

We compute

$$
\sum_{k=1}^{n} \frac{\partial}{\partial\bar{\zeta}_k}\left[\frac{\bar{\zeta}_k - \bar{z}_k}{\|\zeta - z\|^{2n}}\right] = \sum_{k=1}^{n}\left(\frac{1}{\|\zeta - z\|^{2n}} - n\frac{|\zeta_k - z_k|^2}{\|\zeta - z\|^{2n+2}}\right) = 0.
$$

Therefore, $d\big(f(\zeta)\omega(\zeta, z)\big) = \bar{\partial}f(\zeta) \wedge \omega(\zeta, z)$. We apply Stokes:

$$
\begin{aligned}
\int_{\partial U} f(\zeta)\omega(\zeta, z) - \int_{\partial B_r(z)} f(\zeta)\omega(\zeta, z) &= \int_{U\setminus\overline{B_r(z)}} d\big(f(\zeta)\omega(\zeta, z)\big) \\
&= \int_{U\setminus\overline{B_r(z)}} \bar{\partial}f(\zeta) \wedge \omega(\zeta, z).
\end{aligned}
$$

Again, due to the integrability, which you showed in an exercise above, the right-hand side converges to the integral over U as $r \to 0$. Just as for the Cauchy–Pompeiu formula, we now need to show that the integral over $\partial B_r(z)$ goes to $f(z)$ as $r \to 0$. So

$$
\int_{\partial B_r(z)} f(\zeta)\omega(\zeta, z) = f(z)\int_{\partial B_r(z)} \omega(\zeta, z) + \int_{\partial B_r(z)} \big(f(\zeta) - f(z)\big)\omega(\zeta, z).
$$

To finish the proof, we will show that $\int_{\partial B_r(z)} \omega(\zeta, z) = 1$, and that the second term goes to zero. We apply Stokes again and note that the volume of $B_r(z)$ is $\frac{\pi^n}{n!} r^{2n}$.

$$\int_{\partial B_r(z)} \omega(\zeta, z)$$

$$= \int_{\partial B_r(z)} \frac{(n-1)!}{(2\pi i)^n} \sum_{k=1}^{n} \frac{\bar{\zeta}_k - \bar{z}_k}{\|\zeta - z\|^{2n}} \, d\bar{\zeta}_1 \wedge d\zeta_1 \wedge \cdots \wedge \widehat{d\bar{\zeta}_k} \wedge d\zeta_k \wedge \cdots \wedge d\bar{\zeta}_n \wedge d\zeta_n$$

$$= \frac{(n-1)!}{(2\pi i)^n} \frac{1}{r^{2n}} \int_{\partial B_r(z)} \sum_{k=1}^{n} (\bar{\zeta}_k - \bar{z}_k) d\bar{\zeta}_1 \wedge d\zeta_1 \wedge \cdots \wedge \widehat{d\bar{\zeta}_k} \wedge d\zeta_k \wedge \cdots \wedge d\bar{\zeta}_n \wedge d\zeta_n$$

$$= \frac{(n-1)!}{(2\pi i)^n} \frac{1}{r^{2n}} \int_{B_r(z)} d \left(\sum_{k=1}^{n} (\bar{\zeta}_k - \bar{z}_k) d\bar{\zeta}_1 \wedge d\zeta_1 \wedge \cdots \wedge \widehat{d\bar{\zeta}_k} \wedge d\zeta_k \wedge \cdots \wedge d\bar{\zeta}_n \wedge d\zeta_n \right)$$

$$= \frac{(n-1)!}{(2\pi i)^n} \frac{1}{r^{2n}} \int_{B_r(z)} n \, d\bar{\zeta}_1 \wedge d\zeta_1 \wedge \cdots \wedge d\bar{\zeta}_n \wedge d\zeta_n$$

$$= \frac{(n-1)!}{(2\pi i)^n} \frac{1}{r^{2n}} \int_{B_r(z)} n(2i)^n \, dV = 1.$$

Next, we tackle the second term. Via the same computation as above we find

$$\int_{\partial B_r(z)} \big(f(\zeta) - f(z) \big) \omega(\zeta, z)$$

$$= \frac{(n-1)!}{(2\pi i)^n} \frac{1}{r^{2n}} \left(\int_{B_r(z)} \big(f(\zeta) - f(z) \big) n \, d\bar{\zeta}_1 \wedge d\zeta_1 \wedge \cdots \wedge d\bar{\zeta}_n \wedge d\zeta_n \right.$$

$$\left. + \int_{B_r(z)} \sum_{k=1}^{n} \frac{\partial f}{\partial \bar{\zeta}_k}(\zeta)(\bar{\zeta}_k - \bar{z}_k) \, d\bar{\zeta}_1 \wedge d\zeta_1 \wedge \cdots \wedge d\bar{\zeta}_n \wedge d\zeta_n \right).$$

As U is bounded, $\big| f(\zeta) - f(z) \big| \le M\|\zeta - z\|$ and $\left| \frac{\partial f}{\partial \bar{\zeta}_k}(\zeta)(\bar{\zeta}_k - \bar{z}_k) \right| \le M\|\zeta - z\|$ for some M. So for all $\zeta \in \partial B_r(z)$, we have $\big| f(\zeta) - f(z) \big| \le Mr$ and $\left| \frac{\partial f}{\partial \bar{\zeta}_k}(\zeta)(\bar{\zeta}_k - \bar{z}_k) \right| \le Mr$. Hence

$$\left| \int_{\partial B_r(z)} \big(f(\zeta) - f(z) \big) \omega(\zeta, z) \right|$$

$$\le \frac{(n-1)!}{(2\pi)^n} \frac{1}{r^{2n}} \left(\int_{B_r(z)} n 2^n Mr \, dV + \int_{B_r(z)} n 2^n Mr \, dV \right) = 2Mr.$$

Therefore, this term goes to zero as $r \to 0$. $\qquad\square$

One drawback of the Bochner–Martinelli formula $\int_{\partial U} f(\zeta)\omega(\zeta, z)$ is that the kernel is not holomorphic in z unless $n = 1$. It does not simply produce holomorphic functions. If we differentiate in $\bar{z}$ underneath the ∂U integral, we do not necessarily obtain zero. On the other hand, we have an explicit formula and this formula does not depend on U. This is not the case for the Bergman and Szegő kernels, which we will see next, although those are holomorphic in the right way.

Exercise 5.1.3: *Prove that if $z \notin \overline{U}$, then rather than $f(z)$ in the formula you obtain*

$$\int_{\partial U} f(\zeta)\omega(\zeta, z) - \int_U \bar{\partial} f(\zeta) \wedge \omega(\zeta, z) = 0.$$

Exercise 5.1.4: *Suppose f is holomorphic on a neighborhood of $\overline{B_r(z)}$.*
  a) Using the Bochner–Martinelli formula, prove that

$$f(z) = \frac{1}{V(B_r(z))} \int_{B_r(z)} f(\zeta)\, dV(\zeta),$$

  where $V(B_r(z))$ is the volume of $B_r(z)$.
  b) Use part a) to prove the maximum principle for holomorphic functions.

Exercise 5.1.5: *Use Bochner–Martinelli for the solution of $\bar{\partial}$ with compact support. That is, suppose $g = g_1 d\bar{z}_1 + \cdots + g_n d\bar{z}_n$ is a smooth compactly supported $(0,1)$-form on $\mathbb{C}^n$, $n \geq 2$, and $\frac{\partial g_k}{\partial \bar{z}_\ell} = \frac{\partial g_\ell}{\partial \bar{z}_k}$ for all k, ℓ. Prove that*

$$\psi(z) = -\int_{\mathbb{C}^n} g(\zeta) \wedge \omega(\zeta, z)$$

is a compactly supported smooth solution to $\bar{\partial}\psi = g$. Hint: Look at the previous proof.

5.2 The Bergman kernel

Let $U \subset \mathbb{C}^n$ be a domain. Define the *Bergman space* of U:

$$A^2(U) \overset{\text{def}}{=} \mathcal{O}(U) \cap L^2(U).$$

That is, $A^2(U)$ denotes the space of holomorphic functions $f \in \mathcal{O}(U)$ such that

$$\|f\|_{A^2(U)}^2 \overset{\text{def}}{=} \|f\|_{L^2(U)}^2 = \int_U |f(z)|^2 dV < \infty.$$

The space $A^2(U)$ is an inner product space with the $L^2(U)$ inner product

$$\langle f, g \rangle \overset{\text{def}}{=} \int_U f(z)\overline{g(z)}\, dV.$$

We will prove that $A^2(U)$ is complete; in other words, it is a Hilbert space. We first prove that the $A^2(U)$ norm bounds the uniform norm on compact sets.

Lemma 5.2.1. *Let $U \subset \mathbb{C}^n$ be a domain and $K \subset\subset U$ compact. Then there exists a constant C_K, such that*

$$\|f\|_K = \sup_{z \in K} |f(z)| \leq C_K \|f\|_{A^2(U)} \qquad \text{for all } f \in A^2(U).$$

Consequently, $A^2(U)$ is complete.

Proof. As K is compact there exists an $r > 0$ such that $\overline{\Delta_r(z)} \subset U$ for all $z \in K$. Take any $z \in K$, and apply Exercise 1.2.10 and Cauchy–Schwarz:

$$|f(z)| = \left| \frac{1}{V(\Delta_r(z))} \int_{\Delta_r(z)} f(\xi)\, dV(\xi) \right|$$

$$\leq \frac{1}{\pi^n r^{2n}} \sqrt{\int_{\Delta_r(z)} 1^2 \, dV(\xi)} \sqrt{\int_{\Delta_r(z)} |f(\xi)|^2 \, dV(\xi)}$$

$$= \frac{1}{\pi^{n/2} r^n} \|f\|_{A^2(\Delta_r(z))} \leq \frac{1}{\pi^{n/2} r^n} \|f\|_{A^2(U)}.$$

Taking the supremum over $z \in K$ proves the estimate. Therefore, if $\{f_\ell\}$ is a sequence of functions in $A^2(U)$ converging in $L^2(U)$ to some $f \in L^2(U)$, then it converges uniformly on compact sets, and so $f \in \mathcal{O}(U)$. Consequently, $A^2(U)$ is a closed subspace of $L^2(U)$, and hence complete. $\qquad \square$

For a bounded domain, $A^2(U)$ is always infinite-dimensional, see exercise below. There exist unbounded domains for which either $A^2(U)$ is trivial (just the zero function, e.g. $U = \mathbb{C}^n$) or even finite-dimensional. When $n = 1$, $A^2(U)$ is either trivial, or infinite-dimensional.[*]

Exercise 5.2.1: Show that if a domain $U \subset \mathbb{C}^n$ is bounded, then $A^2(U)$ is infinite-dimensional.

Exercise 5.2.2:
   a) Show that $A^2(\mathbb{C}^n)$ is trivial (it is just the zero function).
   b) Show that $A^2(\mathbb{D} \times \mathbb{C})$ is trivial.
   c) Find an example of an unbounded domain U for which $A^2(U)$ is infinite-dimensional. Hint: Think in one dimension for simplicity.

Exercise 5.2.3:
   a) Show that $A^2(\mathbb{D})$ can be identified with $A^2(\mathbb{D} \setminus \{0\})$, that is, every function in the latter can be extended to a function in the former.
   b) Let $U \subset \mathbb{C}^n$ be a domain, $f \in \mathcal{O}(U)$, f not identically zero, and $X = f^{-1}(0)$. Show that every function in $A^2(U \setminus X)$ is a restriction of a function in $A^2(U)$, that is, $A^2(U) \cong A^2(U \setminus X)$.

---

[*]For examples of nontrivial finite dimensional $A^2(U)$ for $n \geq 2$ as well as the result in $n = 1$, see: Wiegerinck, J.J.O.O. *Domains with finite dimensional Bergman space.* Math. Z. **187** (1984), 559–562.

The lemma says that point evaluation is a bounded linear functional. That is, for a fixed $z \in U$, taking $K = \{z\}$, the linear operator

$$f \mapsto f(z)$$

is a bounded linear functional. By the Riesz–Fisher theorem, there exists a $k_z \in A^2(U)$, such that

$$f(z) = \langle f, k_z \rangle.$$

Define the *Bergman kernel* for U as

$$K_U(z, \bar{\zeta}) \overset{\text{def}}{=} \overline{k_z(\zeta)}.$$

The function K_U is defined as $(z, \bar{\zeta})$ vary over $U \times U^*$, where we write

$$U^* = \{\zeta \in \mathbb{C}^n : \bar{\zeta} \in U\}.$$

Then for all $f \in A^2(U)$, we have

$$f(z) = \int_U f(\zeta) K_U(z, \bar{\zeta}) \, dV(\zeta). \tag{5.1}$$

This last equation is sometimes called the *reproducing property* of the kernel.

Note that the Bergman kernel depends on U, which is why we write it as $K_U(z, \bar{\zeta})$.

Proposition 5.2.2. *The Bergman kernel $K_U(z, \bar{\zeta})$ is holomorphic in z, antiholomorphic in ζ, and*

$$\overline{K_U(z, \bar{\zeta})} = K_U(\zeta, \bar{z}).$$

Proof. As each k_z is in $A^2(U)$, it is holomorphic in ζ. Hence, K_U is antiholomorphic in ζ. If we prove $\overline{K_U(z, \bar{\zeta})} = K_U(\zeta, \bar{z})$, then we prove K_U is holomorphic in z.

As $\overline{K_U(z, \bar{\zeta})} = k_z(\zeta)$ is in $A^2(U)$, we have

$$\overline{K_U(z, \bar{\zeta})} = \int_U \overline{K_U(z, \bar{w})} K_U(\zeta, \bar{w}) dV(w)$$

$$= \overline{\left(\int_U \overline{K_U(\zeta, \bar{w})} K_U(z, \bar{w}) dV(w) \right)} = \overline{\overline{K_U(\zeta, \bar{z})}} = K_U(\zeta, \bar{z}). \qquad \square$$

Therefore, thinking of $\bar{\zeta}$ as the variable, K_U is a holomorphic function of $2n$ variables.

Example 5.2.3: Let us compute the Bergman kernel (and the Szegö kernel of the next section while we're at it) explicitly for the unit disc $\mathbb{D} \subset \mathbb{C}$. Let $f \in \mathcal{O}(\mathbb{D}) \cap C(\overline{\mathbb{D}})$, that is, f is holomorphic in $\mathbb{D}$ and continuous up to the boundary. Let $z \in \mathbb{D}$. Then

$$f(z) = \frac{1}{2\pi i} \int_{\partial \mathbb{D}} \frac{f(\zeta)}{\zeta - z} \, d\zeta.$$

On the unit circle, $\zeta\bar{\zeta} = 1$. Let ds be the arc-length measure on the circle, parametrized as $\zeta = e^{is}$. Then $d\zeta = ie^{is}\,ds$ or $\bar{\zeta}\,d\zeta = i\,ds$, and

$$f(z) = \frac{1}{2\pi i}\int_{\partial\mathbb{D}}\frac{f(\zeta)}{\zeta - z}\,d\zeta = \frac{1}{2\pi i}\int_{\partial\mathbb{D}}\frac{f(\zeta)}{1 - z\bar{\zeta}}\bar{\zeta}\,d\zeta = \frac{1}{2\pi}\int_{\partial\mathbb{D}}\frac{f(\zeta)}{1 - z\bar{\zeta}}\,ds.$$

The integral is now a regular line integral of a function whose singularity, which used to be inside the unit disc, disappeared (we "reflected it" to the outside). The kernel $\frac{1}{2\pi}\frac{1}{1-z\bar{\zeta}}$ is called the *Szegö kernel*, which we will briefly mention next. We apply Stokes to the second integral above:

$$\frac{1}{2\pi i}\int_{\partial\mathbb{D}}\frac{f(\zeta)}{1 - z\bar{\zeta}}\bar{\zeta}\,d\zeta = \frac{1}{2\pi i}\int_{\mathbb{D}}f(\zeta)\frac{\partial}{\partial\bar{\zeta}}\left[\frac{\bar{\zeta}}{1 - z\bar{\zeta}}\right]d\bar{\zeta}\wedge d\zeta$$

$$= \frac{1}{\pi}\int_{\mathbb{D}}\frac{f(\zeta)}{(1 - z\bar{\zeta})^2}\,dA(\zeta).$$

We therefore have a candidate for the Bergman kernel in the unit disc:

$$K_{\mathbb{D}}(z,\bar{\zeta}) = \frac{1}{\pi}\frac{1}{(1 - z\bar{\zeta})^2}.$$

That this function really is the Bergman kernel follows from Exercise 5.2.5. That is, $K_{\mathbb{D}}$ is the unique conjugate symmetric reproducing function that is in $A^2(\mathbb{D})$ for a fixed ζ. We have only shown the formula for functions continuous up to the boundary, but those are dense in $A^2(\mathbb{D})$.

Example 5.2.4: In an exercise you found that $A^2(\mathbb{C}^n) = \{0\}$. Therefore, $K_{\mathbb{C}^n}(z,\bar{\zeta}) \equiv 0$.

The Bergman kernel for a more general domain is difficult (usually impossible) to compute explicitly. We do have the following formula however.

Proposition 5.2.5. *Suppose $U \subset \mathbb{C}^n$ is a domain, and $\{\varphi_\ell(z)\}_{\ell\in I}$ is a complete orthonormal system for $A^2(U)$. Then*

$$K_U(z,\bar{\zeta}) = \sum_{\ell\in I}\varphi_\ell(z)\overline{\varphi_\ell(\zeta)},$$

with uniform convergence on compact subsets of $U \times U^*$.

Proof. For a fixed $\zeta \in U$, the function $z \mapsto K_U(z,\bar{\zeta})$ is in $A^2(U)$. Expand this function in terms of the basis and use the reproducing property of K_U:

$$K_U(z,\bar{\zeta}) = \sum_{\ell\in I}\left(\int_U K_U(w,\bar{\zeta})\overline{\varphi_\ell(w)}\,dV(w)\right)\varphi_\ell(z) = \sum_{\ell\in I}\overline{\varphi_\ell(\zeta)}\varphi_\ell(z).$$

The convergence is in L^2 as a function of z, for a fixed ζ. Let $K \subset\subset U$ be a compact set. Via Lemma 5.2.1, L^2 convergence in $A^2(U)$ is uniform convergence on compact sets. Therefore, for a fixed ζ the convergence is uniform in $z \in K$. In particular, we get pointwise convergence. So,

$$\sum_{\ell \in I} |\varphi_\ell(z)|^2 = \sum_{\ell \in I} \varphi_\ell(z)\overline{\varphi_\ell(z)} = K_U(z, \bar{z}) \leq C_K < \infty,$$

where C_K is the supremum of $K_U(z, \bar{\zeta})$ on $K \times K^*$. Hence for $(z, \bar{\zeta}) \in K \times K^*$,

$$\sum_{\ell \in I} |\varphi_\ell(z)\overline{\varphi_\ell(\zeta)}| \leq \sqrt{\sum_{\ell \in I} |\varphi_\ell(z)|^2} \sqrt{\sum_{\ell \in I} |\varphi_\ell(\zeta)|^2} \leq C_K < \infty.$$

So the convergence is uniform on $K \times K^*$.   □

Exercise 5.2.4:
  a) Show that if $U \subset \mathbb{C}^n$ is bounded, then $K_U(z, \bar{z}) > 0$ for all $z \in U$.
  b) Why can this fail if U is unbounded? Find a (trivial) counterexample.

Exercise 5.2.5: Show that given a domain $U \subset \mathbb{C}^n$, the Bergman kernel is the unique function $K_U(z, \bar{\zeta})$ such that
  1) for a fixed ζ, $K_U(z, \bar{\zeta})$ is in $A^2(U)$,
  2) $\overline{K_U(z, \bar{\zeta})} = K_U(\zeta, \bar{z})$,
  3) the reproducing property (5.1) holds.

Exercise 5.2.6: Let $U \subset \mathbb{C}^n$ be either the unit ball or the unit polydisc. Show that $A^2(U) \cap C(\overline{U})$ is dense in $A^2(U)$. In particular, this exercise says we only need to check the reproducing property on functions continuous up to the boundary to show we have the Bergman kernel.

Exercise 5.2.7: Let $U, V \subset \mathbb{C}^n$ be two domains and $f : U \to V$ a biholomorphism. Prove

$$K_U(z, \bar{\zeta}) = \det Df(z) \overline{\det Df(\zeta)} K_V(f(z), \overline{f(\zeta)}).$$

Exercise 5.2.8: Show that the Bergman kernel for the polydisc is

$$K_{\mathbb{D}^n}(z, \bar{\zeta}) = \frac{1}{\pi^n} \prod_{\ell=1}^{n} \frac{1}{(1 - z_\ell \bar{\zeta}_\ell)^2}.$$

Exercise 5.2.9 (Hard): *Show that for some constants c_α, the set of all monomials $\frac{z^\alpha}{c_\alpha}$ gives a complete orthonormal system of $A^2(\mathbb{B}_n)$. Hint: To show orthogonality, compute the integral using polar coordinates in each variable separately, that is, let $z_\ell = r_\ell e^{i\theta_\ell}$ where $\theta \in [0, 2\pi]^n$ and $\sum_\ell r_\ell^2 < 1$. Then show completeness by showing that if $f \in A^2(\mathbb{B}_n)$ is orthogonal to all z^α, then $f = 0$. Note that explicitly finding $c_\alpha = \sqrt{\frac{\pi^n \alpha!}{(n+|\alpha|)!}}$ requires the classical β function of special function theory.*

Exercise 5.2.10: *Using the previous exercise, show that the Bergman kernel for the unit ball is*

$$K_{\mathbb{B}_n}(z, \bar{\zeta}) = \frac{n!}{\pi^n} \frac{1}{(1 - \langle z, \zeta \rangle)^{n+1}},$$

where $\langle z, \zeta \rangle$ is the standard inner product on $\mathbb{C}^n$.

5.3   The Szegö kernel

We use the same technique to create a reproducing kernel on the boundary by starting with $L^2(\partial U, d\sigma)$ instead of $L^2(U)$. We obtain a kernel where we integrate over the boundary rather than the domain itself. Let us give a quick overview, but let us not get into the details.

Let $U \subset \mathbb{C}^n$ be a bounded domain with smooth boundary. Let $C(\overline{U}) \cap \mathcal{O}(U)$ be the set of holomorphic functions in U continuous up to the boundary. The restriction of $f \in C(\overline{U}) \cap \mathcal{O}(U)$ to ∂U is a continuous function, and hence $f|_{\partial U}$ is in $L^2(\partial U, d\sigma)$, where $d\sigma$ is the surface measure on ∂U. Taking the closure of these restrictions in $L^2(\partial U)$ obtains the Hilbert space $H^2(\partial U)$, which is called the *Hardy space*. The inner product on $H^2(\partial U)$ is the $L^2(\partial U, d\sigma)$ inner product:

$$\langle f, g \rangle \overset{\text{def}}{=} \int_{\partial U} f(z)\overline{g(z)} \, d\sigma(z).$$

Exercise 5.3.1: *Show that monomials z^α are a complete orthogonal system in $H^2(\partial \mathbb{B}_n)$.*

Exercise 5.3.2: *Let $U \subset \mathbb{C}^n$ be a bounded domain with smooth boundary. Prove that $H^2(\partial U)$ is infinite-dimensional.*

Given an $f \in H^2(\partial U)$, write the Poisson integral

$$Pf(z) = \int_{\partial U} f(\zeta) P(z, \zeta) \, d\sigma(\zeta),$$

where $P(z, \zeta)$ is the Poisson kernel. The Poisson integral reproduces harmonic functions. As holomorphic functions are harmonic, we find that if $f \in C(\overline{U}) \cap \mathcal{O}(U)$, then $Pf = f$.

Although $f \in H^2(\partial U)$ is only defined on the boundary, through the Poisson integral, we have the values $Pf(z)$ for $z \in U$. For each $z \in U$,

$$f \mapsto Pf(z)$$

defines a continuous linear functional. Again we find a $s_z \in H^2(\partial U)$ such that

$$Pf(z) = \langle f, s_z \rangle.$$

For $z \in U$ and $\zeta \in \partial U$, define

$$S_U(z, \bar{\zeta}) \stackrel{\text{def}}{=} \overline{s_z(\zeta)},$$

although for a fixed z this is a function only defined almost everywhere as it is an element of $L^2(\partial U, d\sigma)$. The function S_U is the *Szegö kernel*. If $f \in H^2(\partial U)$, then

$$Pf(z) = \int_{\partial U} f(\zeta) \, S_U(z, \bar{\zeta}) \, d\sigma(\zeta).$$

Functions in $H^2(\partial U)$ extend to $\overline{U}$, so $f \in H^2(\partial U)$ may be considered a function on $\overline{U}$, where values in U are given by Pf. Of course, since we are extending an L^2 function, the values on the boundary are only defined almost everywhere. Similarly, we extend $S_U(z, \bar{\zeta})$ to a function on $U \times \overline{U}^*$. We state without proof that if $\{\varphi_\ell\}_{\ell \in I}$ is a complete orthonormal system for $H^2(\partial U)$, then

$$S_U(z, \bar{\zeta}) = \sum_{\ell \in I} \varphi_\ell(z) \overline{\varphi_\ell(\zeta)} \tag{5.2}$$

for $(z, \bar{\zeta}) \in U \times U^*$, converging uniformly on compact subsets. As before, this formula shows that S_U is conjugate symmetric, and so it extends to $(U \times \overline{U}^*) \cup (\overline{U} \times U^*)$.

Example 5.3.1: In Example 5.2.3, we computed that if $f \in C(\overline{\mathbb{D}}) \cap \mathcal{O}(\mathbb{D})$, then

$$f(z) = \frac{1}{2\pi} \int_{\partial \mathbb{D}} \frac{f(\zeta)}{1 - z\bar{\zeta}} \, ds.$$

In other words, $S_{\mathbb{D}}(z, \bar{\zeta}) = \frac{1}{2\pi} \frac{1}{1 - z\bar{\zeta}}$.

Exercise 5.3.3: Using the formula (5.2), compute $S_{\mathbb{B}_n}$.

6   Complex Analytic Varieties

6.1   The ring of germs

Definition 6.1.1. Let p be a point in a topological space X. Let Y be a set and $U, V \subset X$ be open neighborhoods of p. Say that two functions $f : U \to Y$ and $g : V \to Y$ are equivalent if there exists a neighborhood W of p such that $f|_W = g|_W$. An equivalence class of functions under this relation defined in neighborhoods of p is called a *germ of a function*. The germ is denoted by (f, p), but we may say f when the context is clear. We usually restrict the functions to a certain category: smooth, holomorphic, etc.

The set of germs of complex-valued functions forms a commutative ring, see exercise below to check the details. For example, to multiply (f, p) and (g, p), take two representatives f and g defined on a common neighborhood, multiply them, and then consider the germ (fg, p). Similarly, $(f, p) + (g, p)$ is defined as $(f + g, p)$. It is easy to check that these operations are well-defined.

Exercise 6.1.1: Let X be a topological space and $p \in X$. Let $\mathcal{F}$ be a class of complex-valued functions defined on open subsets of X such that whenever $f : U \to \mathbb{C}$ is in $\mathcal{F}$ and $W \subset U$ is open, then $f|_W \in \mathcal{F}$, and such that whenever f and g are two functions in $\mathcal{F}$, and W is an open set where both are defined, then $fg|_W$ and $(f + g)|_W$ are also in $\mathcal{F}$. Assume that all constant functions are in $\mathcal{F}$. Show that the ring operations defined above on a set of germs at p of functions from $\mathcal{F}$ are well-defined, and that the set of germs at p of functions from $\mathcal{F}$ is a commutative ring.

Exercise 6.1.2: Let $X = Y = \mathbb{R}$ and $p = 0$. Consider the ring of germs of continuous functions (or smooth functions). Show that for every continuous $f : \mathbb{R} \to \mathbb{R}$ and every neighborhood W of 0, there exists a $g : \mathbb{R} \to \mathbb{R}$ such that $(f, 0) = (g, 0)$, but $g|_W \neq f|_W$.

Germs are particularly useful for holomorphic functions because of the identity theorem. In particular, the behavior of Exercise 6.1.2 does not happen for holomorphic functions. Furthermore, for holomorphic functions, the ring of germs is the same as the ring of convergent power series, see exercise below. No similar result is true for smooth functions.

Definition 6.1.2. Let $p \in \mathbb{C}^n$. Write $_n\mathcal{O}_p = \mathcal{O}_p$ as the ring of germs at p of holomorphic functions.

The ring of germs $\mathcal{O}_p$ has many nice properties, and it is generally a "nicer" ring than the ring $\mathcal{O}(U)$ for some open U, and so it is easier to work with if we are interested in local properties and not in the geometry of U.

> **Exercise 6.1.3:**
> a) Show that $\mathcal{O}_p$ is an integral domain (has no zero divisors).
> b) Prove the ring of germs at $0 \in \mathbb{R}$ of smooth real-valued functions is not an integral domain.

> **Exercise 6.1.4:** Show that the units (elements with multiplicative inverse) of $\mathcal{O}_p$ are the germs of functions which do not vanish at p.

> **Exercise 6.1.5:**
> a) (easy) Show that given a germ $(f, p) \in \mathcal{O}_p$, there exists a fixed open neighborhood U of p and a representative $f \colon U \to \mathbb{C}$ such that any other representative g can be analytically continued from p to a holomorphic function on U.
> b) (easy) Given two representatives $f \colon U \to \mathbb{C}$ and $g \colon V \to \mathbb{C}$ of a germ $(f, p) \in \mathcal{O}_p$, let W be the connected component of $U \cap V$ that contains p. Prove that $f|_W = g|_W$.
> c) Find a germ $(f, p) \in \mathcal{O}_p$, such that for every representative $f \colon U \to \mathbb{C}$, we can find another representative $g \colon V \to \mathbb{C}$ of that same germ such that $g|_{U \cap V} \neq f|_{U \cap V}$. Hint: $n = 1$ is sufficient.

> **Exercise 6.1.6:** Show that $\mathcal{O}_p$ is isomorphic to the ring of convergent power series.

Definition 6.1.3. Let p be a point in a topological space X. Say that sets $A, B \subset X$ are equivalent if there exists a neighborhood W of p such that $A \cap W = B \cap W$. An equivalence class of sets under this relation is called a *germ of a set* at p. It is denoted by (A, p), but we may write A when the context is clear.

The concept of $(A, p) \subset (B, p)$ is defined in an obvious manner, that is, there exist representatives A and B, and a neighborhood W of p such that $A \cap W \subset B \cap W$. Similarly, if $(A, p), (B, p)$ are germs and A, B are some representatives of these germs, then the intersection $(A, p) \cap (B, p)$ is the germ $(A \cap B, p)$, the union $(A, p) \cup (B, p)$ is the germ $(A \cup B, p)$, and the complement $(A, p)^c$ is the germ (A^c, p).

> **Exercise 6.1.7:** Check that the definition of subset, union, intersection, and complement of germs of sets is well-defined.

Let R be some ring of germs of complex-valued functions at $p \in X$ for some topological space X. If f is a complex-valued function, let Z_f be the zero set of f, that is $f^{-1}(0)$. When $(f, p) \in R$ is a germ of a function it makes sense to talk about the germ (Z_f, p). We take the zero set of some representative and look at its germ at p.

Exercise 6.1.8: *Suppose f and g are two representatives of a germ (f, p). Show that the germs (Z_f, p) and (Z_g, p) are the same.*

Exercise 6.1.9: *Show that if (f, p) and (g, p) are in R and f and g are some representatives, then $(Z_f, p) \cup (Z_g, p) = (Z_{fg}, p)$.*

6.2 \ Weierstrass preparation and division theorems

Suppose f is (a germ of) a holomorphic function at a point $p \in \mathbb{C}^n$. Write

$$f(z) = \sum_{k=0}^{\infty} f_k(z - p),$$

where f_k is a homogeneous polynomial of degree k, that is, $f_k(tz) = t^k f_k(z)$.

Definition 6.2.1. Let $p \in \mathbb{C}^n$ and f be a function holomorphic in a neighborhood of p. If f is not identically zero, define

$$\operatorname{ord}_p f \overset{\text{def}}{=} \min\{k \in \mathbb{N}_0 : f_k \not\equiv 0\}.$$

If $f \equiv 0$, define $\operatorname{ord}_p f = \infty$. We call the number $\operatorname{ord}_p f$ the *order of vanishing* of f at p.

In other words, the order of vanishing of f at p is k whenever all partial derivatives of order less than k vanish at p, and there exists at least one derivative of order k that does not vanish at p.

In one complex variable, a holomorphic function f with $\operatorname{ord}_0 f = k$ can be written (locally) as $f(z) = z^k u(z)$ for a nonvanishing holomorphic u. Such a u is a *unit* in the ring $\mathcal{O}_0$, that is, an element with a multiplicative inverse. In several variables, there is a similar theorem, or in fact a pair of theorems, the so-called Weierstrass preparation and division theorems. We first need to replace z^k with something.

Definition 6.2.2. Let $U \subset \mathbb{C}^{n-1}$ be open, $0 \in U$, and let $z' \in \mathbb{C}^{n-1}$ denote the coordinates. Suppose $P \in \mathcal{O}(U)[z_n]$ is a monic polynomial of degree $k \geq 0$,

$$P(z', z_n) = z_n^k + \sum_{\ell=0}^{k-1} c_\ell(z') z_n^\ell,$$

where c_ℓ are holomorphic functions defined on U such that $c_\ell(0) = 0$ for all ℓ. Then P is called a *Weierstrass polynomial* of degree k. If the c_ℓ are germs in $\mathcal{O}_0 = {}_{n-1}\mathcal{O}_0$, then $P \in \mathcal{O}_0[z_n]$ and P is called a *germ of a Weierstrass polynomial*.

The definition (and the theorem that follows) still holds for $n = 1$. If you read the definition carefully, you will find that if $n = 1$, then the only Weierstrass polynomial of degree k is z^k. Note that for any n, if $k = 0$, then $P = 1$.

The purpose of this section is to show that every holomorphic function in $\mathcal{O}_0$ is, up to a unit and a possible small rotation, a Weierstrass polynomial, which carries the zeros of f. Consequently the algebraic and geometric properties of $_n\mathcal{O}_0$ can be understood via algebraic and geometric properties of $_{n-1}\mathcal{O}_0[z_n]$.

Theorem 6.2.3 (Weierstrass preparation theorem). *Suppose $f \in \mathcal{O}(U)$ for an open $U \subset \mathbb{C}^{n-1} \times \mathbb{C}$, where $0 \in U$, and $f(0) = 0$. Suppose $z_n \mapsto f(0, z_n)$ is not identically zero near the origin and its order of vanishing at the origin is $k \geq 1$.*

Then there exists an open polydisc $V = V' \times D \subset \mathbb{C}^{n-1} \times \mathbb{C}$ with $0 \in V \subset U$, a unique $u \in \mathcal{O}(V)$, $u(z) \neq 0$ for all $z \in V$, and a unique Weierstrass polynomial P of degree k with coefficients holomorphic in V' such that

$$f(z', z_n) = u(z', z_n) P(z', z_n),$$

and such that all k zeros (counting multiplicity) of $z_n \mapsto P(z', z_n)$ lie in D for all $z' \in V'$

Proof. There exists a small disc $D \subset \mathbb{C}$ centered at zero such that $\{0\} \times \overline{D} \subset U$ and such that $f(0, z_n) \neq 0$ for $z_n \in \overline{D} \setminus \{0\}$. By continuity of f, there is a small polydisc $V = V' \times D$ such that $\overline{V} \subset U$ and f is not zero on $V' \times \partial D$. See Figure 6.1 for the setup. We will consider the zeros of $z_n \mapsto f(z', z_n)$ for $z' \in V'$. See Figure 6.2.

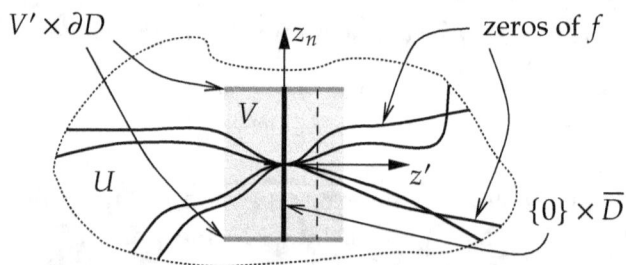

Figure 6.1: Setting up the neighborhood V. The two discs in Figure 6.2 are the vertical thick black line and the thin dashed line.

By the one-variable argument principle (Theorem B.25) the number of zeros (with multiplicity) of $z_n \mapsto f(z', z_n)$ in D is

$$\frac{1}{2\pi i} \int_{\partial D} \frac{\frac{\partial f}{\partial z_n}(z', \zeta)}{f(z', \zeta)} \, d\zeta.$$

As $f(z', \zeta)$ does not vanish when $z' \in V'$ and $\zeta \in \partial D$, the expression above is a continuous integer-valued function of $z' \in V'$. The expression is equal to k when

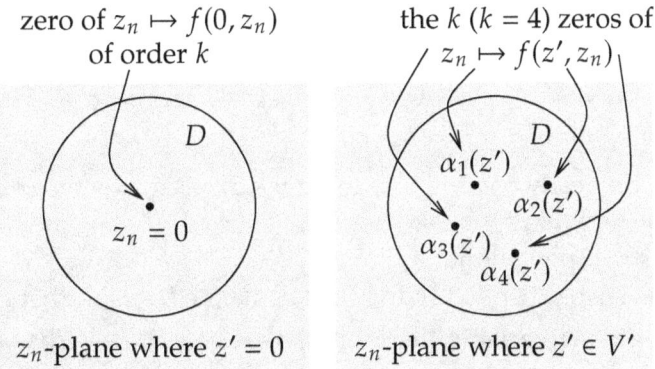

Figure 6.2: The zeros of $z_n \mapsto f(z', z_n)$.

$z' = 0$, and so it is equal to k for all $z' \in V'$. Write the zeros of $z_n \mapsto f(z', z_n)$ as $\alpha_1(z'), \ldots, \alpha_k(z')$, including multiplicity. The zeros are not ordered in any particular way—pick *some* ordering for every z'. Write

$$P(z', z_n) = \prod_{\ell=1}^{k} (z_n - \alpha_\ell(z')) = z_n^k + c_{k-1}(z') z_n^{k-1} + \cdots + c_0(z').$$

For a fixed z', P (and thus the coefficients $c_0, \ldots, c_{k-1}$) is uniquely defined as its definition is independent of the ordering of the zeros. That $c_j(0) = 0$ for all j follows as $\alpha_\ell(0) = 0$ for all ℓ. As the above is the unique way to define a monic polynomial with these zeros (Exercise 6.2.6), the uniqueness part of the theorem follows. We need to show that the coefficients are holomorphic functions on V', and that u is a holomorphic function on V.

No matter how you ordered the zeros for each z', the functions α_ℓ may not be continuous in general (see Example 6.2.4). However, we will prove that the functions c_ℓ are holomorphic. The functions c_ℓ are (up to sign) the *elementary symmetric functions* of $\alpha_1, \ldots, \alpha_k$ (see below). A standard theorem in algebra (Newton's identities, see Exercise 6.2.1) says that the elementary symmetric functions are polynomials in the so-called *power sum* functions in the α_ℓs:

$$s_m(z') = \sum_{\ell=1}^{k} \alpha_\ell(z')^m, \qquad m = 1, \ldots, k.$$

So if the power sums s_m are holomorphic on V', then c_ℓ are holomorphic on V'.

A refinement of the argument principle (see Theorem B.25) says: If h and g are holomorphic functions on a disc D, continuous on $\overline{D}$, such that g has no zeros on ∂D, and $\alpha_1, \ldots, \alpha_k$ are the zeros of g in D, then

$$\frac{1}{2\pi i} \int_{\partial D} h(\zeta) \frac{g'(\zeta)}{g(\zeta)} \, d\zeta = \sum_{\ell=1}^{k} h(\alpha_\ell).$$

With $h(\zeta) = \zeta^m$ and $g(\zeta) = f(z', \zeta)$, the theorem says

$$s_m(z') = \sum_{\ell=1}^{k} \alpha_\ell(z')^m = \frac{1}{2\pi i} \int_{\partial D} \zeta^m \frac{\frac{\partial f}{\partial z_n}(z', \zeta)}{f(z', \zeta)} \, d\zeta.$$

The function s_m is clearly continuous, and if we differentiate under the integral with $\frac{\partial}{\partial \bar{z}_1}, \ldots, \frac{\partial}{\partial \bar{z}_{n-1}}$ we find that s_m is holomorphic. Thus $c_0, \ldots, c_{k-1}$ are holomorphic, and so P is a Weierstrass polynomial.

Finally, we wish to show that P divides f as claimed, that is, that u is holomorphic. For each fixed z', one variable theory says that $z_n \mapsto \frac{f(z', z_n)}{P(z', z_n)}$ has only removable singularities, and in fact, it has no zeros as we defined P to exactly cancel them all out. The Cauchy formula on f/P then says that the function

$$u(z', z_n) = \frac{1}{2\pi i} \int_{\partial D} \frac{f(z', \zeta)}{P(z', \zeta)(\zeta - z_n)} \, d\zeta$$

is equal to $\frac{f(z', z_n)}{P(z', z_n)}$. The function u is clearly continuous and holomorphic in z_n for each fixed z'. Differentiating under the integral shows it is also holomorphic in z'. $\square$

Example 6.2.4: Consider the zero set of $f(z_1, z_2) = z_2^2 - z_1$, a Weierstrass polynomial in z_2 of degree $k = 2$. So $z' = z_1$. For all z_1 except the origin there are two zeros, $\pm\sqrt{z_1}$. Call one of them $\alpha_1(z_1)$ and the other $\alpha_2(z_1)$. Recall there is no continuous choice of a square root that works for all z_1, so no matter how you choose, α_1 and α_2 will not be continuous. At the origin there is one zero of multiplicity two, so $\alpha_1(0) = \alpha_2(0) = 0$. On the other hand, the symmetric functions $c_1(z_1) = -\alpha_1(z_1) - \alpha_2(z_1) = 0$ and $c_0(z_1) = \alpha_1(z_1)\alpha_2(z_1) = -z_1$ are holomorphic. See Figure 6.3.

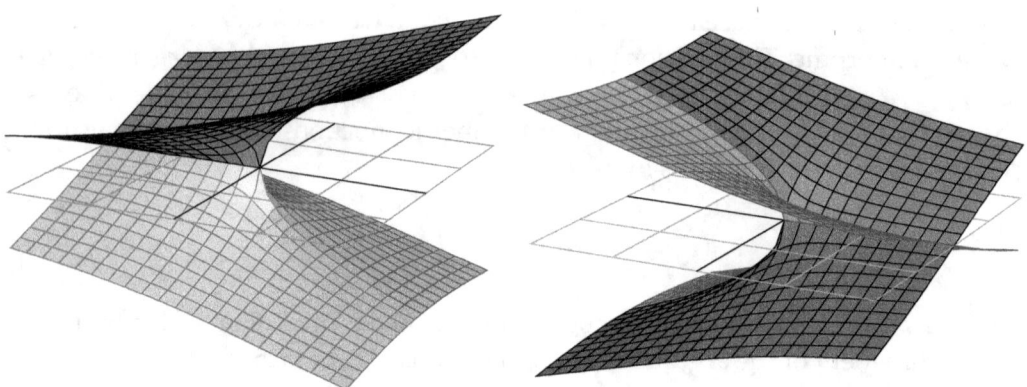

Figure 6.3: Graphs of the real and imaginary parts of both branches $\pm\sqrt{z_1}$. A possible choice of branch $\alpha_1(z_1)$ is drawn darker; note the discontinuity of its imaginary part. We remark that the surface $z_2^2 - z_1 = 0$ does not cross itself in $\mathbb{C}^2$.

The k depends on the coordinates chosen. Consider $g(z_1, z_2) = -f(z_2, z_1) = z_2 - z_1^2$, which is a Weierstrass polynomial in z_2 of degree $k = 1$. In these coordinates, there is only one zero for each z', $\alpha_1(z_1) = z_1^2$, and so $c_0(z_1) = -z_1^2$.

A function $f(z_1, \ldots, z_n)$ is *symmetric* if $f = f \circ p$ for all permutations of the variables p. The *elementary symmetric functions* of $\alpha_1, \ldots, \alpha_k$ are the coefficients $\sigma_1, \ldots, \sigma_k$ of the polynomial

$$\prod_{\ell=1}^{k} (t + \alpha_\ell) = t^k + \sigma_1 t^{k-1} + \cdots + \sigma_{k-2} t^2 + \sigma_{k-1} t + \sigma_k.$$

So,

$$\sigma_1 = \alpha_1 + \alpha_2 + \cdots + \alpha_k,$$
$$\sigma_2 = \alpha_1\alpha_2 + \alpha_1\alpha_3 + \cdots + \alpha_{k-1}\alpha_k,$$
$$\vdots$$
$$\sigma_{k-1} = \alpha_2\alpha_3 \cdots \alpha_k + \alpha_1\alpha_3\alpha_4 \cdots \alpha_k + \cdots + \alpha_1\alpha_2 \cdots \alpha_{k-1},$$
$$\sigma_k = \alpha_1\alpha_2 \cdots \alpha_k.$$

For example, when $k = 2$, then $\sigma_2 = \alpha_1\alpha_2$ and $\sigma_1 = \alpha_1 + \alpha_2$. The function σ_1 happens to already be a power sum. We can write σ_2 as a polynomial in the power sums:

$$\sigma_2 = \frac{1}{2} \left((\alpha_1 + \alpha_2)^2 - (\alpha_1^2 + \alpha_2^2) \right).$$

In general, as we said we can write any σ_ℓ in terms of the power sums of the α_js. The formulas for this are called the *Newton's identities* or *Girard–Newton formulas*, although we will avoid writing these down explicitly, and we leave finding them (or just proving that they exist) as an exercise.

Exercise 6.2.1: Show that elementary symmetric functions are polynomials in the power sums. Equivalently, show that the elementary symmetric functions σ_ℓ can be found in terms of the power sums of the α_js.

Exercise 6.2.2: Prove the fundamental theorem of symmetric polynomials: *Every symmetric polynomial can be written as a polynomial in the elementary symmetric functions. Use the following procedure. Using double induction, suppose the theorem is true if the number of variables is less than k, and the theorem is true in k variables for degree less than d. Consider a symmetric $P(z_1, \ldots, z_k)$ of degree d. Write $P(z_1, \ldots, z_{k-1}, 0)$ by induction hypothesis as a polynomial in the elementary symmetric functions of one less variable. Use the same coefficients, but plug in the elementary symmetric functions of k variables except the symmetric polynomial in k variables of degree k, that is, except $z_1 z_2 \cdots z_k$. You will obtain a symmetric function $L(z_1, \ldots, z_k)$ and you need to show $L(z_1, \ldots, z_{k-1}, 0) = P(z_1, \ldots, z_{k-1}, 0)$. Now use symmetry to prove that*

$$P(z_1, \ldots, z_k) = L(z_1, \ldots, z_k) + z_1 z_2 \cdots z_k Q(z_1, \ldots, z_k).$$

Then note that Q has lower degree and finish by induction.

Exercise **6.2.3:** *Extend the previous exercise to power series. Suppose $f(z_1, \ldots, z_k)$ is a convergent symmetric power series at 0, show that f can be written as a convergent power series in the elementary symmetric functions.*

Exercise **6.2.4:** *Suppose $P(z', z_n)$ is a Weierstrass polynomial of degree k, and write the zeros as $\alpha_1(z'), \ldots, \alpha_k(z')$. These are not holomorphic functions, but suppose that f is a symmetric convergent power series at the origin in k variables. Show that $f\big(\alpha_1(z'), \ldots, \alpha_k(z')\big)$ is a holomorphic function of z' near the origin.*

The hypotheses of the preparation theorem are not an obstacle. If a holomorphic function f is such that $z_n \mapsto f(0, z_n)$ vanishes identically, then we can make a small linear change of coordinates L (L can be a matrix arbitrarily close to the identity) such that $f \circ L$ satisfies the hypotheses of the theorem. For example, $f(z_1, z_2, z_3) = z_1 z_3 + z_2 z_3$ does not satisfy the hypotheses of the theorem as $f(0, 0, z_3) \equiv 0$. But for an arbitrarily small $\epsilon \neq 0$, replacing z_2 with $z_2 + \epsilon z_3$ leads to $\tilde{f}(z_1, z_2, z_3) = f(z_1, z_2 + \epsilon z_3, z_3) = z_1 z_3 + z_2 z_3 + \epsilon z_3^2$, and $\tilde{f}(0, 0, z_3) = \epsilon z_3^2$. Thence, $\tilde{f}$ satisfies the hypotheses of the theorem.

Exercise **6.2.5:** *Prove the fact above about the existence of L arbitrarily close to the identity.*

Exercise **6.2.6:** *Prove that a monic polynomial $P(\zeta)$ of one variable is uniquely determined by its zeros up to multiplicity: If P and Q are two monic polynomials with the same zeros up to multiplicity, then $P = Q$. That proves the uniqueness of the Weierstrass polynomial.*

Exercise **6.2.7:** *Suppose $D \subset \mathbb{C}$ is a bounded domain, $0 \in D$, $U' \subset \mathbb{C}^{n-1}$ is a domain, $0 \in U'$, and $P \in \mathcal{O}(U')[z_n]$ is a Weierstrass polynomial such that $P(z', z_n)$ is not zero on $U' \times \partial D$. Show that for every $z' \in U$, all zeros of $z_n \mapsto P(z', z_n)$ are in D.*

Exercise **6.2.8:** *Let $D \subset \mathbb{C}$ be a bounded domain, and $U' \subset \mathbb{C}^{n-1}$ a domain. Suppose f is a continuous function on $U' \times \overline{D}$ holomorphic on $U' \times D$, where f is zero on at least one point of $U' \times D$, and f is never zero on $U' \times \partial D$. Prove that $z_n \mapsto f(z', z_n)$ has at least one zero in D for every $z' \in U'$.*

The order of vanishing of f at the origin is a lower bound on the number k in the theorem. The order of vanishing for a certain variable may be larger than this lower bound. If $f(z_1, z_2) = z_1^2 + z_2^3$, then the k we get is 3, but $\mathrm{ord}_0 f = 2$. We can make a small linear change of coordinates to ensure $k = \mathrm{ord}_0 f$. With the f as above, $f(z_1 + \epsilon z_2, z_2)$ gives $k = 2$ as expected.

When $k = 1$ in the Weierstrass preparation theorem, we obtain the Weierstrass polynomial $z_n + c_0(z')$. That is, the zero set of f is a graph of the holomorphic function $-c_0$. Therefore, the Weierstrass theorem is a generalization of the implicit function theorem to the case when $\frac{\partial f}{\partial z_n}$ is zero. In such a case, we can still "solve" for z_n, but we find a k-valued solution given by the zeros of the obtained Weierstrass polynomial.

There is an obvious statement of the preparation theorem for germs.

> *Exercise 6.2.9:* *State and prove a germ version of the preparation theorem.*

The next theorem is rather trivial in one variable. Let f be any holomorphic function near the origin in $\mathbb{C}$ and take any $k \in \mathbb{N}$. Let r be the Taylor polynomial for f at 0 of degree $k - 1$. Then $f - r$ is divisible by z^k, in other words, $f = qz^k + r$. In several variables, we replace z^k with a Weierstrass polynomial and we still have this division theorem.

Theorem 6.2.5 (Weierstrass division theorem). *Suppose f is holomorphic near the origin, and suppose P is a Weierstrass polynomial of degree $k \geq 1$ in z_n. Then there exists a neighborhood V of the origin and unique $q, r \in \mathcal{O}(V)$, where r is a polynomial in z_n of degree less than k, and on V,*

$$f = qP + r.$$

Note that r need not be a Weierstrass polynomial; it need not be monic nor do the coefficients need to vanish at the origin. It is simply a polynomial in z_n with coefficients that are holomorphic functions of the first $n - 1$ variables.

Proof. Uniqueness is left as an exercise. There exists a connected neighborhood $V = V' \times D$ of the origin, where D is a disc, f and P are continuous on $V' \times \overline{D}$, and P is not zero on $V' \times \partial D$. Let

$$q(z', z_n) = \frac{1}{2\pi i} \int_{\partial D} \frac{f(z', \zeta)}{P(z', \zeta)(\zeta - z_n)} \, d\zeta.$$

As P is not zero on $V' \times \partial D$, the function q is holomorphic in V (differentiate under the integral). If P did divide f, then q would really be f/P. But if P does not divide f, then the Cauchy integral formula does not apply and q is not equal to f/P. Interestingly, the expression does give the quotient in the division with remainder.

Write f using the Cauchy integral formula in z_n and subtract qP to obtain r:

$$r(z', z_n) = f(z', z_n) - q(z', z_n)P(z', z_n)$$

$$= \frac{1}{2\pi i} \int_{\partial D} \frac{f(z', \zeta)P(z', \zeta) - f(z', \zeta)P(z', z_n)}{P(z', \zeta)(\zeta - z_n)} \, d\zeta.$$

We need to show r is a polynomial in z_n of degree less than k. In the expression inside the integral, the numerator is of the form $\sum_{\ell=1}^{k} h_\ell(z', \zeta)(\zeta^\ell - z_n^\ell)$ and is therefore divisible by $(\zeta - z_n)$. The numerator is a polynomial of degree k in z_n. After dividing by $(\zeta - z_n)$, the integrand becomes a polynomial in z_n of degree $k - 1$. Use linearity of the integral to integrate the coefficients of the polynomial. Each coefficient is a holomorphic function in V' and the proof is finished. Some coefficients may have integrated to zero, so we can only say that r is a polynomial of degree $k - 1$ or less. $\square$

For example, let $f(z, w) = e^z + z^4 e^w + zw^2 e^w + zw$ and $P(z, w) = w^2 + z^3$. Then P is a Weierstrass polynomial in w of degree $k = 2$. A bit of computation shows

$$\frac{1}{2\pi i} \int_{\partial \mathbb{D}} \frac{e^z + z^4 e^\zeta + z\zeta^2 e^\zeta + z\zeta}{(\zeta^2 + z^3)(\zeta - w)} d\zeta = ze^w, \quad \text{so} \quad f(z, w) = \underbrace{(ze^w)}_{q} \underbrace{(w^2 + z^3)}_{P} + \underbrace{zw + e^z}_{r}.$$

Notice that r is a polynomial of degree 1 in w, but it is neither monic, nor do the coefficients vanish at 0.

> **Exercise 6.2.10:** *Prove the uniqueness part of the theorem. Hint: Don't forget that we defined V to be connected.*
>
> **Exercise 6.2.11:** *State and prove a germ version of the division theorem.*

The Weierstrass division theorem is a generalization of the division algorithm for polynomials with coefficients in a field, such as the complex numbers: If $f(\zeta)$ is a polynomial, and $P(\zeta)$ is a nonzero polynomial of degree k, then there exist polynomials $q(\zeta)$ and $r(\zeta)$ with degree of r less than k such that $f = qP + r$. If the coefficients are in a commutative ring, we can divide as long as P is monic. The Weierstrass division theorem says that we can divide by a monic $P \in {}_{n-1}\mathcal{O}_p[z_n]$, even if f is a holomorphic function (a "polynomial of infinite degree").

Remark 6.2.6. Despite what it looks like given our proofs, the preparation and division theorems are really theorems about power series. They also hold for formal power series, that is, power series which do not necessarily converge. Another standard way to prove the theorems is to prove the formal version and then to prove that in case we stick in convergent power series, the series we obtain back are also convergent.

6.3 \ The dependence of zeros on parameters

Let us prove that the zeros change holomorphically as long as they do not come together. We will prove shortly that the zeros come together only on a small set—a zero set of a certain holomorphic function called the discriminant.

Zeros of a function of one variable that are distinct points of $\mathbb{C}$ are said to be *geometrically distinct*. If the set of zeros is a single complex number, we say it is a *geometrically unique* zero. For example, $(\zeta - 1)^2$ has a geometrically unique zero at 1, and $(\zeta - 1)^2(\zeta + 1)$ has two geometrically distinct zeros, 1 and −1.

Proposition 6.3.1. *Let $U' \subset \mathbb{C}^{n-1}$ and $D \subset \mathbb{C}$ be domains, and $f \in \mathcal{O}(U' \times D)$. Suppose that for each fixed $z' \in U'$ the function $z_n \mapsto f(z', z_n)$ has a geometrically unique zero $\alpha(z') \in D$. Then α is holomorphic in U'.*

The proposition shows that the regularity conclusion of the holomorphic implicit function theorem holds under the hypothesis that there exists some locally unique solution for z_n, regardless of the derivative vanishing or not. Such a result holds only for holomorphic functions and not for real-analytic or smooth functions. For example, $x^2 - y^3 = 0$ has a unique real solution $y = x^{2/3}$, but $x^{2/3}$ is not even differentiable.

Proof. We must show that α is holomorphic near any point, which, without loss of generality, is the origin and $\alpha(0) = 0$. Apply the preparation theorem to find $f = uP$, where P is a Weierstrass polynomial in $\mathcal{O}(V')[z_n]$ for some $V' \subset U'$ and all zeros of $z_n \mapsto P(z', z_n)$ are in D. As α is a geometrically unique zero in D,

$$P(z', z_n) = \big(z_n - \alpha(z')\big)^k = z_n^k - k\alpha(z')z_n^{k-1} + \cdots$$

The coefficients of P are holomorphic, so α is holomorphic. $\qquad\square$

Proposition 6.3.2. *Let $U' \subset \mathbb{C}^{n-1}$ and $D \subset \mathbb{C}$ be domains, and $f \in \mathcal{O}(U' \times D)$. Let $m \in \mathbb{N}$ be such that for each $z' \in U'$, the function $z_n \mapsto f(z', z_n)$ has precisely m geometrically distinct zeros. Then locally near each point in U' there exist m holomorphic functions $\alpha_1(z'), \ldots, \alpha_m(z')$, positive integers $k_1, \ldots, k_m$, and a nonvanishing holomorphic function u such that*

$$f(z', z_n) = u(z', z_n) \prod_{\ell=1}^{m} \big(z_n - \alpha_\ell(z')\big)^{k_\ell}.$$

The proof is left as an exercise. We can only define α_1 through α_m locally (on a smaller domain) as we cannot consistently order α_1 through α_m as we move around U' if it is not simply connected. If U' is simply connected, then the functions can be defined globally by analytic continuation. For an example where U' is not simply connected, recall Example 6.2.4. Consider $U' = \mathbb{C} \setminus \{0\}$ and think $D = \mathbb{C}$ rather than a disc for simplicity. Then U' is not simply connected, and there do not exist continuous functions $\alpha_1(z_1)$ and $\alpha_2(z_1)$ defined in U' that are zeros of the Weierstrass polynomial, that is $z_2^2 - z_1 = \big(z_2 - \alpha_1(z_1)\big)\big(z_2 - \alpha_2(z_1)\big)$. These would be the two square roots of z_1, and there is no continuous (let alone holomorphic) square root defined in $\mathbb{C} \setminus \{0\}$. Such roots can be chosen to be holomorphic on any smaller simply connected open subset of U', for example, on any disc $\Delta \subset U'$.

Exercise 6.3.1: Let $D \subset \mathbb{C}$ be a bounded domain, $U' \subset \mathbb{C}^{n-1}$ a domain, f a continuous function on $U' \times \overline{D}$ holomorphic on $U' \times D$, where f is zero on at least one point of $U' \times D$, and f is never zero on $U' \times \partial D$. Suppose that for each $z' \in U'$, the function $z_n \mapsto f(z', z_n)$ has no more than one geometrically distinct zero in D. Prove that for each $z' \in U'$, $z_n \mapsto f(z', z_n)$ has exactly one geometrically unique zero in D. Note: By Proposition 6.3.1, that zero is a holomorphic function.

Exercise 6.3.2: Prove Proposition 6.3.2. See the exercise above and Proposition 6.3.1.

Theorem 6.3.3. *Let $D \subset \mathbb{C}$ be a bounded domain, $U' \subset \mathbb{C}^{n-1}$ a domain, and $f \in \mathcal{O}(U' \times D)$. Suppose the zero set $f^{-1}(0)$ has no limit points on $U' \times \partial D$. Then there exists an $m \in \mathbb{N}$ and a holomorphic function $\Delta \colon U' \to \mathbb{C}$, not identically zero, such that for every $z' \in U' \setminus E$, where $E = \Delta^{-1}(0)$, $z_n \mapsto f(z', z_n)$ has exactly m geometrically distinct zeros in D, and $z_n \mapsto f(z', z_n)$ has strictly fewer than m geometrically distinct zeros for $z' \in E$.*

The complement of a zero set of a holomorphic function is connected, open, and dense. We call Δ the *discriminant function* and its zero set E the *discriminant set*. For the quadratic equation $a(z')z_n^2 + b(z')z_n + c(z') = 0$ (assuming a never zero and 2 distinct zeros at some z'), Δ is, up to a unit, the discriminant we learned about in high school: $b^2 - 4ac$. To be precise, for the Δ from the proof below, $-a^2\Delta = b^2 - 4ac$.

Proof. The zeros of $z_n \mapsto f(z', z_n)$ are isolated, and there are finitely many for every z' as D is bounded and $f^{-1}(0)$ has no limit points on $U' \times \partial D$. For any $p' \in U'$, we define two useful paths. Let γ be the union of nonintersecting small simple closed curves around small nonintersecting discs in D, one around each geometrically distinct zero of $z_n \mapsto f(p', z_n)$. Let λ be a large closed path in D going exactly once around all the zeros and such that the interior of λ is in D. Suppose γ and λ intersect no zeros. See Figure 6.4. By continuity, the curves γ and λ do not intersect any zeros for z' near p'. Since the set $f^{-1}(0)$ is closed and the zeros do not accumulate on $U' \times \partial D$, for z' near p', the zeros stay a positive distance away from the boundary. So λ can be picked to go around all the zeros of $z_n \mapsto f(z', z_n)$ exactly once for z' near p'.

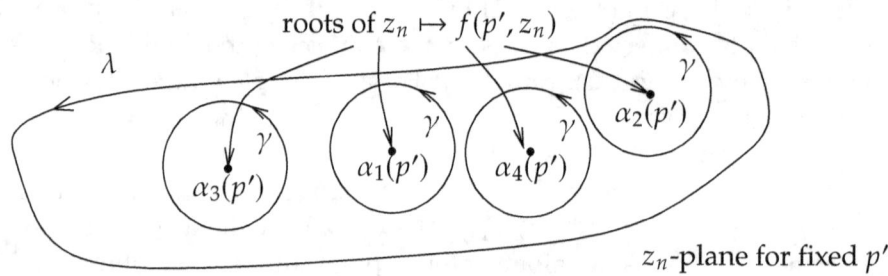

Figure 6.4: Curve around each zero.

Let $M(z')$ be the number of zeros (counting multiplicity) of $z_n \mapsto f(z', z_n)$. Given any p' pick a λ as above that contains all zeros of $z_n \mapsto f(z', z_n)$ for all z' in some neighborhood of p'. The argument principle shows that

$$M(z') = \frac{1}{2\pi i} \int_\lambda \frac{\frac{\partial f}{\partial z_n}(z', \zeta)}{f(z', \zeta)} d\zeta,$$

and so M is constant in this neighborhood (it is an integer-valued continuous function). So M is a locally constant function on U', which is connected and so M is constant. The number of geometrically distinct zeros at any z' is bounded by M, although the number of geometrically distinct zeros may not be constant. Let m be the maximal number of geometrically distinct zeros and suppose that at some point in U', there are exactly m geometrically distinct zeros.

Let $U'_m \subset U'$ be the set of $z' \in U'$ for which $z_n \mapsto f(z', z_n)$ has exactly m geometrically distinct zeros. Write U' as a union of disjoint sets $U' = U'_m \cup E$, where $E = U' \setminus U'_m$. By the definition of m, U'_m is nonempty. Suppose $p' \in U'_m$ and γ goes around the zeros as above. Let γ_j be a single component curve of the path γ going around one of the zeros. The argument principle with respect to γ_j says that γ_j must contain at least one zero for all z' near p'. There are finitely many components of γ, so for z' in some neighborhood of p', $z_n \mapsto f(z', z_n)$ has at least m zeros in γ (at least one in each component), and as m is the maximum, it has exactly m zeros. In other words, U'_m is open.

Locally on U'_m, there exist m holomorphic functions $\alpha_1, \ldots, \alpha_m$ giving the zeros by the previous proposition. We cannot define these on all of U'_m as we do not know how they are ordered. The function

$$\Delta(z') = \prod_{j \neq k} \big(\alpha_j(z') - \alpha_k(z')\big)$$

defined for $z' \in U'_m$ does not depend on the order. That means Δ is well-defined as a function on the open set U'_m, and since α_k can locally be picked to be holomorphic, Δ is holomorphic.

Let $p' \in E \cap \overline{U'_m}$, so there are fewer than m zeros at p'. Suppose γ and λ are as above, so there are fewer than m components of γ. In each component γ_j of γ, there is at least one zero for all z' near p' by the same argument as above. The path λ goes around all the zeros of $z_n \mapsto f(z', z_n)$ for z' near p'. The number of zeros between λ and γ at z' is

$$\frac{1}{2\pi i} \int_{\lambda - \gamma} \frac{\frac{\partial f}{\partial z_n}(z', \zeta)}{f(z', \zeta)} d\zeta,$$

which is a continuous integer-valued function that is zero at $z' = p'$ and so it is zero in a neighborhood. Thus there are no zeros between γ and λ. As λ goes around all the zeros for z' near p', all zeros of $z_n \mapsto f(z', z_n)$ lie inside γ for z' near p'. There exist such $z' \in U'_m$ arbitrarily near p', in which case, by pigeonhole principle, some component γ_j contains at least two geometrically distinct zeros of $z_n \mapsto f(z', z_n)$. Let $\{z'_\ell\}$ be an arbitrary sequence of points in U'_m going to p'. As the number of components of γ is finite, we pass to a subsequence so that there is some fixed component γ_j of γ where $z_n \mapsto f(z'_\ell, z_n)$ has at least two distinct zeros in γ_j for every z'_ℓ. Label the two distinct zeros as $\alpha_1(z'_\ell)$ and $\alpha_2(z'_\ell)$. At p' there is only a geometrically unique zero in γ_j, let us name it $\alpha_1(p')$. As $f^{-1}(0)$ is closed, $\alpha_1(z'_\ell)$ and $\alpha_2(z'_\ell)$ both approach $\alpha_1(p')$ as $\ell \to \infty$. The zeros are bounded, so $\lim_{\ell \to \infty} \Delta(z'_\ell) = 0$. As the limit is zero for a subsequence of an arbitrary sequence,

$$\lim_{z' \in U'_m \to p'} \Delta(z') = 0.$$

We have already defined Δ on U'_m, where it is nonzero, so set $\Delta(z') = 0$ for $z' \in E$. The function Δ is continuous on U' and is zero precisely on E and holomorphic on U'_m. Radó's theorem (Theorem 2.4.12) says that Δ is holomorphic in U'.   □

The discriminant given above is really the discriminant of the set $f^{-1}(0)$ rather than of the corresponding function, which we can, via the preparation theorem, assume is a Weierstrass polynomial. For Weierstrass polynomials, the discriminant is often defined as $\prod_{j \neq k} \big(\alpha_j(z') - \alpha_k(z')\big)$ taking multiple zeros into account, and therefore the "discriminant" could be identically zero. It will be clear from upcoming exercises that if the Weierstrass polynomial is irreducible, then the two notions do in fact coincide.

Exercise 6.3.3: Prove that if $f \in \mathcal{O}(U)$, then $U \setminus f^{-1}(0)$ is not simply connected if $f^{-1}(0)$ is nonempty. In particular, in the theorem, $U' \setminus E$ is not simply connected if $E \neq \emptyset$.

Exercise 6.3.4: Let $D \subset \mathbb{C}$ be a bounded domain and $U' \subset \mathbb{C}^{n-1}$ a domain. Suppose f is a continuous function on $U' \times \overline{D}$ holomorphic on $U' \times D$, and f is never zero on $U' \times \partial D$. Suppose $\gamma \colon [0,1] \to U'$ is continuous and $f\big(\gamma(0), c\big) = 0$ for some $c \in D$. Prove that there exists a continuous $\alpha \colon [0,1] \to \mathbb{C}$ such that $\alpha(0) = c$ and $f\big(\gamma(t), \alpha(t)\big) = 0$ for all $t \in [0,1]$. Hint: Start with a path arbitrarily close to γ that misses the discriminant.

6.4 \ Properties of the ring of germs

Given a commutative ring R, an *ideal* $I \subset R$ is a nonempty subset such that firstly, $fg \in I$ whenever $f \in R$ and $g \in I$, and secondly, $g + h \in I$ whenever $g, h \in I$. An intersection of ideals is again an ideal, and hence it makes sense to talk about the smallest ideal containing a set of elements. An ideal I is generated by $f_1, \ldots, f_k$ if I is the smallest ideal containing $\{f_1, \ldots, f_k\}$. We then write $I = (f_1, \ldots, f_k)$. Every element in I can be written as $c_1 f_1 + \cdots + c_k f_k$ where $c_1, \ldots, c_k \in R$. A *principal ideal* is an ideal generated by a single element, that is, (f).

For convenience, when talking about germs of functions, we often identify a representative with the germ when the context is clear. So by abuse of notation, we may write $f \in \mathcal{O}_p$ instead of $(f, p) \in \mathcal{O}_p$ and $(f_1, \ldots, f_k)$ instead of $((f_1, p), \ldots, (f_k, p))$, as in the following exercises.

Exercise 6.4.1:
  a) Suppose $f \in \mathcal{O}_p$, $f(p) \neq 0$, and (f) is the ideal generated by f. Prove $(f) = \mathcal{O}_p$.
  b) Let $\mathfrak{m}_p = (z_1 - p_1, \ldots, z_n - p_n) \subset \mathcal{O}_p$ be the ideal generated by the coordinate functions. Show that if $f(p) = 0$, then $f \in \mathfrak{m}_p$.
  c) Show that if $I \subsetneq \mathcal{O}_p$ is a proper ideal (ideal such that $I \neq \mathcal{O}_0$), then $I \subset \mathfrak{m}_p$, that is, $\mathfrak{m}_p$ is a maximal ideal.

Exercise **6.4.2:** *Suppose $n = 1$. Show that ${}_1\mathcal{O}_p$ is a principal ideal domain (PID), that is, every ideal is a principal ideal. More precisely, show that given an ideal $I \subset {}_1\mathcal{O}_p$, $I \neq \{0\}$, there exists a $k = 0, 1, 2, \ldots$, such that $I = ((z - p)^k)$.*

Exercise **6.4.3:** *Suppose $U, V \subset \mathbb{C}^n$ are two neighborhoods of p and $h \colon U \to V$ is a biholomorphism. First prove that it makes sense to talk about $f \circ h$ for any $(f, p) \in \mathcal{O}_p$. Then prove that $f \mapsto f \circ h$ is a ring isomorphism.*

A commutative ring R is *Noetherian* if every ideal in R is finitely generated. That is, for every ideal $I \subset R$ there exist finitely many generators $f_1, \ldots, f_k \in I$: Every $g \in I$ can be written as $g = c_1 f_1 + \cdots + c_k f_k$, for some $c_1, \ldots, c_k \in R$. In an exercise above, you proved ${}_1\mathcal{O}_p$ is a PID. So ${}_1\mathcal{O}_p$ is Noetherian. In higher dimensions, the ring of germs may not be a PID, but it is Noetherian.

Theorem 6.4.1. *$\mathcal{O}_p$ is Noetherian.*

Proof. Without loss of generality, $p = 0$. The proof is by induction on dimension. By Exercise 6.4.2, ${}_1\mathcal{O}_0$ is Noetherian. By Exercise 6.4.3, we are allowed a biholomorphic change of coordinates near the origin.

For induction, suppose ${}_{n-1}\mathcal{O}_0$ is Noetherian and let $I \subset {}_n\mathcal{O}_0$ be an ideal. If $I = \{0\}$ or $I = {}_n\mathcal{O}_0$, then the assertion is obvious. Therefore, assume that all elements of I vanish at the origin ($I \neq {}_n\mathcal{O}_0$), and that there exist elements that are not identically zero ($I \neq \{0\}$). Let g be such an element. After perhaps a linear change of coordinates, assume g is a Weierstrass polynomial in z_n by the preparation theorem.

The ring ${}_{n-1}\mathcal{O}_0[z_n]$ is a subring of ${}_n\mathcal{O}_0$. The set $J = {}_{n-1}\mathcal{O}_0[z_n] \cap I$ is an ideal in the ring ${}_{n-1}\mathcal{O}_0[z_n]$. By the Hilbert basis theorem (see Theorem D.4 in the appendix for a proof), as ${}_{n-1}\mathcal{O}_0$ is Noetherian, the ring ${}_{n-1}\mathcal{O}_0[z_n]$ is also Noetherian. Thus J has finitely many generators, that is, $J = (h_1, \ldots, h_k)$ in the ring ${}_{n-1}\mathcal{O}_0[z_n]$.

By the division theorem, every $f \in I$ is of the form $f = qg + r$, where $r \in {}_{n-1}\mathcal{O}_0[z_n]$ and $q \in {}_n\mathcal{O}_0$. As f and g are in I, so is r. As g and r are in ${}_{n-1}\mathcal{O}_0[z_n]$, they are both in J. Write $g = c_1 h_1 + \cdots + c_k h_k$ and $r = d_1 h_1 + \cdots + d_k h_k$. Then $f = (qc_1 + d_1)h_1 + \cdots + (qc_k + d_k)h_k$. So $h_1, \ldots, h_k$ also generate I in ${}_n\mathcal{O}_0$. $\square$

Exercise **6.4.4:** *Prove that after possibly a linear change of coordinates, every proper ideal $I \subset \mathcal{O}_0$ where $I \neq \{0\}$ is generated by Weierstrass polynomials. As a technicality, note that a Weierstrass polynomial of degree 0 is just 1, so it works for $I = \mathcal{O}_0$.*

Exercise **6.4.5:** *We saw above that ${}_1\mathcal{O}_p$ is a PID. Prove that if $n > 1$, then ${}_n\mathcal{O}_p$ is not a PID.*

In a commutative ring R, $f \in R$ is *irreducible* if f is not a unit and whenever $f = gh$, either g or h is a unit.

Theorem 6.4.2. *$\mathcal{O}_p$ is a unique factorization domain (UFD). That is, up to multiplication by a unit, every nonzero nonunit has a unique factorization into irreducible elements of $\mathcal{O}_p$.*

Proof. Again assume $p = 0$ and induct on the dimension. The one-dimensional statement is an exercise below. If $_{n-1}\mathcal{O}_0$ is a UFD, then $_{n-1}\mathcal{O}_0[z_n]$ is a UFD by the Gauss lemma (see Theorem D.6).

Take $f \in {_n\mathcal{O}_0}$ such that $f(0) = 0$ and $f \not\equiv 0$. After perhaps a linear change of coordinates $f = qP$, for q a unit in $_n\mathcal{O}_0$, and P a Weierstrass polynomial in z_n. As $_{n-1}\mathcal{O}_0[z_n]$ is a UFD, P has a unique factorization in $_{n-1}\mathcal{O}_0[z_n]$ into $P = P_1 P_2 \cdots P_k$. So $f = qP_1P_2 \cdots P_k$. We choose $P_1, \ldots, P_k$ to be monic. As P has a geometrically unique zero at $z' = 0$, $P_1, \ldots, P_k$ do as well, and so they are Weierstrass polynomials. That P_ℓ are irreducible in $_n\mathcal{O}_0$ is left as an exercise.

Suppose $f = \tilde{q}g_1g_2 \cdots g_m$ is another factorization. The preparation theorem applies to each g_ℓ. Therefore, write $g_\ell = u_\ell \widetilde{P}_\ell$ for a unit u_ℓ and a Weierstrass polynomial $\widetilde{P}_\ell$. We obtain $f = u\widetilde{P}_1\widetilde{P}_2 \cdots \widetilde{P}_m$ for a unit u. By the uniqueness part of the preparation theorem we obtain $P = \widetilde{P}_1\widetilde{P}_2 \cdots \widetilde{P}_m$. The conclusion is obtained by noting that $_{n-1}\mathcal{O}_0[z_n]$ is a UFD.   □

Exercise 6.4.6: *Prove that $_1\mathcal{O}_p$ is a UFD.*

Exercise 6.4.7: *Show that an irreducible Weierstrass polynomial in $_{n-1}\mathcal{O}_0[z_n]$ is irreducible in $_n\mathcal{O}_0$.*

6.5 \ Varieties

As before, if $f: U \to \mathbb{C}$ is a function, let $Z_f = f^{-1}(0) \subset U$ denote the zero set of f.

Definition 6.5.1. Let $U \subset \mathbb{C}^n$ be an open set. Let $X \subset U$ be a set such that near each point $p \in U$, there exists a neighborhood W of p and a family of holomorphic functions $\mathcal{F}$ defined on W such that

$$W \cap X = \{z \in W : f(z) = 0 \text{ for all } f \in \mathcal{F}\} = \bigcap_{f \in \mathcal{F}} Z_f.$$

Then X is called a (*complex* or *complex-analytic*) *variety* or a *subvariety* of U. Sometimes X is called an *analytic set*. We say $X \subset U$ is a proper subvariety if $\emptyset \neq X \subsetneq U$.

We generally leave out the "complex" from "complex subvariety" as it is clear from context. But you should know that there are other types of subvarieties, namely real subvarieties given by real-analytic functions. We will not cover those in this book.

Example 6.5.2: The set $X = \{0\} \subset \mathbb{C}^n$ is a subvariety as it is the only common vanishing point of the functions $\mathcal{F} = \{z_1, \ldots, z_n\}$. Similarly, $X = \mathbb{C}^n$ is a subvariety of $\mathbb{C}^n$, where we let $\mathcal{F} = \emptyset$.

Example 6.5.3: The set defined by $z_2 = e^{1/z_1}$ is a subvariety of $U = \{z \in \mathbb{C}^2 : z_1 \neq 0\}$. It is not a subvariety of any open set larger than U.

It is useful to note what happens when we replace "near each point $p \in U$" with "near each point $p \in X$." We get a slightly different concept, and X is said to be a *local variety*. A local variety X is a subvariety of some neighborhood of X, but it is not necessarily closed in U. As a simple example, the set $X = \{z \in \mathbb{C}^2 : z_1 = 0, |z_2| < 1\}$ is a local variety, but not a subvariety of $\mathbb{C}^2$. On the other hand, X is a subvariety of the unit ball $\{z \in \mathbb{C}^2 : \|z\| < 1\}$.

Note that $\mathscr{F}$ depends on p and near each point may have a different set of functions. Clearly the family $\mathscr{F}$ is not unique. A priori, we let $\mathscr{F}$ be infinite, but let us note why it would be sufficient to restrict to finite families $\mathscr{F}$.

We work with germs of functions. Recall that when (f, p) is a germ of a function the germ (Z_f, p) is the germ of the zero set of some representative. Let

$$I_p(X) \overset{\text{def}}{=} \{(f, p) \in \mathcal{O}_p : (X, p) \subset (Z_f, p)\}.$$

That is, $I_p(X)$ is the set of germs of holomorphic functions vanishing on X near p. The sum of two functions that vanish on X also vanishes on X, and if a function vanishes on X, then any multiple of it also vanishes on X. So $I_p(X)$ is an ideal. Really $I_p(X)$ depends only on the germ of X at p, so define $I_p((X, p)) = I_p(X)$.

As $\mathcal{O}_p$ is Noetherian, every ideal in $\mathcal{O}_p$ is finitely generated. Let $I \subset \mathcal{O}_p$ be an ideal generated by $f_1, f_2, \ldots, f_k$. Write

$$V_p(I) \overset{\text{def}}{=} (Z_{f_1}, p) \cap (Z_{f_2}, p) \cap \cdots \cap (Z_{f_k}, p).$$

That is, $V_p(I)$ is the germ of the subvariety "cut out" by the elements of I, since every element of I vanishes on the points where all the generators vanish. Suppose representatives $f_1, \ldots, f_k$ of the generators are defined in some neighborhood W of p, and a germ $(g, p) \in I$ has a representative g defined in W such that $g = c_1 f_1 + \cdots + c_k f_k$, where c_j are also holomorphic functions on W. If $q \in Z_{f_1} \cap \cdots \cap Z_{f_k}$, then $g(q) = 0$. Thus, $Z_{f_1} \cap \cdots \cap Z_{f_k} \subset Z_g$, or in terms of germs, $V_p(I) \subset (Z_g, p)$. The reason why we did not define $V_p(I)$ to be the intersection of zero sets of all germs in I is that this would be an infinite intersection, and we did not define such an object for germs.

Exercise 6.5.1: *Show that $V_p(I)$ is independent of the choice of generators.*

Exercise 6.5.2: *Suppose $I_p(X)$ is generated by the functions $f_1, f_2, \ldots, f_k$. Prove*

$$(X, p) = (Z_{f_1}, p) \cap (Z_{f_2}, p) \cap \cdots \cap (Z_{f_k}, p).$$

Exercise 6.5.3: *Given a germ (X, p) of a subvariety at p, show $V(I_p(X)) = (X, p)$ (see above). Then given an ideal $I \subset \mathcal{O}_p$, show $I_p(V_p(I)) \supset I$.*

Exercise 6.5.4: *Let $X \subset \mathbb{C}^n$ be a subvariety that is a complex cone, in other words, if $z \in X$, then $\lambda z \in X$ for all $\lambda \in \mathbb{C}$. Prove that $I_0(X)$ is generated by finitely many homogeneous polynomials. Hint: Given any f holomorphic near 0 that vanishes on X, write $f = \sum_k f_k(z)$ where f_k are homogeneous polynomials. Show that f_k vanish on X. Use that Hilbert basis theorem applies and so the ring of polynomials is Noetherian.*

Exercise 6.5.5: Suppose $\Omega \subset \mathbb{C}^n$ is a domain, $U \subset \mathbb{C}^k$ open, $f \colon \Omega \to U$ holomorphic, and $X \subset U$ is a subvariety. Suppose that there exists a nonempty open subset $W \subset \Omega$ such that $f(W) \subset X$. Prove that $f(\Omega) \subset X$.

The ideal $I_p(X)$ is finitely generated. Near each point p only finitely many functions are necessary to define a subvariety, that is, by an exercise above, those functions "cut out" the subvariety. When one says *defining functions* for a germ of a subvariety, one generally means that those functions generate the ideal, not just that their common zero set happens to be the subvariety. A theorem that we will not prove here in full generality, the *Nullstellensatz*, says that if we take the germ of a subvariety defined by functions in an ideal $I \subset \mathcal{O}_p$, and look at the ideal given by that subvariety, we obtain the radical of I. The *radical* of I is defined as $\sqrt{I} \overset{\text{def}}{=} \{f : f^m \in I \text{ for some } m\}$. In more concise language, the Nullstellensatz says $I_p(V_p(I)) = \sqrt{I}$. Germs of subvarieties are in one-to-one correspondence with radical ideals of $\mathcal{O}_p$.

Example 6.5.4: The subvariety $X = \{0\} \subset \mathbb{C}^2$ can be given by $\mathcal{F} = \{z_1^2, z_2^2\}$. If $I = (z_1^2, z_2^2) \subset \mathcal{O}_0$ is the ideal of germs generated by these two functions, then $I_0(X) \neq I$. We have seen that the ideal $I_0(X)$ is the maximal ideal $\mathfrak{m}_0 = (z_1, z_2)$. As $I \subset (z_1, z_2) = \mathfrak{m}_0$ and the square of z_1 and z_2 are both in I, we find $\sqrt{I} = (z_1, z_2) = \mathfrak{m}_0$.

The local properties of a subvariety at p are encoded in the properties of the ideal $I_p(X)$. Therefore, the study of subvarieties often involves the study of the various algebraic properties of the ideals of $\mathcal{O}_p$. Let us also mention in passing that the other object that is studied is the so-called *coordinate ring* $\mathcal{O}_p / I_p(X)$, which represents the functions on (X, p). That is, we identify two functions if they differ by something in the ideal, since then they are equal on X.

At most points a subvariety is like a piece of $\mathbb{C}^k$, more precisely like a graph over $\mathbb{C}^k$. A graph of $f \colon U' \subset \mathbb{C}^k \to \mathbb{C}^{n-k}$ is the set $\Gamma_f \subset U' \times \mathbb{C}^{n-k} \subset \mathbb{C}^k \times \mathbb{C}^{n-k}$ defined by

$$\Gamma_f \overset{\text{def}}{=} \{(z, w) \in U' \times \mathbb{C}^{n-k} : w = f(z)\}.$$

Definition 6.5.5. Let $U \subset \mathbb{C}^n$ be open and $X \subset U$ a subvariety. Let $p \in X$ be a point where after a permutation of coordinates, the set X is a graph of a holomorphic mapping near p. That is, after relabeling coordinates, there is a neighborhood $U' \times U'' \subset \mathbb{C}^k \times \mathbb{C}^{n-k}$ of p, for some $k = 0, 1, \ldots, n$, and a holomorphic $f \colon U' \to \mathbb{C}^{n-k}$ such that

$$X \cap (U' \times U'') = \Gamma_f.$$

Then p is a *regular point* (or a *simple point*) of X and the (complex) *dimension* of X at p is k. We write $\dim_p X = k$. As the ambient* dimension is n (X is a subvariety of U), we say X is of *codimension* $n - k$ at p. If all points of X are regular points of dimension k, then X is a *complex manifold*, or a *complex submanifold*, of (complex) dimension k.

---

*The word *ambient* is used often to mean the set that contains whatever object we are talking about.

The set of regular points of X is denoted by X_{reg}. Any point that is not regular is *singular*. The set of singular points of X is denoted by X_{sing}.

A couple of remarks are in order. A subvariety X can have regular points of several different dimensions, although if a point is a regular point of dimension k, then all nearby points are regular points of dimension k as the same U' and U'' works. In particular, X_{reg} is an open subset of X. An isolated point of X is automatically a regular point of dimension 0. Sometimes the empty set is considered a complex manifold of dimension -1 (or $-\infty$). Although it may not perhaps be immediately clear, it is not difficult to show that the definition of a regular point is invariant under biholomorphic changes of coordinates (Exercise 6.5.8). Finally, we remark that dimension is well-defined (Exercise 6.5.6).

Example 6.5.6: The set $U = \mathbb{C}^n$ is a complex submanifold of dimension n (codimension 0). In particular, $U_{reg} = U$ and $U_{sing} = \emptyset$.

The set $M = \{z \in \mathbb{C}^3 : z_3 = z_1^2 - z_2^2\}$ is a complex submanifold of dimension 2 (codimension 1). Again, $M_{reg} = M$ and $M_{sing} = \emptyset$.

On the other hand, the so-called *cusp*, $C = \{z \in \mathbb{C}^2 : z_1^3 - z_2^2 = 0\}$ is not a complex submanifold. The origin is a singular point of C (see exercise below). At every other point, we can write $z_2 = \pm z_1^{3/2}$, so $C_{reg} = C \setminus \{0\}$, and so $C_{sing} = \{0\}$. The dimension at every regular point is 1. See Figure 6.5 for a plot of $C \cap \mathbb{R}^2$.

Another type of singularity could be where two complex manifolds intersect. For example, $X = \{z \in \mathbb{C}^2 : z_1^2 - z_2^2 = 0\}$ is the union of the two complex manifolds of dimension 1 given by $z_1 + z_2 = 0$ and $z_1 - z_2 = 0$. In this case $X_{sing} = \{0\}$ and $X_{reg} = X \setminus \{0\}$. See Figure 6.5 for a plot of $X \cap \mathbb{R}^2$.

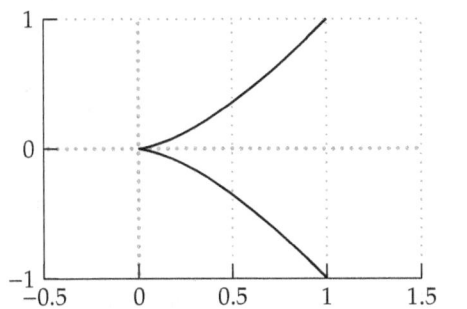 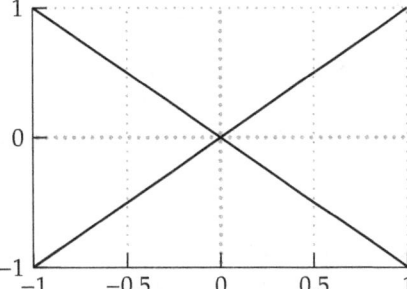

Figure 6.5: The cusp C (left), and the intersecting manifolds X (right).

Exercise 6.5.6: Prove that if p is a regular point of a subvariety $X \subset U \subset \mathbb{C}^n$ of a domain U, then the dimension at p is well-defined. Hint: If there were two possible U' of different dimension (possibly different affine coordinates), construct a map from one such U' to another such U' with nonvanishing derivative.

Exercise 6.5.7: *Consider the cusp* $C = \{z \in \mathbb{C}^2 : z_1^3 - z_2^2 = 0\}$. *Prove that the origin is not a regular point of* C.

Exercise 6.5.8: *Show that* p *is a regular point of dimension* k *of a subvariety* X *if and only if there exists a local biholomorphic change of coordinates that puts* p *to the origin and near* 0, X *is given by* $w = 0$, *where* $(z, w) \in \mathbb{C}^k \times \mathbb{C}^{n-k}$. *In other words, if we allow a biholomorphic change of coordinates instead of just reordering of coordinates, we can let* $f = 0$ *in the definition.*

We also define dimension at a singular point. Below, we will prove that the set of regular points is not only nonempty, but also dense in any subvariety, and so arbitrarily near to every point, there is a regular point.

Definition 6.5.7. Let $X \subset U \subset \mathbb{C}^n$ be a subvariety of U. Let $p \in X$ be a point. We define the (complex) *dimension* of X at p to be

$$\dim_p X \overset{\text{def}}{=} \max\{k \in \mathbb{N}_0 : \forall \text{ open } W \ni p, \exists q \in W \cap X_{reg} \text{ with } \dim_q X = k\}.$$

If (X, p) is a germ and X a representative, the *dimension* of (X, p) is the dimension of X at p. The dimension of the entire subvariety X is defined to be

$$\dim X \overset{\text{def}}{=} \max_{p \in X_{reg}} \dim_p X.$$

We say that X is of *pure dimension* k if at all regular points p, dimension of X at p is k. We say a germ (X, p) is of pure dimension k if there exists a representative of X that is of pure dimension k. We define the word *codimension* as before, that is, the ambient dimension minus the dimension of X.

Example 6.5.8: The cusp $C = \{z \in \mathbb{C}^2 : z_1^3 - z_2^2 = 0\}$ is of dimension 1 at all the regular points, and the only singular point is the origin. Hence $\dim_0 C = 1$, and so $\dim C = 1$. The subvariety C is of pure dimension 1.

Let us restate Theorem 1.6.2 we proved in section 1.6 in the language of varieties.

Theorem 6.5.9. *Let* $U \subset \mathbb{C}^n$ *be a domain and* $f \in \mathcal{O}(U)$. *Then* Z_f *is empty,* Z_f *is a subvariety of pure codimension* 1, *or* $Z_f = U$. *Furthermore,* $(Z_f)_{reg}$ *is open and dense in* Z_f.

Let us improve on this for arbitrary varieties.

Lemma 6.5.10. *Let* $U \subset \mathbb{C}^n$ *be open and let* $X \subset U$ *be a subvariety, then* X_{reg} *is nonempty. Consequently,* X_{reg} *is open and dense in* X.

Proof. In the same way that we proved Theorem 1.6.2, by intersecting X with arbitrary neighborhoods, it is sufficient to show the existence of one regular point.

We will induct on dimension. For $n = 1$, the only subvarieties are either isolated points or entire components of U, so all points are regular. Suppose the lemma is true in dimension $n - 1$. It is enough to find one regular point on X as that means there is a regular point on any neighborhood of any point of X. Suppose $0 \in X$ for simplicity. Either X contains a whole neighborhood of 0, in which case 0 is a regular point, or there is some nontrivil holomorphic function f near 0 that vanishes on X. After a small linear change of coordinates, the Weierstrass preparation theorem applies and we can assume that f is a Weierstrass polynomial. Write the variables as $(z_1, \ldots, z_{n-1}, z_n) = (z', z_n)$. In some neighborhood V of the origin (where $V \subset U$), Theorem 6.3.3 applies to f, so let $\Delta(z')$ denote the discriminant function. Thinking of Δ as a function on V, suppose that $\Delta(q) \neq 0$ for some $q \in X \cap V$. Then there is some neighborhood W of q, such that $Z_f \cap W$ is a graph of a holomorphic function and $X \cap W \subset Z_f \cap W$. After a local biholomorphic change of variables, we can assume that $Z_f \cap W$ an open subset of $\mathbb{C}^{n-1} \times \{0\}$, so $X \cap W$ is really contained in $\mathbb{C}^{n-1}$ and we apply induction.

So suppose Δ vanishes on X. Note that Δ is a function of $z' = (z_1, \ldots, z_{n-1})$. After a linear change of variables in z', we apply the preparation theorem in z' with respect to z_{n-1} and get $\Delta = uP$ in some neighborhood of 0, where u is a unit and P is a Weierstrass polynomial in z_{n-1} with coefficients that only depend on $z_1, \ldots, z_{n-2}$. Let Δ' be the discriminant for P and repeat the procedure above. Either Δ' is nonzero at some point of X, in which case we apply the induction hypothesis as above, or near the origin, X is contained in the zero set of Δ'. If X is contained in the zero set of Δ', we again apply the preparation theorem and get a polynomial in z_{n-2} with coefficients depending only on $z_1, \ldots, z_{n-3}$. Rinse and repeat. Either at some point we could apply the induction hypothesis, or we end after n steps with X being in the zero set of the Weierstrass polynomial in one variable, that is, X is locally near the origin contained in the set where $z_1 = 0$, and we can again apply the induction. $\square$

We will state without proof that we actually have that the singular set is a subvariety, that is, we have the following theorem. We will prove it for varieties of pure codimension 1 in the next section.

Theorem 6.5.11. *Let $U \subset \mathbb{C}^n$ be open and let $X \subset U$ be a subvariety, then $X_{sing} \subset X$ is a subvariety, which is nowhere dense in X and $\dim X_{sing} < \dim X$.*

Exercise 6.5.9: Suppose that $X \subset U \subset \mathbb{C}^n$ is a subvariety of a domain U, such that X_{reg} is connected. Show that X is of pure dimension.

6.6 \ Hypervarieties

Pure codimension-1 subvarieties are particularly nice. Sometimes pure codimension-1 subvarieties are called *hypervarieties*. We will prove two things for hypervarieties. First we will prove that locally, a hypervariety can be defined via a single function, and second, we will prove that the singular set of a hypervariety is a subvariety.

Theorem 6.6.1. *If (X, p) is a germ of a pure codimension-1 subvariety, then there is a germ of a holomorphic function f at p such that $(Z_f, p) = (X, p)$ and $I_p(X)$ is generated by (f, p).*

Proof. We need to find a function that vanishes on (X, p) and divides every other function that vanishes there. There must exist at least one germ of a function that vanishes on X near p (although it could vanish on a larger set). Without loss of generality, assume $p = 0$ and after a linear change of coordinates the Weierstrass preparation theorem applies. More precisely, suppose X is a pure codimension-1 subvariety of a small enough polydisc $U' \times D \subset \mathbb{C}^{n-1} \times \mathbb{C}$ centered at the origin, and the function that vanishes on X is a Weierstrass polynomial $P(z', z_n)$ defined for $z' \in U'$, and all zeros of $z_n \mapsto P(z', z_n)$ are in D for $z' \in U$. Theorem 6.3.3 applies. Let $E \subset U'$ be the discriminant set, a zero set of a holomorphic function. On $U' \setminus E$, there are a certain number of geometrically distinct zeros of $z_n \mapsto P(z', z_n)$.

Let X' be a topological component of $X \setminus (E \times D)$. Above each $z' \in U' \setminus E$, let $\alpha_1(z'), \dots, \alpha_k(z')$ denote the distinct zeros that are in X', that is, $(z', \alpha_\ell(z')) \in X'$. Near each point X' is a graph of a holomorphic function over $U' \setminus E$, and so we can locally choose $\alpha_1, \dots, \alpha_k$ to be holomorphic. Furthermore, this means that the set X' contains only regular points of X, which are of dimension $n - 1$. The number of such geometrically distinct zeros in X' above each point in $U' \setminus E$ is locally constant. As $U' \setminus E$ is connected (Exercise 1.6.5), there exists a unique k. Take

$$F(z', z_n) = \prod_{\ell=1}^{k} (z_n - \alpha_\ell(z')) = z_n^k + \sum_{\ell=0}^{k-1} g_\ell(z') z_n^\ell.$$

The coefficients g_ℓ are well-defined for $z \in U' \setminus E$ as they are independent of how $\alpha_1, \dots, \alpha_k$ are ordered. The g_ℓ are holomorphic for $z \in U' \setminus E$ as locally we can ensure that each α_ℓ is holomorphic. The coefficients g_ℓ are bounded on U' and so extend to holomorphic functions of U' via the Riemann extension theorem. Hence, F is a polynomial in $\mathcal{O}(U')[z_n]$. The zeros of F above $z' \in U' \setminus E$ are simple and give precisely X'. The zeros of F above $z' \in E$ must be limits of zeros above points of $U' \setminus E$ by the argument principle. Consequently, the zero set of F is the closure of X' in $U' \times D$. It is left to the reader to check that all the functions g_ℓ vanish at the origin and F is a Weierstrass polynomial, a fact that will be useful in the exercises below.

If the polynomial $P(z', z_n)$ is of degree m, then $z' \mapsto P(z', z_n)$ has at most m zeros. Together with the fact that $U' \setminus E$ is connected, this means that $X \setminus (E \times D)$ has at most finitely many components (at most m). We find an F for every topological component of $X \setminus (E \times D)$ and we multiply those functions together to get f. No open piece $M \subset X_{reg}$ can lie completely in $E \times D$, as otherwise an open subset of M would also be an open piece of $E \times D$, see Exercise 6.6.4, but we know that P must vanish on M, which is impossible as it only vanishes at finitely many points for each fixed z'. Therefore, as X_{reg} is dense in X (Lemma 6.5.10), the closure of $X \setminus (E \times D)$ contains X and so $Z_f = X$.

The fact that this f generates $I_p(X)$ is left as Exercise 6.6.3. $\qquad\square$

In other words, local properties of a codimension 1 subvariety can be studied by studying the zero set of a single Weierstrass polynomial.

Example 6.6.2: It is not true that a subvariety in $\mathbb{C}^n$ of dimension $n - k$ (codimension k) has k holomorphic functions that "cut it out." That only works for $k = 1$. The set defined by

$$\text{rank} \begin{bmatrix} z_1 & z_2 & z_3 \\ z_4 & z_5 & z_6 \end{bmatrix} < 2$$

is a pure 4-dimensional subvariety of $\mathbb{C}^6$, so of codimension 2, and the defining equations are $z_1 z_5 - z_2 z_4 = 0$, $z_1 z_6 - z_3 z_4 = 0$, and $z_2 z_6 - z_3 z_5 = 0$. We state without proof that the unique singular point is the origin and there exist no 2 holomorphic functions near the origin that define this subvariety. In more technical language, the subvariety is not a *complete intersection*.

Interestingly, a small refinement of the proof of the theorem above gives the following. The same result holds for higher codimension, but it is harder to prove.

Corollary 6.6.3. *Let (X, p) be a germ of a subvariety of pure codimension 1. Then there exists a neighborhood U of p, a representative $X \subset U$ of (X, p) and subvarieties $X_1, \dots, X_k \subset U$ of pure codimension 1 such that $(X_\ell)_{reg}$ is connected for every ℓ, and $X = X_1 \cup \dots \cup X_k$.*

Proof. A particular X_ℓ is defined by considering a topological component of $X \setminus (E \times D)$ as in the proof of Theorem 6.6.1, getting the F, and setting $X_\ell = Z_F$. The topological component is a connected set and it is dense in $(X_\ell)_{reg}$, which proves the corollary. $\quad\square$

Exercise 6.6.1: Suppose $p(z', z_n)$ is a Weierstrass polynomial of degree k such that for an open dense set of z' near the origin $z_n \mapsto p(z', z_n)$ has geometrically k zeros, and such that the set of regular points of Z_p is connected. Show that p is irreducible in the sense that if $p = rs$ for two Weierstrass polynomials r and s, then either $r = 1$ or $s = 1$.

Exercise 6.6.2: Suppose f is a function holomorphic in a neighborhood of the origin with $z_n \mapsto f(0, z_n)$ being of finite order. Show that

$$f = u p_1^{d_1} p_2^{d_2} \cdots p_\ell^{d_\ell},$$

where p_k are Weierstrass polynomials of degree μ_k such that for an open dense set of z', $z_n \mapsto f(z', z_n)$ has μ_k geometrically distinct zeros (no multiple zeros), the set of regular points of Z_{p_k} is connected, and u is a nonzero holomorphic function near 0. Note: In the next section, these polynomials will be the irreducible factors in the factorization of f.

Exercise 6.6.3: Prove the last part of Theorem 6.6.1: Show that if (X, p) is a germ of a pure codimension-1 subvariety, then the ideal $I_p(X)$ is a principal ideal (has a single generator).

Exercise 6.6.4: *Suppose $U \subset \mathbb{C}^n$ is open and $X \subset U$ is a subvariety of dimension $n - 1$. Suppose M is a small piece of a complex submanifold of dimension $n - 1$ such that $M \subset X$. Prove that M agrees with X_{reg} on a dense open set, that is, for each p a dense open subset of M, there is a neighborhood W of p such that $M \cap W = X_{reg} \cap W$. Hint: Consider coordinates where M is a graph and Theorem 6.3.3 applies to X.*

Exercise 6.6.5: *Suppose $I \subset \mathcal{O}_p$ is a principal ideal. Prove the* Nullstellensatz *for hypervarieties: $I_p(V_p(I)) = \sqrt{I}$. That is, show that if $(f, p) \in I_p(V_p(I))$, then $(f^k, p) \in I$ for some integer k.*

Exercise 6.6.6: *Suppose $X \subset U$ is a subvariety of pure codimension 1 for an open set $U \subset \mathbb{C}^n$. Let X' be a topological component of X_{reg}. Prove that the closure $\overline{X'}$ is a subvariety of U of pure codimension 1.*

Example 6.6.4: If X is a hypervariety, the preparation theorem applies, and E the corresponding discriminant set, it is tempting to say that the singular set of X is the set $X \cap (E \times \mathbb{C})$, which is a codimension-2 subvariety. It is true that $X \cap (E \times \mathbb{C})$ will contain the singular set, but in general the singular set is smaller. A simple example of this behavior is the set defined by $z_2^2 - z_1 = 0$. The defining function is a Weierstrass polynomial in z_2 and the discriminant set is given by $z_1 = 0$. However, the subvariety has no singular points as it is the graph of z_1 over z_2. See Figure 6.6.

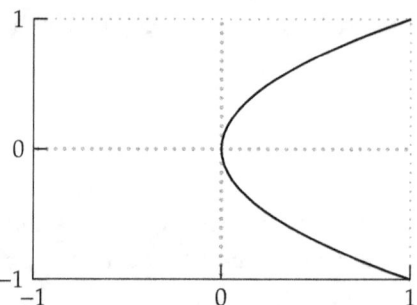

Figure 6.6: The sideways parabola $z_2^2 - z_1 = 0$ for real z_1 and z_2. For each nonzero z_1, there are two solutions (for negative z_1 these are obviously not real). But for each fixed z_2, there is exactly one solution.

A less trivial example $z_1^2 + \cdots + z_k^2 = 0$, where the singular set is of any given dimension $n - k$ is given in Exercise 6.6.8.

Let us prove the version of Theorem 6.5.11 for hypervarieties. We now know that locally hypervarieties are defined by a single function, so we can use the discriminant to locate singularities, but we must allow infinitely many linear changes of coordinates.

Theorem 6.6.5. *Let $U \subset \mathbb{C}^n$ be open and $X \subset U$ a subvariety of pure codimension 1 (a hypervariety). Then X_{sing} is a subvariety of dimension less than or equal to $n - 2$.*

Proof. It is sufficient to consider a certain fixed point $p \in X$ and prove the result locally near p. At p, there is a single holomorphic function f that defines the germ of (X, p) meaning that its zero set is equal to X near p. Without loss of generality, assume that $p = 0$, and after a linear change of coordinates, assume that we can apply the preparation theorem and Theorem 6.3.3 near the origin with respect to each variable. Then we can assume that f is holomorphic in some neighborhood W, $Z_f = X \cap W$, and there exists an open neighborhood V of the origin so that for every variable z_k, $k = 1, \ldots, n$, there is a polydisc $D = D_1 \times \cdots \times D_n$ centered at the origin with $\overline{V} \subset D$ and $\overline{D} \subset W$, where Theorem 6.3.3 applies with respect to z_k, that is, $Z_f \cap (D_1 \times \cdots \times \partial D_k \times \cdots \times D_n) = \emptyset$.

Consider a $q \in X_{reg} \cap V$. By definition, X is a graph near q, so after reordering variables, we assume it is a graph of z_n over $z' = (z_1, \ldots, z_{n-1})$. Let D be the corresponding polydisc and write $D = D' \times D_n \subset \mathbb{C}^{n-1} \times \mathbb{C}$. Let $E \subset D'$ be the discriminant set given by the function $\Delta \in \mathcal{O}(D')$. We think of Δ as a function in $\mathcal{O}(D)$. If $q \notin E \times D_n$, then $\Delta(q) \neq 0$, so we have found a function holomorphic in V that is nonzero at q. Let us start a collection $\mathcal{F}$ of holomorphic functions on V, one for each $q \in X_{reg} \cap V$, and we put Δ in $\mathcal{F}$.

Suppose that $\Delta(q) = 0$. We may assume that the f is the Weierstrass polynomial in z_n we found in the proof of Theorem 6.6.1. In particular, it is a Weierstrass polynomial of degree m and for a generic z' (outside of the discriminant set), the function $z_n \mapsto f(z', z_n)$ has m geometrically distinct roots (so m roots up to multiplicity as well). Write $q = (q', q_n)$. The zero of $z_n \mapsto f(q', z_n)$ at $z_n = q_n$ is simple, but the others are not all simple as q' is in the discriminant. We will change variables so that the new vertical line through q intersects X only at simple zeros. The root at q is already simple and so we will rotate the line around q.

We will change variables to $\tilde{z} = (\tilde{z}', z_n)$ where $(z', z_n) = (\tilde{z}' + (q_n - z_n)\epsilon', z_n)$ and where $\epsilon' \in \mathbb{C}^{n-1}$ is small, so that the chosen line becomes the vertical $\{\tilde{z}' = q'\}$. For small ϵ' and $\tilde{z}'$ near q', the function $z_n \mapsto f(\tilde{z}' + (q_n - z_n)\epsilon', z_n)$ still has exactly m roots up to multiplicity via the argument principle. If $z_n \mapsto f(q' + (q_n - z_n)\epsilon', z_n)$ has m geometrically distinct roots, then so does $z_n \mapsto f(\tilde{z}' + (q_n - z_n)\epsilon', z_n)$ for $\tilde{z}'$ near q' via the same argument with the m small discs and the argument principle as in the proof of Theorem 6.3.3. The problem of finding arbitrarily small ϵ' that do the trick is left as an exercise. It can be done one intersection of the line $z_n \mapsto (q' + (q_n - z_n)\epsilon', z_n)$ with X at a time, that is, if we have an intersection of multiplicity k, a small generic change in ϵ' will give us k distinct intersections nearby. See Exercise 6.6.7.

Take a slightly smaller polydisc $\widetilde{D} = \widetilde{D}' \times D_n \subset D$ in the $\tilde{z}$ variables such that still $V \subset \widetilde{D}$ (we may need to pick ϵ' small enough to arrange this) and Theorem 6.3.3 applies in $\widetilde{D}$. As the number of distinct zeros of $f(\tilde{z}' + (q_n - z_n)\epsilon', z_n)$ is m for all $\tilde{z}'$ near q' including $\tilde{z}' = q'$, the discriminant $\widetilde{\Delta}$ in these variables does not vanish at q'.

We again consider $\widetilde{\Delta}$ as a function on V and as it does not vanish at q, we add $\widetilde{\Delta}$ to $\mathcal{F}$. See Figure 6.7 for the setup. We define $\mathcal{F}$ by repeating for each $q \in X_{reg} \cap V$.

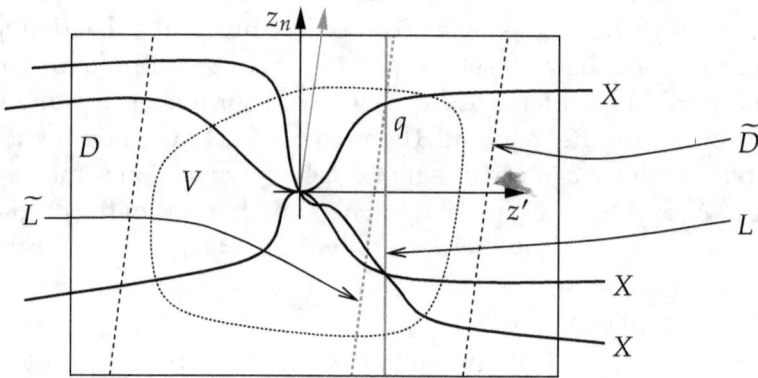

Figure 6.7: Changing variables to make the discriminant not vanish at q, where L is the line $\{z' = q'\}$ while $\widetilde{L}$ is the tilted line $\{\widetilde{z}' = q'\}$.

Next, if $q \in X \cap V$ is singular, then every discriminant function used above must be zero at q; outside of the discriminant set, all points of X are graphs of the zeros and hence nonsingular. That is, $\varphi(q) = 0$ for every $\varphi \in \mathcal{F}$. Thus the common zero set of all the functions in $\mathcal{F}$ intersected with $X \cap V$ gives us precisely $X_{sing} \cap V$, so X_{sing} is a subvariety. It cannot be of dimension $n - 1$ as if it were, it would be a complex submanifold of dimension $n - 1$ near some point and then not all of those points would be singular for X, see Exercise 6.6.4. □

We used infinitely many functions in $\mathcal{F}$ to define X_{sing}. It is possible to use finitely many near each point as the ideals $I_p(X_{sing})$ are Noetherian, but despite the temptation, it is not possible to do a single generic linear change of variables and use just the n discriminant functions, one for each variable. More than n functions may be necessary as situations like the one depicted in Figure 6.7 may occur for some q, no matter how we change variables to start with.

Exercise 6.6.7:
  a) *Suppose $D \subset \mathbb{C}^n$ is a polydisc, $0 \in D$, $q = (0, 1) \in \mathbb{C}^{n-1} \times \mathbb{C}$, and P is a Weierstrass polynomial of degree k such that for a generic z' (not in the discriminant set), $z_n \mapsto P(z', z_n)$ has k simple zeros. Prove that there exists a ball $B \subset \mathbb{C}^{n-1}$ centered at the origin and a dense open set $W \subset B$ such that for every $\epsilon' \in W$, the function $z_n \mapsto \big((1 - z_n)\epsilon', z_n\big)$ has exactly k geometrically distinct zeros. Hint: Change coordinates near the origin to make all these lines vertical.*
  b) *Show that part a) proves the claim in the proof of theorem.*

Exercise 6.6.8:
a) Prove that the hypervariety in $\mathbb{C}^n$, $n \geq 2$, given by $z_1^2 + z_2^2 + \cdots + z_n^2 = 0$ has an isolated singularity at the origin (that is, the origin is the only singular point).
b) For every $0 \leq k \leq n - 2$, find a hypervariety X of $\mathbb{C}^n$ whose set of singular points is a subvariety of dimension k.

6.7 \ Irreducibility, local parametrization, and Puiseux

Definition 6.7.1. A germ of a subvariety $(X, p) \subset (\mathbb{C}^n, p)$ is *reducible* at p if there exist two germs (X_1, p) and (X_2, p) with $(X_1, p) \not\subset (X_2, p)$ and $(X_2, p) \not\subset (X_1, p)$ such that $(X, p) = (X_1, p) \cup (X_2, p)$. Otherwise, the germ (X, p) is *irreducible* at p.

Similarly globally, a subvariety $X \subset U$ is *reducible* in U if there exist two subvarieties X_1 and X_2 of U with $X_1 \not\subset X_2$ and $X_2 \not\subset X_1$ such that $X = X_1 \cup X_2$. Otherwise, the subvariety X is *irreducible* in U.

Example 6.7.2: Local and global reducibility are different. The subvariety given by

$$z_2^2 = z_1(z_1 - 1)^2$$

is irreducible in $\mathbb{C}^2$ (the set of regular points is connected), but locally at the point $(1, 0)$ it is reducible. There, the subvariety is a union of two graphs: $z_2 = \pm\sqrt{z_1}(z_1 - 1)$. See Figure 6.8 for a plot in two real dimensions.

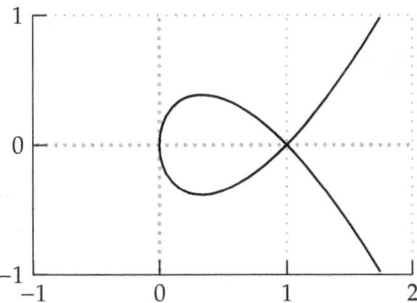

Figure 6.8: Locally reducible curve.

Exercise 6.7.1: *Prove a germ of a subvariety* (X, p) *is irreducible if and only if* $I_p(X)$ *is a prime ideal. Recall an ideal* I *is* prime *if* $ab \in I$ *implies either* $a \in I$ *or* $b \in I$.

Exercise 6.7.2: *Suppose a germ of a subvariety* (X, p) *is of pure codimension 1. Prove* (X, p) *is irreducible if and only if there exists a representative of* X *where* X_{reg} *is connected.*

Exercise 6.7.3: *Let* $X \subset U$ *be a subvariety of pure codimension 1 of a domain* $U \subset \mathbb{C}^n$. *Prove* X *is irreducible if and only if* X_{reg} *is connected. Hint: See previous exercise.*

For complex subvarieties, a subvariety is irreducible if and only if the set of regular points is connected. We omit the proof in the general case, and for hypervarieties it is an exercise above. It then makes sense that we can split a subvariety into its irreducible parts.

Proposition 6.7.3. *Let $(X, p) \subset (\mathbb{C}^n, p)$ be a germ of a subvariety. Then there exist finitely many irreducible subvarieties $(X_1, p), \ldots, (X_k, p)$ such that $(X_1, p) \cup \ldots \cup (X_k, p) = (X, p)$ and such that $(X_m, p) \not\subset (X_\ell, p)$ for all m and ℓ.*

Proof. Suppose (X, p) is reducible. Find (Y_1, p) and (Y_1, p) such that $(Y_1, p) \not\subset (Y_2, p)$, $(Y_2, p) \not\subset (Y_1, p)$, and $(Y_1, p) \cup (Y_2, p) = (X, p)$. As $(Y_\ell, p) \subsetneq (X, p)$, we find $I_p(Y_\ell) \supsetneq I_p(X)$ for both ℓ. If both (Y_1, p) and (Y_2, p) are irreducible, then stop, we are done. Otherwise apply the same reasoning to whichever (or both) (Y_ℓ, p) that was reducible. After finitely many steps you must come to a stop as you cannot have an infinite ascending chain of ideals since $\mathcal{O}_p$ is Noetherian. $\square$

These $(X_1, p), \ldots, (X_k, p)$ are the *irreducible components*. We omit the proof that they are unique in general. For a germ of a hypervariety, the UFD property of $_n\mathcal{O}_p$ gives the irreducible components. You found this factorization in an exercise above, and so this factorization is unique.

Each irreducible component has the following structure. We give the theorem without proof in the general case, although we have essentially proved it already for pure codimension 1 (to put it together is left as an exercise).

Theorem 6.7.4 (Local parametrization theorem). *Let $(X, 0)$ be an irreducible germ of a subvariety of dimension k in $\mathbb{C}^n$. Let X denote a representative of the germ. Then after a possible linear change of coordinates, letting $\pi \colon \mathbb{C}^n \to \mathbb{C}^k$ be the projection onto the first k components, there exists a neighborhood $U \subset \mathbb{C}^n$ of the origin, and a proper subvariety $E \subset \pi(U)$ (the discriminant set) such that*

(i) *$X' = X \cap U \setminus \pi^{-1}(E)$ is a connected k-dimensional complex manifold that is dense in $X \cap U$.*

(ii) *$\pi \colon X' \to \pi(U) \setminus E$ is an m-sheeted covering map for some integer m.*

(iii) *$\pi \colon X \cap U \to \pi(U)$ is a proper mapping.*

The m-sheeted covering map in this case is a local biholomorphism that is an m-to-1 map.

> *Exercise 6.7.4:* Use Theorem 6.3.3 to prove the parametrization theorem if $(X, 0)$ is of pure codimension 1.

Let $(z_1, \ldots, z_n)$ be the coordinates. The linear change of coordinates needed in the theorem is to ensure that the set defined by $z_1 = z_2 = \cdots = z_k = 0$ intersected with

X is an isolated point at the origin. This is precisely the same condition needed to apply Weierstrass preparation theorem in the case when X is the zero set of a single function.

We saw that hypersurfaces are the simpler cases of subvarieties. At the other end of the spectrum, subvarieties of dimension 1 are also reasonably simple for different reasons. Locally, subvarieties of dimension 1 are analytic discs. Moreover, these discs can be chosen to be one-to-one, and so such subvarieties have a natural topological manifold structure even at singular points.

Example 6.7.5: The image of the holomorphic map $\xi \mapsto (\xi^2, \xi^3)$ is the cusp subvariety defined by $z_1^3 - z_2^2 = 0$ in $\mathbb{C}^2$.

The following theorem is often stated only in $\mathbb{C}^2$ for zero sets of a single function although it follows in the same way from the local parametrization theorem in higher-dimensional spaces. Of course, we only proved that theorem (or in fact you the reader did so in an exercise), for codimension-1 subvarieties, and therefore, we also only have a complete proof of the following in $\mathbb{C}^2$.

Theorem 6.7.6 (Puiseux). *Let $(z, w) \in \mathbb{C} \times \mathbb{C}^{n-1}$ be coordinates. Suppose $f \colon U \subset \mathbb{C} \times \mathbb{C}^{n-1} \to \mathbb{C}^\ell$ is a holomorphic map such that $f(z, w) = 0$ defines a one-dimensional subvariety X of U, $0 \in X$, and $w \mapsto f(0, w)$ has an isolated zero at the origin. Then there exists an integer k and a holomorphic $\mathbb{C}^{n-1}$-valued g defined near the origin in $\mathbb{C}$ such that for all ξ near the origin*

$$f\left(\xi^k, g(\xi)\right) = 0.$$

Proof. Without loss of generality assume $(X, 0)$ is irreducible, so that the local parametrization theorem applies. We work in a small disc $D \subset \mathbb{C}$ centered at the origin, so that the origin is the unique point of the discriminant set (the subvariety E). Let $N = \{z \in D : \operatorname{Im} z = 0, \operatorname{Re} z \leq 0\}$. As $D \setminus N$ is simply connected, we have the well-defined functions $\alpha_1(z), \ldots, \alpha_m(z)$ holomorphic on $D \setminus N$ that are solutions to $f(z, \alpha_j(z)) = 0$. These functions continue analytically across N away from the origin. The continuation equals one of the zeros, e.g. $\alpha_j(z)$ becomes $\alpha_\ell(z)$ (and by continuity it is the same zero along the entire N). So there is a permutation σ on m elements such that as z moves counter-clockwise around the origin from the upper half-plane across N to the lower half-plane, $\alpha_j(z)$ is continued as $\alpha_{\sigma(j)}(z)$. There exists some number k (e.g. $k = m!$) such that σ^k is the identity. As ξ goes around a circle around the origin, ξ^k goes around the origin k times. Start at a positive real ξ and start defining a function $g(\xi)$ as $\alpha_1(\xi^k)$. Move ξ around the origin counter-clockwise continuing g analytically. Divide the disc into sectors of angle $2\pi/k$, whose boundaries are where $\xi^k \in N$. Transition to $\alpha_{\sigma(1)}(\xi^k)$ after we reach the boundary of the first sector, then to $\alpha_{\sigma(\sigma(1))}(\xi^k)$ after we reach the boundary of the next sector, and so on. After k steps, that is, as ξ moved all the way around the circle, we are back at $\alpha_1(\xi^k)$, because σ^k is the identity. So $g(\xi)$ is a well-defined holomorphic function outside the origin. Let $g(0) = 0$, and g is holomorphic at 0 by the Riemann extension theorem. See Figure 6.9 for an example. $\qquad \square$

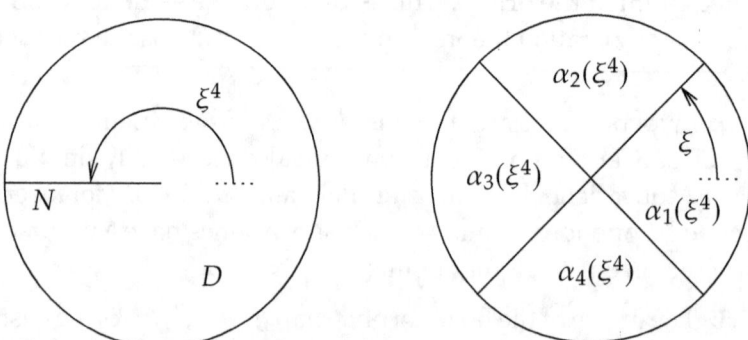

Figure 6.9: Proving Puiseux with $m = k = 4$. The permutation σ takes 1 to 2, 2 to 3, 3 to 4, and 4 to 1. As ξ moves along the short circular arrow on the right, ξ^4 moves along the long circular arrow on the left. The definition of g is given in the right-hand diagram.

Exercise 6.7.5: *Consider an irreducible germ $(X,0) \subset (\mathbb{C}^2,0)$ defined by an irreducible Weierstrass polynomial $f(z,w) = 0$ (polynomial in w) of degree k. Prove there exists a holomorphic g such that $f\left(\xi^k, g(\xi)\right) = 0$ and $\xi \mapsto \left(\xi^k, g(\xi)\right)$ is one-to-one and onto a neighborhood of 0 in X.*

Exercise 6.7.6: *Suppose $(X,0) \subset (\mathbb{C}^2,0)$ is a germ of a one-dimensional subvariety. Show that after a possible linear change of coordinates, there are natural numbers $d_1, \ldots, d_k$ and holomorphic functions $c_1(\xi), \ldots, c_k(\xi)$ vanishing at 0, such that X is given near 0 by*

$$\prod_{\ell=1}^{k} \prod_{m=0}^{d_\ell-1} \left(w - c_\ell\left(e^{m2\pi i/d_\ell} z^{1/d_\ell}\right)\right) = 0.$$

Exercise 6.7.7: *Using the local parametrization theorem, prove that if (X,p) is an irreducible germ of a subvariety of dimension greater than 1, then there exists a neighborhood U of p and a closed subvariety $X \subset U$ (whose germ at p is (X,p)), such that for every $q \in X$ there exists an irreducible subvariety $Y \subset X$ of dimension 1 such that $p \in Y$ and $q \in Y$.*

Exercise 6.7.8: *Prove a stronger version of the exercise above. Show that not only is there a Y, but an analytic disc $\varphi \colon \mathbb{D} \to U$ such that $\varphi(\mathbb{D}) \subset X$, $\varphi(0) = p$ and $\varphi(1/2) = q$.*

Exercise 6.7.9: *Suppose $X \subset U$ is a subvariety of a domain $U \subset \mathbb{C}^n$. Show that X is irreducible if and only if for every pair of points $p, q \in X$ there exists a finite sequence of points $p_0 = p, p_1, \ldots, p_k = q$ in X, and a finite sequence of analytic discs $\Delta_\ell \subset X$ such that p_ℓ and $p_{\ell-1}$ are in Δ_ℓ.*

Exercise 6.7.10: *Prove a* maximum principle *for subvarieties* using the exercises *above: Suppose $X \subset U$ is an irreducible subvariety of an open set U, and suppose $f \colon U \to \mathbb{R} \cup \{-\infty\}$ is a plurisubharmonic function. If the restriction $f|_X$ achieves a maximum at some point $p \in X$, then the restriction $f|_X$ is constant.*

> **Exercise 6.7.11:** *Prove that an analytic disc in $\mathbb{C}^2$ is locally a one-dimensional local variety in the following sense: If $f \colon \mathbb{D} \to \mathbb{C}^2$ is nonconstant and holomorphic and $p \in \mathbb{D}$, then there is some neighbourhood Δ of p so that $f(\Delta)$ is a one-dimensional local variety.*

Using the Puiseux theorem, we often simply parametrize germs of complex one-dimensional subvarieties. And for larger-dimensional varieties, we can find enough one-dimensional curves through any point and parametrize those.

It is not true that every irreducible subvariety is locally an injective image of a piece of $\mathbb{C}^k$ via a holomorphic map, but it is a very deep theorem, the resolution of singularities, that says you can do so if you allow some points where the function is not one-to-one.

6.8 \ Segre varieties and CR geometry

The existence of analytic discs (or subvarieties) in boundaries of domains says a lot about the geometry of the boundary.

Example 6.8.1: Let $M \subset \mathbb{C}^n$ be a smooth real hypersurface containing a complex hypersurface X (zero set of a holomorphic function with nonzero derivative), at $p \in X \subset M$. Apply a local biholomorphic change of coordinates at p, so that in the new coordinates $(z, w) \in \mathbb{C}^{n-1} \times \mathbb{C}$, X is given by $w = 0$, and p is the origin. The tangent hyperplane to M at 0 contains $\{w = 0\}$. By rotating the w coordinate (multiplying it by $e^{i\theta}$), we assume M is tangent to the set $\{(z, w) : \operatorname{Im} w = 0\}$. In other words, M is given by

$$\operatorname{Im} w = \rho(z, \bar{z}, \operatorname{Re} w),$$

where $d\rho = 0$. As $w = 0$ on M, we find $\rho = 0$ when $\operatorname{Re} w = 0$. That is, ρ is divisible by $\operatorname{Re} w$. So M is defined by

$$\operatorname{Im} w = (\operatorname{Re} w)\widetilde{\rho}(z, \bar{z}, \operatorname{Re} w),$$

for a smooth function $\widetilde{\rho}$. The Levi form at the origin vanishes. As $p = 0$ was an arbitrary point on $M \cap X$, the Levi form of M vanishes on $M \cap X$.

Example 6.8.2: The vanishing of the Levi form is not necessary if the complex varieties in M are of low enough dimension. Consider $M \subset \mathbb{C}^3$ with a nondegenerate (but not definite) Levi form:

$$\operatorname{Im} w = |z_1|^2 - |z_2|^2 .$$

For every $\theta \in \mathbb{R}$, M contains the complex line L_θ, given by $z_1 = e^{i\theta}z_2$ and $w = 0$. The union $\bigcup_\theta L_\theta$ of those complex lines is not contained in some single unique complex subvariety inside M. Any complex subvariety that contains all the lines L_θ must contain the entire complex hypersurface given by $w = 0$, which is not contained in M. See Figure 6.10.

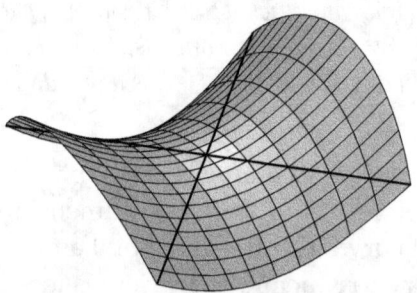

Figure 6.10: A trace of the hypersurface in the $(\operatorname{Re} z_1, \operatorname{Re} z_2, \operatorname{Im} w)$ space. The traces of the two complex lines L_0 and L_π in the plane $w = 0$ corresponding to $z_1 = z_2$ and $z_1 = -z_2$ are visible.

Exercise 6.8.1: Let $M \subset \mathbb{C}^n$ be a smooth real hypersurface. Show that if M at p contains a complex submanifold of (complex) dimension more than $\frac{n-1}{2}$, then the Levi form must be degenerate, that is, it must have at least one zero eigenvalue.

Exercise 6.8.2: Let $M \subset \mathbb{C}^n$ be a smooth pseudoconvex real hypersurface (one side of M is pseudoconvex). Suppose M at p contains a dimension k complex submanifold X. Show that the Levi form has at least k zero eigenvalues.

Exercise 6.8.3: Find an example of a smooth real hypersurface $M \subset \mathbb{C}^n$ that contains a germ of a singular complex-analytic subvariety (X, p) through a point p, which is unique in the sense that if (Y, p) is another germ of a complex analytic subvariety in M, then $(Y, p) \subset (X, p)$.

Let us discuss a tool, the *Segre variety*, that allows us to find such complex subvarieties inside M, and much more. Segre varieties only work in the real-analytic setting and rely on complexification.

Let $M \subset \mathbb{C}^n$ be a real-analytic hypersurface and $p \in M$. Suppose $M \subset U$, where $U \subset \mathbb{C}^n$ is a small domain with a defining function $r \colon U \to \mathbb{R}$ for M. That is, r is a real-analytic function in U such that $M = r^{-1}(0)$, but $dr \neq 0$ on M. Define

$$U^* = \left\{ z \in \mathbb{C}^n : \bar{z} \in U \right\}.$$

Suppose U is small enough so that the Taylor series for r converges in $U \times U^*$ when treating z and $\bar{z}$ as separate variables. That is, $r(z, \zeta)$ is a well-defined function on $U \times U^*$, and $r(z, \zeta) = 0$ defines a complexification $\mathcal{M}$ in $U \times U^*$. Assume also that U is small enough that the derivative dr of the complexified r does not vanish on $\mathcal{M}$ and that $\mathcal{M}$ is connected. See also Proposition 3.2.8.

Given $q \in U$, define the *Segre variety*

$$\Sigma_q(U, r) = \left\{ z \in U : r(z, \bar{q}) = 0 \right\} = \left\{ z \in U : (z, \bar{q}) \in \mathcal{M} \right\}.$$

See a diagram in Figure 6.11. A priori, the subvariety Σ_p depends on U and r. However, if $\tilde{r}$ is a real-analytic function that complexifies to $U \times U^*$ and vanishes on M, it must also vanish on the complexification $\mathcal{M}$. If $\tilde{r}$ is a defining function as above, that is, $d\tilde{r}$ does not vanish on its zero set and the zero set of the complexified $\tilde{r}$ is connected in $U \times U^*$, then $\tilde{r}(z, \zeta) = 0$ also defines $\mathcal{M}$. Hence the actual r does not matter. As long as $q \in M$, we have $q \in \Sigma_q(U, r)$, and furthermore the Segre variety is a complex hypersurface for every q. It is not hard to see that if $\tilde{U}$ is a small neighborhood of q, the same r is a defining function in $\tilde{U}$, and we get the same complexification in $\tilde{U} \times \tilde{U}^*$. So the germ at $q \in U$ is well-defined, and we write

$$\Sigma_q = \left(\Sigma_q(U, r), q\right).$$

The Segre variety is well-defined as a germ, and so often when one talks about Σ_q without mentioning the U or r, then one means some small enough representative of a Segre variety or the germ itself.

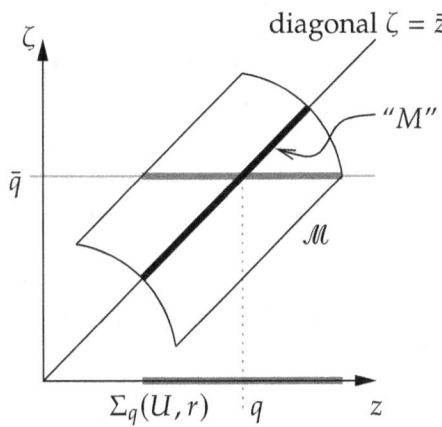

Figure 6.11: Diagram of $\Sigma_q(U, r)$. The "M" is in quotation marks as it is really in the z space not in the diagonal, but we identify it with a subset of the diagonal in this picture.

Exercise 6.8.4: Let $r \colon U \to \mathbb{R}$ be a real-valued real-analytic function that complexifies to $U \times U^*$. Show that $r(z, \bar{\zeta}) = 0$ if and only if $r(\zeta, \bar{z}) = 0$. In other words, $z \in \Sigma_\zeta(U, r)$ if and only if $\zeta \in \Sigma_z(U, r)$.

Example 6.8.3: Consider a real-analytic hypersurface M given by

$$\operatorname{Im} w = (\operatorname{Re} w)\rho(z, \bar{z}, \operatorname{Re} w),$$

with ρ vanishing at the origin. The hypersurface given by $w = 0$ lies in M. Rewrite in terms of w and $\bar{w}$,

$$\frac{w - \bar{w}}{2i} = \left(\frac{w + \bar{w}}{2}\right)\rho\left(z, \bar{z}, \frac{w + \bar{w}}{2}\right).$$

Set $\bar{z} = \bar{w} = 0$ to find
$$\frac{w}{2i} = \left(\frac{w}{2}\right) \rho\left(z, 0, \frac{w}{2}\right).$$

The function ρ vanishes at the origin, so near the origin, the equation defines the complex hypersurface given by $w = 0$. That is, Σ_0 is defined by $w = 0$. Hence, Σ_0 is precisely the complex hypersurface that lies inside M.

The last example is not a fluke. The most important property of Segre varieties is that it locates complex subvarieties in a real-analytic submanifold. We will phrase it in terms of analytic discs, which is enough as complex subvarieties can be filled with analytic discs, as we have seen.

Proposition 6.8.4. *Let $M \subset \mathbb{C}^n$ be a real-analytic hypersurface and $p \in M$. Suppose $\Delta \subset M$ is an analytic disc through p. Then as germs $(\Delta, p) \subset \Sigma_p$.*

Proof. Let U be a neighborhood of p where a representative of Σ_p is defined, that is, assume that Σ_p is a closed subset of U, and suppose $r(z, \bar{z})$ is the corresponding defining function. Let $\varphi \colon \mathbb{D} \to \mathbb{C}^n$ be the parametrization of Δ with $\varphi(0) = p$. We can restrict φ to a smaller disc around the origin, and since we are only interested in the germ of Δ at p this is sufficient (if there are multiple points of $\mathbb{D}$ that go to p, we repeat the argument for each one). So let us assume without loss of generality that $\varphi(\mathbb{D}) = \Delta \subset U$. Since $\Delta \subset M$, we have
$$r\left(\varphi(\xi), \overline{\varphi(\xi)}\right) = r\left(\varphi(\xi), \bar{\varphi}(\bar{\xi})\right) = 0.$$

The function $\xi \mapsto r\left(\varphi(\xi), \bar{\varphi}(\bar{\xi})\right)$ is a real-analytic function of ξ, and therefore for some small neighborhood of the origin, it complexifies. In fact, it complexifies to $\mathbb{D} \times \mathbb{D}$ as $\varphi(\xi) \in U$ for all $\xi \in \mathbb{D}$. So we can treat ξ and $\bar{\xi}$ as separate variables. By complexification, the equation holds for all such independent ξ and $\bar{\xi}$. Set $\bar{\xi} = 0$ to obtain
$$0 = r\left(\varphi(\xi), \bar{\varphi}(0)\right) = r\left(\varphi(\xi), \bar{p}\right) \qquad \text{for all } \xi \in \mathbb{D}.$$

In particular, $\varphi(\mathbb{D}) \subset \Sigma_p$ and the result follows. $\square$

Exercise 6.8.5: *Show that if a real-analytic real hypersurface $M \subset \mathbb{C}^n$ is strongly pseudoconvex at $p \in M$ (one side of M is strongly pseudoconvex at p), then $\Sigma_p \cap (M, p) = \{p\}$ (as germs).*

Exercise 6.8.6: *Use the proposition and the exercise above to show that if a real-analytic real hypersurface M is strongly pseudoconvex, then M contains no analytic discs.*

We end our discussion of Segre varieties by its perhaps most well-known application, the so-called Diederich–Fornæss lemma. Although we state and prove it only for real-analytic hypersurfaces, it works in greater generality. There are two parts to it, although it is generally the corollary that is called the *Diederich–Fornæss lemma*.

First, for real-analytic hypersurfaces each point has a fixed neighborhood such that germs of complex subvarieties contained in the hypersurface extend to said fixed neighborhood.

Theorem 6.8.5 (Diederich–Fornæss). *Suppose $M \subset \mathbb{C}^n$ is a real-analytic hypersurface. For every $p \in M$ there exists a neighborhood U of p with the following property: If $q \in M \cap U$ and (X, q) is a germ of a complex subvariety such that $(X, q) \subset (M, q)$, then there exists a complex subvariety $Y \subset U$ (in particular a closed subset of U) such that $Y \subset M$ and $(X, q) \subset (Y, q)$.*

Proof. Suppose U is a polydisc centered at p, small enough so that the defining function r of M complexifies to $U \times U^*$ as above. Suppose $q \in M \cap U$ is a point such that (X, q) is a germ of a positive-dimensional complex subvariety with $(X, q) \subset (M, q)$. Most points of a subvariety are regular, so without loss of generality assume q is a regular point, that is, (X, q) is a germ of a complex submanifold. Let X be a representative of the germ (X, q) such that $X \subset M$, and $X \subset U$, although we do not assume it is closed.

Assume X is an image of an open subset $V \subset \mathbb{C}^k$ via a holomorphic surjective mapping $\varphi \colon V \to X$. Since $r\big(\varphi(\xi), \overline{\varphi(\xi)}\big) = 0$ for all $\xi \in V$, then we may treat ξ and $\bar{\xi}$ separately. In particular, $r(z, \bar{\zeta}) = 0$ for all $z, \zeta \in X$.

Define complex subvarieties $Y', Y \subset U$ (closed in U) by

$$Y' = \bigcap_{a \in X} \Sigma_a(U, r) \qquad \text{and} \qquad Y = \bigcap_{a \in Y'} \Sigma_a(U, r).$$

If $a \in Y'$ and $b \in X$, then $r(a, \bar{b}) = 0$. Because r is real-valued, $r(b, \bar{a}) = 0$. Therefore, $X \subset Y \subset Y'$. Furthermore, $r(z, \bar{z}) = 0$ for all $z \in Y$, and so $Y \subset M$. $\qquad \square$

Corollary 6.8.6 (Diederich–Fornæss). *Suppose $M \subset \mathbb{C}^n$ is a compact real-analytic hypersurface. Then there does not exist any point $q \in M$ such that there exists a germ of a positive-dimensional complex subvariety (X, q) such that $(X, q) \subset (M, q)$.*

Proof. Let $S \subset M$ be the set of points through which there exists a germ of a positive-dimensional complex subvariety contained in M. As M, and hence $\bar{S}$, is compact, there must exist a point $p \in \bar{S}$ that is furthest from the origin. After a rotation by a unitary and rescaling assume $p = (1, 0, \ldots, 0)$. Let U be the neighborhood from the previous theorem around p. There exist germs of varieties in M through points arbitrarily close to p. So for any distance $\epsilon > 0$, there exists a subvariety $Y \subset U$ (in particular, Y closed in U) of positive dimension with $Y \subset M$ that contains points ϵ close to p. Consider the function $\operatorname{Re} z_1$, which attains a strict maximum on $\bar{S}$ at p. Because $\operatorname{Re} z_1$ achieves a maximum strictly smaller than 1 on $\partial U \cap \bar{S}$, for a small enough ϵ, we would obtain a pluriharmonic function with a strict maximum on Y, which is impossible by the maximum principle for varieties that you proved in Exercise 6.7.10. The picture would look as in Figure 6.12. $\qquad \square$

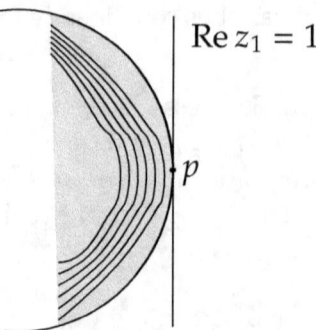

Figure 6.12: Contradicting the maximum principle at p.

Example 6.8.7: The results above do not work in the smooth setting. Let us disprove the theorem in the smooth setting. Disproving the corollary is an exercise. Let $g \colon \mathbb{R} \to \mathbb{R}$ be a smooth function that is strictly positive for $|t| > 1$, and $g(t) = 0$ for all $|t| \leq 1$. Define M in $(z, w) \in \mathbb{C}^{n-1} \times \mathbb{C}$ by

$$\operatorname{Im} w = g\left(\|z\|^2 + (\operatorname{Re} w)^2\right).$$

The M is a smooth real hypersurface. Consider $p = (0, \ldots, 0, 1) \in M$. For every $0 < s < 1$, let $q_s = (0, \ldots, 0, s) \in M$ and $X_s = \{(z, w) \in M : w = s\}$. Each X_s is the closure of a local complex subvariety of dimension $n - 1$ and $(X_s, q_s) \subset (M, q_s)$. The size (diameter) of X_s goes to zero as $s \to 1$ and X_s cannot extend to a larger complex subvariety inside M. So, no neighborhood U at p (as in the theorem) exists.

Exercise 6.8.7: Find a compact smooth real hypersurface $M \subset \mathbb{C}^n$ that contains a germ of a positive dimensional complex subvariety.

. . . and that is how using sheep's bladders can prevent earthquakes!

A Basic Notation and Terminology

We quickly review some basic notation used in this book that is perhaps not described elsewhere. We use $\mathbb{C}$, $\mathbb{R}$ for complex and real numbers, and i for imaginary unit (a square root of -1). We use $\mathbb{N} = \{1, 2, 3, \ldots\}$ for the natural numbers, $\mathbb{N}_0 = \{0, 1, 2, 3, \ldots\}$ for the zero-based natural numbers, and $\mathbb{Z}$ for all integers. When we write $\mathbb{C}^n$ or $\mathbb{R}^n$, we implicitly mean that $n \geq 1$, unless otherwise stated.

We denote set subtraction by $A \setminus B$, meaning all elements of A that are not in B. We denote complement of a set by X^c. The ambient set should be clear. So, for example, if $X \subset \mathbb{C}$ naturally, then $X^c = \mathbb{C} \setminus X$. Topological closure of a set S is denoted by $\overline{S}$, its boundary is denoted by ∂S. If S is a relatively compact subset of X (its closure in X is compact), we write $S \subset\subset X$.

We denote a function with domain X and codomain Y by $f \colon X \to Y$. The direct image of S by f is $f(S)$. We write f^{-1} for the inverse image of sets and single points. When f is bijective (one-to-one and onto), we use f^{-1} for the inverse mapping. So $f^{-1}(T)$ for a set $T \subset Y$ denotes the points of X that f maps to T. For a point q, $f^{-1}(q)$ denotes the points that map to q, but if the mapping is bijective, then it means the unique point mapping to q. To define a function without giving it a name, we use $x \mapsto F(x)$, where $F(x)$ is some formula giving the output. The notation $f|_S$ is the restriction of f to S: a function $f|_S \colon S \to Y$ such that $f|_S(x) = f(x)$ for all $x \in S$. A function $f \colon U \to \mathbb{C}$ is *compactly supported* if the *support*, that is the set $\overline{\{p \in U : f(p) \neq 0\}}$ (closure in U), is compact. If $f(x) = g(x)$ for all x in the domain, we write $f \equiv g$, and we say that f and g are identically equal. The notation $f \circ g$ denotes the composition defined by $x \mapsto f\big(g(x)\big)$.

To define X to be Y rather than just show equality, we write $X \overset{\text{def}}{=} Y$.

B Results from One Complex Variable

We review some results from one complex variable useful for reading this book. The reader should first look through section 0.1 for basic notation and motivation, although we review some of the results again here. Let $U \subset \mathbb{C}$ be open. A function $f : U \to \mathbb{C}$ is *holomorphic* if it is complex differentiable at every point, that is,

$$f'(z) = \lim_{h \in \mathbb{C} \to 0} \frac{f(z+h) - f(z)}{h}$$

exists for all $z \in U$. For example, polynomials and rational functions in z are holomorphic. Perhaps the most important holomorphic function is the solution to the differential equation $f'(z) = f(z)$, $f(0) = 1$, the complex exponential,

$$f(z) = e^z = e^{x+iy} = e^x \big(\cos y + i \sin(y) \big).$$

A piecewise-C^1 *path* (or curve) in $\mathbb{C}$ is a continuous $\gamma : [a,b] \to \mathbb{C}$, continuously differentiable except at finitely many points, such that one-sided limits of $\gamma'(t)$ exist at all $t \in [a,b]$ and such that γ' (or its one-sided limits) is never zero. By abuse of notation, when γ is used as a set, we mean the image $\gamma([a,b])$. For a continuous $f : \gamma \to \mathbb{C}$, define

$$\int_\gamma f(z)\, dz \overset{\text{def}}{=} \int_a^b f\big(\gamma(t)\big) \gamma'(t)\, dt.$$

As γ' is continuous at all but finitely many points, the integral is well-defined. Similarly, one defines the more general path integral in $dz = dx + i\, dy$ and $d\bar{z} = dx - i\, dy$. Let $z = \gamma(t) = \gamma_1(t) + i\gamma_2(t) = x + i\, y$ parametrize the path. Then

$$\int_\gamma f(z)\, dz + g(z)\, d\bar{z} = \int_\gamma \Big(f(x+iy) + g(x+iy) \Big) dx + i \Big(f(x+iy) - g(x+iy) \Big) dy$$

$$= \int_a^b \Big(f\big(\gamma(t)\big)\gamma'(t) + g\big(\gamma(t)\big)\overline{\gamma'(t)} \Big) dt$$

$$= \int_a^b \Big(\big(f\big(\gamma(t)\big) + g\big(\gamma(t)\big) \big) \gamma_1'(t) + i \big(f\big(\gamma(t)\big) - g\big(\gamma(t)\big) \big) \gamma_2'(t) \Big) dt.$$

A path is *closed* if $\gamma(a) = \gamma(b)$, and a path is *simple* if $\gamma|_{(a,b]}$ is one-to-one with the possible exception of $\gamma(a) = \gamma(b)$.

An open $U \subset \mathbb{C}$ has *piecewise-C^1 boundary* if for each $p \in \partial U$ there is an open neighborhood W of p such that $\partial U \cap W = \gamma((a, b))$ where $\gamma: [a, b] \to \mathbb{C}$ is an injective piecewise-C^1 path, and such that each $p \in \partial U$ is in the closure of $\mathbb{C} \setminus \overline{U}$. Intuitively, the boundary is locally a piecewise-C^1 curve that locally cuts the plane into two open pieces. If at each point where the parametrization of ∂U is differentiable the domain is on the left ($\gamma'(t)$ rotated by $\frac{\pi}{2}$ points into the domain), then the boundary is *positively oriented*. As in the introduction, we have the following version of Cauchy integral formula.

Theorem B.1 (Cauchy integral formula). *Let $U \subset \mathbb{C}$ be a bounded open set with piecewise-C^1 boundary ∂U oriented positively, and let $f: \overline{U} \to \mathbb{C}$ be a continuous function holomorphic in U. Then for $z \in U$,*

$$f(z) = \frac{1}{2\pi i} \int_{\partial U} \frac{f(\zeta)}{\zeta - z} \, d\zeta.$$

Usually the theorem is stated with *winding numbers*. The winding number is the number of times a closed path γ "goes around" a point $p \notin \gamma$. More precisely it is defined by

$$n(\gamma; p) \overset{\text{def}}{=} \frac{1}{2\pi i} \int_\gamma \frac{1}{z - p} \, dz.$$

It is easy to show that $n(\gamma; p)$ is always an integer and it is constant on the components of $\mathbb{C} \setminus \gamma$. It is also defined on *cycles*, which are just formal sums of closed paths $\Gamma = \gamma_1 + \gamma_2 + \cdots + \gamma_n$, by simply summing the corresponding integrals.

A common statement of the Cauchy integral formula with winding numbers is that if U is open, $f: U \to \mathbb{C}$ holomorphic, and γ is a closed piecewise C^1 path (or cycle) in U, such that $n(\gamma; p) = 0$ for all $p \notin U$, then

$$n(\gamma; z)f(z) = \frac{1}{2\pi i} \int_\gamma \frac{f(\zeta)}{\zeta - z} \, d\zeta.$$

By the *Jordan curve theorem*, a simple closed path divides the plane into two components, one bounded and one unbounded. The bounded component is called the *interior* of γ, and the unbounded component is called the *exterior*. It can be shown that for a piecewise-C^1 path γ, $n(\gamma; p) = \pm 1$ for p in the interior of γ and $n(\gamma; p) = 0$ for p in the exterior of γ. We say γ is oriented positively if $n(\gamma; p) = 1$ on the interior.

More generally, if U is a bounded open set with piecewise-C^1 boundary ∂U oriented positively, then one can show that ∂U is composed of finitely many simple closed paths oriented in such a way that $n(\partial U; p) = 1$ for $p \in U$ and $n(\partial U; p) = 0$ for $p \in \mathbb{C} \setminus \overline{U}$.

One way to get at the Cauchy integral formula is via Green's theorem, which is the Stokes' theorem in two dimensions. In the versions we state, one needs to approximate the open set by smaller open sets from the inside to ensure the partial derivatives are bounded. See Theorem 4.1.1. Let us state Green's theorem using the dz and $d\bar{z}$ for completeness. See appendix C for an overview of differential forms.

Theorem B.2 (Green's theorem). *Let $U \subset \mathbb{C}$ be a bounded open set with piecewise-C^1 boundary ∂U oriented positively, and let $f: \bar{U} \to \mathbb{C}$ and $g: \bar{U} \to \mathbb{C}$ be continuous with bounded continuous partial derivatives in U. Then*

$$\int_{\partial U} f(z)\,dz + g(z)\,d\bar{z} = \int_U d\Big(f(z)\,dz + g(z)\,d\bar{z}\Big) = \int_U \left(\frac{\partial g}{\partial z} - \frac{\partial f}{\partial \bar{z}}\right) dz \wedge d\bar{z}$$

$$= (-2i)\int_U \left(\frac{\partial g}{\partial z} - \frac{\partial f}{\partial \bar{z}}\right) dx \wedge dy = (-2i)\int_U \left(\frac{\partial g}{\partial z} - \frac{\partial f}{\partial \bar{z}}\right) dA.$$

It is sufficient to prove the conclusion of Green's with $g \equiv 0$, and it is sometimes stated that way. The Cauchy integral formula is equivalent to Cauchy's theorem:

Theorem B.3 (Cauchy). *Let $U \subset \mathbb{C}$ be a bounded open set with piecewise-C^1 boundary ∂U oriented positively, and let $f: \bar{U} \to \mathbb{C}$ be a continuous function holomorphic in U. Then*

$$\int_{\partial U} f(z)\,dz = 0.$$

Again, the alternative statement with winding numbers is that if U is open, $f: U \to \mathbb{C}$ holomorphic, and γ is a closed piecewise C^1 path (or cycle) in U, such that $n(\gamma; p) = 0$ for all $p \notin U$, then the integral of f over γ vanishes.

There is a converse to Cauchy. A triangle $T \subset \mathbb{C}$ is the convex hull of the three vertices (we include the inside of the triangle), and ∂T is the boundary of the triangle oriented counter-clockwise. We state the following theorem as an "if and only if," even though, usually it is only the reverse direction that is called Morera's theorem.

Theorem B.4 (Morera). *Suppose $U \subset \mathbb{C}$ is an open set, and $f: U \to \mathbb{C}$ is continuous. Then f is holomorphic if and only if*

$$\int_{\partial T} f(z)\,dz = 0 \qquad \text{for all triangles } T \subset U.$$

As we saw in the introduction, a holomorphic function has a power series.

Proposition B.5. *If $U \subset \mathbb{C}$ is open and $f: U \to \mathbb{C}$ is holomorphic, then f is infinitely differentiable, and if $\Delta_\rho(p) \subset \mathbb{C}$ is a disc, then f has a power series that converges absolutely uniformly on compact subsets of $\Delta_\rho(p)$:*

$$f(z) = \sum_{k=0}^{\infty} c_k(z-p)^k,$$

where given a simple closed (piecewise-C^1) path γ going once counter-clockwise around p inside $\Delta_\rho(p)$,

$$c_k = \frac{f^{(k)}(p)}{k!} = \frac{1}{2\pi i} \int_\gamma \frac{f(\zeta)}{(\zeta - p)^{k+1}}\,d\zeta.$$

Cauchy estimates follow: If M is the maximum of $|f|$ on the circle $\partial \Delta_r(p)$, then

$$|c_k| \leq \frac{M}{r^k}.$$

Conversely, if a power series satisfies such estimates, it converges on $\Delta_r(p)$.

A holomorphic $f \colon \mathbb{C} \to \mathbb{C}$ is said to be *entire*. An immediate application of Cauchy estimates is Liouville's theorem:

Theorem B.6 (Liouville). *If f is entire and bounded, then f is constant.*

And as a holomorphic function has a power series it satisfies the identity theorem:

Theorem B.7 (Identity). *Suppose $U \subset \mathbb{C}$ is a domain and $f \colon U \to \mathbb{C}$ is holomorphic. If the zero set $f^{-1}(0)$ has a limit point in U, then $f \equiv 0$.*

Another consequence of the Cauchy integral formula is that there is a differential equation characterizing holomorphic functions.

Proposition B.8 (Cauchy–Riemann equations). *Let $U \subset \mathbb{C}$ be open. A function $f \colon U \to \mathbb{C}$ is holomorphic if and only if f is continuously differentiable and*

$$\frac{\partial f}{\partial \bar{z}} = \frac{1}{2}\left(\frac{\partial f}{\partial x} + i\frac{\partial f}{\partial y}\right) = 0 \qquad \text{on } U.$$

Yet another consequence of the Cauchy formula (and one can make an argument that everything in this appendix is a consequence of the Cauchy formula) is the open mapping theorem.

Theorem B.9 (Open mapping theorem). *Suppose $U \subset \mathbb{C}$ is a domain and $f \colon U \to \mathbb{C}$ is holomorphic and not constant. Then f is an open mapping, that is, $f(V)$ is open whenever V is open.*

The real and imaginary parts u and v of a holomorphic function $f = u + iv$ are harmonic, that is $\nabla^2 u = \nabla^2 v = 0$, where ∇^2 is the Laplacian. A domain U is *simply connected* if every simple closed path is homotopic in U to a constant, in other words, if the domain has no holes. For example a disc is simply connected.

Proposition B.10. *If $U \subset \mathbb{C}$ is a simply connected domain and $u \colon U \to \mathbb{R}$ is harmonic, then there exists a harmonic function $v \colon U \to \mathbb{R}$ such that $f = u + iv$ is holomorphic.*

The function v is called the *harmonic conjugate* of u. For further review of harmonic functions see section 2.4 on harmonic functions. We have the following versions of the maximum principle.

Theorem B.11 (Maximum principles). *Suppose $U \subset \mathbb{C}$ is a domain.*

(i) *If $f \colon U \to \mathbb{C}$ is holomorphic and $|f|$ achieves a local maximum in U, then f is constant.*

(ii) *If U is bounded and $f \colon \overline{U} \to \mathbb{C}$ is holomorphic in U and continuous, then $|f|$ achieves its maximum on ∂U.*

(iii) *If $f: U \to \mathbb{R}$ is harmonic and achieves a local maximum or a minimum in U, then f is constant.*

(iv) *If U is bounded and $f: \overline{U} \to \mathbb{R}$ is harmonic in U and continuous, then f achieves its maximum and minimum on ∂U.*

The first two items are sometimes called the *maximum modulus principle*. The maximum principle immediately implies the following lemma.

Lemma B.12 (Schwarz's lemma). *Suppose $f: \mathbb{D} \to \mathbb{D}$ is holomorphic and $f(0) = 0$, then*

(i) *$|f(z)| \leq |z|$, and*

(ii) *$|f'(0)| \leq 1$.*

Furthermore, if $|f(z_0)| = |z_0|$ for some $z_0 \in \mathbb{D} \setminus \{0\}$ or $|f'(0)| = 1$, then for some $\theta \in \mathbb{R}$ we have $f(z) = e^{i\theta} z$ for all $z \in \mathbb{D}$.

The theorem above is actually quite general.

Theorem B.13 (Riemann mapping theorem). *If $U \subset \mathbb{C}$ is a nonempty simply connected domain such that $U \neq \mathbb{C}$, then U is biholomorphic to $\mathbb{D}$. Given $z_0 \in U$ there exists a unique biholomorphic $f: U \to \mathbb{D}$ such that $f(z_0) = 0$, $f'(z_0) > 0$, and f maximizes $|f'(z_0)|$ among all injective holomorphic maps to $\mathbb{D}$ such that $f(z_0) = 0$.*

Schwarz's lemma can also be used to classify the automorphisms of the disc (and hence any simply connected domain). Let $\text{Aut}(\mathbb{D})$ denote the group of biholomorphic (both f and f^{-1} are holomorphic) self maps of the disc to itself.

Proposition B.14. *If $f \in \text{Aut}(\mathbb{D})$, then there exists an $a \in \mathbb{D}$ and $\theta \in \mathbb{R}$ such that*

$$f(z) = e^{i\theta} \frac{z - a}{1 - \bar{a}z}.$$

Speaking of automorphisms. We have the following version of inverse function theorem.

Theorem B.15. *Suppose U and V are open subsets of $\mathbb{C}$.*

(i) *If $f: U \to V$ is holomorphic and bijective (one-to-one and onto), then $f'(z) \neq 0$ for all $z \in V$, and $f^{-1}: V \to U$ is holomorphic. If $f(p) = q$, then*

$$\left(f^{-1}\right)'(q) = \frac{1}{f'(p)}.$$

(ii) *If $f: U \to V$ is holomorphic, $f(p) = q$, and $f'(p) \neq 0$, then there exists a neighborhood W of q and a holomorphic function $g: W \to U$ that is one-to-one and $f\big(g(z)\big) = z$ for all $z \in W$.*

The Riemann mapping theorem actually follows from the following theorem about existence of branches of the logarithm.

Theorem B.16. *Suppose $U \subset \mathbb{C}$ is a simply connected domain, and $f \colon U \to \mathbb{C}$ is a holomorphic function without zeros in U. Then there exists a holomorphic function $L \colon U \to \mathbb{C}$ such that*

$$e^L = f.$$

In particular, we can take roots: For every $k \in \mathbb{N}$, there exists a holomorphic function $g \colon U \to \mathbb{C}$ such that

$$g^k = f.$$

In one complex variable, zeros of holomorphic functions can be divided out. Moreover, zeros of holomorphic functions are of finite order unless the function is identically zero.

Proposition B.17. *Suppose $U \subset \mathbb{C}$ is a domain and $f \colon U \to \mathbb{C}$ is holomorphic, not identically zero, and $f(p) = 0$ for some $p \in U$. There exists a $k \in \mathbb{N}$ and a holomorphic function $g \colon U \to \mathbb{C}$, such that $g(p) \neq 0$ and*

$$f(z) = (z - p)^k g(z) \qquad \text{for all } z \in U.$$

The number k above is called the *order* or *multiplicity* of the zero at p. We can use this fact and the existence of roots to show that every holomorphic function is locally like z^k. The function φ below can be thought of as a local change of coordinates.

Proposition B.18. *Suppose $U \subset \mathbb{C}$ is a domain and $f \colon U \to \mathbb{C}$ is holomorphic, not identically zero, and $p \in U$. Then there exists a $k \in \mathbb{N}$, a neighborhood $V \subset U$ of p, and a holomorphic function $\varphi \colon V \to \mathbb{C}$ with $\varphi'(p) \neq 0$, such that*

$$\bigl(\varphi(z)\bigr)^k = f(z) - f(p) \qquad \text{for all } z \in V.$$

Convergence of holomorphic functions is the same as for continuous functions: uniform convergence on compact subsets. Sometimes this is called *normal convergence*.

Proposition B.19. *Suppose $U \subset \mathbb{C}$ is open and $f_k \colon U \to \mathbb{C}$ is a sequence of holomorphic functions which converge uniformly on compact subsets of U to $f \colon U \to \mathbb{C}$. Then f is holomorphic, and every derivative $f_k^{(\ell)}$ converges uniformly on compact subsets to the derivative $f^{(\ell)}$.*

Holomorphic functions satisfy a Heine–Borel-like property:

Theorem B.20 (Montel). *Suppose $U \subset \mathbb{C}$ is open and $f_n \subset U \to \mathbb{C}$ is a sequence of holomorphic functions. If $\{f_n\}$ is uniformly bounded on compact subsets of U, then there exists a subsequence converging uniformly on compact subsets of U.*

A sequence of holomorphic functions cannot create or delete zeros out of thin air:

Theorem B.21 (Hurwitz). *Suppose $U \subset \mathbb{C}$ is a domain and $f_n \subset U \to \mathbb{C}$ is a sequence of holomorphic functions converging uniformly on compact subsets of U to $f : U \to \mathbb{C}$. If f is not identically zero and z_0 is a zero of f, then there exists a disc $\Delta_r(z_0)$ and an N, such that for all $n \geq N$, f_n has the same number of zeros (counting multiplicity) in $\Delta_r(z_0)$ as f (counting multiplicity).*

A common application, and sometimes the way the theorem is stated, is that if f_n have no zeros in U, then either the limit f is identically zero, or it also has no zeros.

If $U \subset \mathbb{C}$ is open, $p \in U$, and $f : U \setminus \{p\} \to \mathbb{C}$ is holomorphic, we say that f has an *isolated singularity* at p. An isolated singularity is *removable* if there exists a holomorphic function $F : U \to \mathbb{C}$ such that $f(z) = F(z)$ for all $z \in U \setminus \{p\}$. An isolated singularity is a *pole* if

$$\lim_{z \to p} f(z) = \infty \qquad \text{(that is, } |f(z)| \to \infty \text{ as } |z - p| \to 0).$$

An isolated singularity that is neither removable nor a pole is said to be *essential*.

At nonessential isolated singularities the function blows up to a finite integral order. The first part of the following proposition is usually called the *Riemann extension theorem*.

Proposition B.22. *Suppose $U \subset \mathbb{C}$ is an open set, $p \in U$, and $f : U \setminus \{p\} \to \mathbb{C}$ holomorphic.*

(i) *If f is bounded (near p is enough), then p is a removable singularity.*

(ii) *If p is a pole, there exists a $k \in \mathbb{N}$ such that*

$$g(z) = (z - p)^k f(z)$$

is bounded near p and hence g has a removable singularity at p.

The number k above is called the *order* of the pole. There is a symmetry between zeros and poles: If f has a zero of order k, then $\frac{1}{f}$ has a pole of order k. If f has a pole of order k, then $\frac{1}{f}$ has a removable singularity, and the extended function has a zero of order k.

Let $\mathbb{P}^1 = \mathbb{C} \cup \{\infty\}$ be the *Riemann sphere*. The topology on $\mathbb{P}^1$ is given by insisting that the function $\frac{1}{z}$ is a homeomorphism of $\mathbb{P}^1$ to itself, where $\frac{1}{\infty} = 0$ and $\frac{1}{0} = \infty$. A function $f : U \to \mathbb{P}^1$ is called *meromorphic*, if it is not identically ∞, is holomorphic on $U \setminus f^{-1}(\infty)$, and has poles at $f^{-1}(\infty)$. A holomorphic function with poles is meromorphic by setting the value to be ∞ at the poles. A meromorphic function is one that can locally be written as a quotient of holomorphic functions.

At an isolated singularity we can expand a holomorphic function via the so-called *Laurent series* by adding all negative powers. The Laurent series also characterizes the type of the singularity.

Proposition B.23. *If $\Delta \subset \mathbb{C}$ is a disc centered at $p \in \mathbb{C}$, and $f \colon \Delta \setminus \{p\} \to \mathbb{C}$ holomorphic, then there exists a double sequence $\{c_k\}_{k=-\infty}^{\infty}$ such that*

$$f(z) = \sum_{k=-\infty}^{\infty} c_k(z-p)^k,$$

converges absolutely uniformly on compact subsets of Δ. If γ is a simple closed piecewise-C^1 path going once counter-clockwise around p in Δ, then

$$c_k = \frac{1}{2\pi i} \int_{\gamma} \frac{f(\zeta)}{(\zeta - z)^{k+1}} \, d\zeta.$$

The singularity at p is

   (i) *removable if $c_k = 0$ for all $k < 0$.*

   (ii) *pole of order $\ell \in \mathbb{N}$ if $c_k = 0$ for all $k < -\ell$ and $c_{-\ell} \neq 0$.*

   (iii) *essential if there exist infinitely many negative k such that $c_k \neq 0$.*

Suppose that f has an isolated singularity at p. The part of the series for negative k, that is, $\sum_{k=-\infty}^{-1} c_k(z-p)^k$, is called the *principal part*. The singularity is removable if the principal part is zero, it is a pole if the principal part is a finite sum, and essential if the principal part is an infinite series.

We call the c_{-1} corresponding to p the *residue* of f at p, and write it as $\mathrm{Res}(f, p)$. The proposition says that for a small γ around p in the positive direction,

$$\mathrm{Res}(f, p) = c_{-1} = \frac{1}{2\pi i} \int_{\gamma} f(z) \, dz.$$

Combining this equation with Cauchy's theorem tells us that to compute integrals of functions with isolated singularities we simply need to find the residues, which tend to be simpler to compute. For example, if p is a simple pole (of order 1), then

$$\mathrm{Res}(f, p) = \lim_{z \to p} (z - p) f(z).$$

Theorem B.24 (Residue theorem). *Suppose $U \subset \mathbb{C}$ is an open set, and γ is a piecewise-C^1 closed path in U such that $n(\gamma; p) = 0$ for all $p \notin U$. Suppose that $f \colon U \setminus S \to \mathbb{C}$ is a holomorphic function with isolated singularities in a finite set S, and suppose S lies in the interior of γ. Then*

$$\int_{\gamma} f(z) \, dz = 2\pi i \sum_{p \in S} n(\gamma; p) \, \mathrm{Res}(f, p).$$

If γ is a simple closed curve positively oriented, then $n(\gamma; p) = 1$ for all p in its interior, and we can replace the hypothesis on γ by requiring that the interior of γ lies in U and hence avoid mentioning winding numbers.

The identity theorem says that zeros of a nonconstant holomorphic f have no limit points; they are isolated points. Since $\frac{1}{f}$ is a meromorphic function with zeros at the poles of f, poles are also isolated. Zeros and poles can be counted fairly easily.

Theorem B.25 (Argument principle). *Suppose $U \subset \mathbb{C}$ is an open set, and γ is a piecewise-C^1 closed path in U such that $n(\gamma; p) = 0$ for all $p \notin U$ and $n(\gamma; p) \in \{0, 1\}$ for all $p \notin \gamma$. Suppose that $f \colon U \to \mathbb{P}^1$ is a meromorphic function with no zeros or poles on γ. Then*

$$\frac{1}{2\pi i} \int_\gamma \frac{f'(z)}{f(z)}\, dz = N - P,$$

where N is the number of zeros of f inside γ and P is the number of poles inside γ, both counted with multiplicity.

Furthermore, suppose $h \colon U \to \mathbb{C}$ is holomorphic. Let $z_1, \ldots, z_N$ be the zeros of f inside γ and $w_1, \ldots, w_P$ be the poles of f inside γ. Then

$$\frac{1}{2\pi i} \int_\gamma h(z)\frac{f'(z)}{f(z)}\, dz = \sum_{k=1}^N h(z_k) \quad - \quad \sum_{k=1}^P h(w_k).$$

Again, if γ is simple closed and positively oriented, the hypothesis on γ could be replaced with the requirement that the interior of γ lies in U. The proof is an immediate application of the residue theorem. Simply compute the residues at the zeros and poles of f. In particular, if f has a zero at p of multiplicity m, then $h(z)\frac{f'(z)}{f(z)}$ has a simple pole at p with residue $m\, h(p)$. Similarly, if f has a pole at p of order m, then $h(z)\frac{f'(z)}{f(z)}$ has a simple pole with residue $-m\, h(p)$ at p.

Another useful way to count zeros is Rouché's theorem.

Theorem B.26 (Rouché). *Suppose $U \subset \mathbb{C}$ is an open set, and γ is a piecewise-C^1 closed path in U such that $n(\gamma; p) = 0$ for all $p \notin U$ and $n(\gamma; p) \in \{0, 1\}$ for all $p \notin \gamma$. Suppose that $f \colon U \to \mathbb{C}$ and $g \colon U \to \mathbb{C}$ are holomorphic functions such that*

$$|f(z) - g(z)| < |f(z)| + |g(z)|$$

for all $z \in \gamma$. Then f and g have the same number of zeros inside γ (up to multiplicity).

In the classical (weaker) statement of the theorem uses the the stronger inequality $|f(z) - g(z)| < |f(z)|$. Notice that either inequality precludes any zeros on γ itself.

A holomorphic function with an essential singularity achieves essentially every value. A weak version of this result (and an easy to prove one) is the *Casorati–Weierstrass theorem*: If a holomorphic f has an essential singularity at p, then for every neighborhood W of p, $f(W \setminus \{p\})$ is dense in $\mathbb{C}$. Let us state the much stronger theorem of Picard: A function with an essential singularity is very wild. It achieves every value (except possibly one) infinitely often.

Theorem B.27 (Picard's big theorem). *Suppose $U \subset \mathbb{C}$ is open, $f \colon U \setminus \{p\} \to \mathbb{C}$ is holomorphic, and f has an essential singularity at p. Then for every neighborhood W of p, $f(W \setminus \{p\})$ is either $\mathbb{C}$ or $\mathbb{C}$ minus a point.*

For example, $e^{1/z}$ has an essential singularity at the origin and the function is never 0. Since we stated the big theorem, let us also state the little theorem.

Theorem B.28 (Picard's little theorem). *If $f : \mathbb{C} \to \mathbb{C}$ is holomorphic and nonconstant, then $f(\mathbb{C})$ is either $\mathbb{C}$ or $\mathbb{C}$ minus a point.*

One theorem from algebra that is important in complex analysis, and becomes perhaps even more important in several variables is the fundamental theorem of algebra. It really is a theorem of complex analysis and its standard proof is via the maximum principle.

Theorem B.29 (Fundamental theorem of algebra). *If $P : \mathbb{C} \to \mathbb{C}$ is a nonzero polynomial of degree k, then P has exactly k zeros (roots) in $\mathbb{C}$ counted with multiplicity.*

The set of rational functions is dense in the space of holomorphic functions, and we even have control over where the poles need to be. Note that a nonconstant polynomial has a "pole at infinity" meaning $P(z) \to \infty$ as $z \to \infty$. Letting $\mathbb{P}^1$ again be the Riemann sphere, we have Runge's approximation theorem.

Theorem B.30 (Runge). *Suppose $U \subset \mathbb{C}$ is an open set and $A \subset \mathbb{P}^1 \setminus U$ is a set containing at least one point from each component of $\mathbb{P}^1 \setminus U$. Suppose $f : U \to \mathbb{C}$ is holomorphic. Then for every $\epsilon > 0$ and every compact $K \subset\subset U$, there exists a rational function R with poles in A such that*

$$|R(z) - f(z)| < \epsilon \qquad \text{for all } z \in K.$$

Perhaps a surprising generalization of the classical Weierstrass approximation theorem, and one of my favorite one-variable theorems, is Mergelyan's theorem. It may be good to note that Mergelyan does not follow from Runge.

Theorem B.31 (Mergelyan). *Suppose $K \subset\subset \mathbb{C}$ is a compact set such that $\mathbb{C} \setminus K$ is connected and $f : K \to \mathbb{C}$ is a continuous function that is holomorphic in the interior K°. Then for every $\epsilon > 0$, there exists a polynomial P such that*

$$|P(z) - f(z)| < \epsilon \qquad \text{for all } z \in K.$$

The reason why the theorem is perhaps surprising is that K may have only a small or no interior. Using a closed interval $K = [a, b]$ of the real line we recover the Weierstrass approximation theorem.

Given an open set $U \subset \mathbb{C}$, we say U is *symmetric with respect to the real axis* if $z \in U$ implies $\bar{z} \in U$. We divide U into three parts

$$U_+ = \{z \in U : \text{Im } z > 0\}, \qquad U_0 = \{z \in U : \text{Im } z = 0\}, \qquad U_- = \{z \in U : \text{Im } z < 0\}.$$

We have the following theorem for extending (reflecting) holomorphic functions past boundaries.

Theorem B.32 (Schwarz reflection principle). *Suppose $U \subset \mathbb{C}$ is a domain symmetric with respect to the real axis, $f : U_+ \cup U_0 \to \mathbb{C}$ a continuous function holomorphic on U_+ and real-valued on U_0. Then the function $g : U \to \mathbb{C}$ defined by*

$$g(z) = f(z) \quad \text{if } z \in U_+ \cup U_0, \qquad g(z) = \overline{f(\bar{z})} \quad \text{if } z \in U_-,$$

is holomorphic on U.

In fact, the reflection is really about harmonic functions.

Theorem B.33 (Schwarz reflection principle for harmonic functions). *Suppose $U \subset \mathbb{C}$ is a domain symmetric with respect to the real axis, $f \colon U_+ \cup U_0 \to \mathbb{R}$ a continuous function harmonic on U_+ and zero on U_0. Then the function $g \colon U \to \mathbb{R}$ defined by*

$$g(z) = f(z) \quad \text{if } z \in U_+ \cup U_0, \qquad g(z) = -f(\bar{z}) \quad \text{if } z \in U_-,$$

is harmonic on U.

Functions may be defined locally, and continued along paths. Suppose p is a point and D is a disc centered at $p \in D$. A holomorphic function $f \colon D \to \mathbb{C}$ can be *analytically continued* along a path $\gamma \colon [0,1] \to \mathbb{C}$, $\gamma(0) = p$, if for every $t \in [0,1]$ there exists a disc D_t centered at $\gamma(t)$, where $D_0 = D$, and a holomorphic function $f_t \colon D_t \to \mathbb{C}$, where $f_0 = f$, and for each $t_0 \in [0,1]$ there is an $\epsilon > 0$ such that if $|t - t_0| < \epsilon$, then $f_t = f_{t_0}$ in $D_t \cap D_{t_0}$. The monodromy theorem says that as long as there are no holes, analytic continuation defines a function uniquely.

Theorem B.34 (Monodromy theorem). *If $U \subset \mathbb{C}$ is a simply connected domain, $D \subset U$ a disc and $f \colon D \to \mathbb{C}$ a holomorphic function that can be analytically continued from $p \in D$ to every $q \in U$ along every path from p to q, then there exists a unique holomorphic function $F \colon U \to \mathbb{C}$ such that $F|_D = f$.*

An interesting and useful theorem getting an inequality in the opposite direction from Schwarz's lemma, and one which is often not covered in a one-variable course is the Koebe $\frac{1}{4}$-theorem. Think of why no such theorem could possibly hold for just smooth functions. At first glance the theorem should seem quite counterintuitive, and at second glance, it should seem outright outrageous.

Theorem B.35 (Koebe quarter theorem). *Suppose $f \colon \mathbb{D} \to \mathbb{C}$ is holomorphic and injective. Then $f(\mathbb{D})$ contains a disc of radius $\frac{|f'(0)|}{4}$ centered at $f(0)$.*

The $\frac{1}{4}$ is sharp, that is, it is the best it can be.

Finally, it is useful to factor out all the zeros of a holomorphic function, not just finitely many. Similarly, we can work with poles.

Theorem B.36 (Weierstrass product theorem). *Suppose $U \subset \mathbb{C}$ is a domain, $\{a_k\}$, $\{b_k\}$ are countable sets in U with no limit points in U, and $\{n_k\}$, $\{m_k\}$ countable sets of natural numbers. Then there exists a meromorphic function f of U whose zeros are exactly at a_k, with orders given by n_k, and poles are exactly at b_k, with orders given by m_k.*

For a more explicit statement, we need infinite products. The product $\prod_{k=1}^{\infty}(1+a_k)$ *converges* if the sequence of partial products $\prod_{k=1}^{n}(1 + a_k)$ converges. We say that the product *converges absolutely* if $\prod_{k=1}^{\infty}(1 + |a_k|)$ converges, which is equivalent to $\sum_{k=1}^{\infty}|a_k|$ converging.

Define

$$E_0(z) = (1 - z), \qquad E_m(z) = (1 - z) \exp\left(z + \frac{z^2}{2} + \cdots + \frac{z^m}{m}\right).$$

The function $E_m(z/a)$ has a zero of order 1 at a.

Theorem B.37 (Weierstrass factorization theorem). *Let f be an entire function with zeros (repeated according to multiplicity) at points of the sequence $\{a_k\}$ except the zero at the origin, whose order is m (possibly $m = 0$). Then there exists an entire function g and a sequence $\{p_k\}$ such that*

$$f(z) = z^m e^{g(z)} \prod_{k=1}^{\infty} E_{p_k}\left(\frac{z}{a_k}\right),$$

converges uniformly absolutely on compact subsets.

The p_k are chosen such that

$$\sum_{k=1}^{\infty} \left|\frac{r}{a_k}\right|^{1+p_k}$$

converges for all $r > 0$.

A companion to the Weierstrass product theorem, which says that you can prescribe zeros, is the Mittag-Leffler theorem, which says that you can prescribe principal parts of poles.

Theorem B.38 (Mittag-Leffler). *Suppose $U \subset \mathbb{C}$ is open, $S \subset U$ is a countable set with no limit point in U, and for every $p \in S$ there is a principal part*

$$P_p(z) = \sum_{n=1}^{k_p} \frac{c_{p,n}}{(z - p)^n}$$

of a pole of order k_p. Then there exists a meromorphic function f in U with poles precisely at points of S, and for each $p \in S$, the principal part of f at p is P_p.

* * *

There are many other useful theorems in one complex variable, and we could spend a lot of time listing them all. However, hopefully the listing above is useful as a refresher for the reader of the most common results, some of which are used in this book, some of which are useful in the exercises, and some of which are just too interesting not to mention.

C Differential Forms and Stokes' Theorem

Differential forms come up quite a bit in this book, especially in chapter 4 and chapter 5. Let us overview their definition and state the general Stokes' theorem. No proofs are given; this appendix is just a bare-bones guide. For a more complete introduction to differential forms, see Rudin [R1].

The short story about differential forms is that a k-form is an object that can be integrated (summed) over a k-dimensional object, taking orientation into account. For simplicity, as in most of this book, everything in this appendix is stated for smooth (C^∞) objects to avoid worrying about how much regularity is needed.

The main point of differential forms is to find the proper context for the Fundamental theorem of calculus,

$$\int_a^b f'(x)\, dx = f(b) - f(a).$$

We interpret both sides as integration. The left-hand side is an integral of the 1-form $f'\, dx$ over the 1-dimensional interval $[a,b]$ and the right-hand side is an integral of the 0-form (a function) f over the 0-dimensional (two-point) set $\{a,b\}$. Both sides consider orientation, $[a,b]$ is integrated from a to b, $\{a\}$ is oriented negatively and $\{b\}$ is oriented positively. The two-point set $\{a,b\}$ is the boundary of $[a,b]$, and the orientation of $\{a,b\}$ is induced by the orientation of $[a,b]$.

Let us define the objects over which we integrate, that is, smooth submanifolds of $\mathbb{R}^n$. Our model for a k-dimensional submanifold-with-boundary is the upper-half-space and its boundary:

$$\mathbb{H}^k \overset{\text{def}}{=} \{x \in \mathbb{R}^k : x_k \geq 0\}, \qquad \partial\mathbb{H}^k \overset{\text{def}}{=} \{x \in \mathbb{R}^k : x_k = 0\},$$

Definition C.1. Let $M \subset \mathbb{R}^n$ have the induced subspace topology. Let $k \in \mathbb{N}_0$. Let M have the property that for each $p \in M$, there exists a neighborhood $W \subset \mathbb{R}^n$ of p, a point $q \in \mathbb{H}^k$, a neighborhood $U \subset \mathbb{H}^k$ of q, and a smooth one-to-one open* mapping $\varphi : U \to M$ such that $\varphi(q) = p$, the derivative $D\varphi$ has rank k at all points,

---

*By open, we mean that $\varphi(V)$ is a relatively open set of M for every open set $V \subset U$.

and $\varphi(U) = M \cap W$. Then M is an *embedded submanifold-with-boundary* of dimension k. The map φ is called a *local parametrization*. If q is such that $q_k = 0$ (the last component is zero), then $p = \varphi(q)$ is a *boundary point*. Let ∂M denote the set of boundary points. If $\partial M = \emptyset$, then we say M is simply an *embedded submanifold*.

The situation for a boundary point and an interior point is depicted in Figure C.1. Note that W is a bigger neighborhood in $\mathbb{R}^n$ than the image $\varphi(U)$.

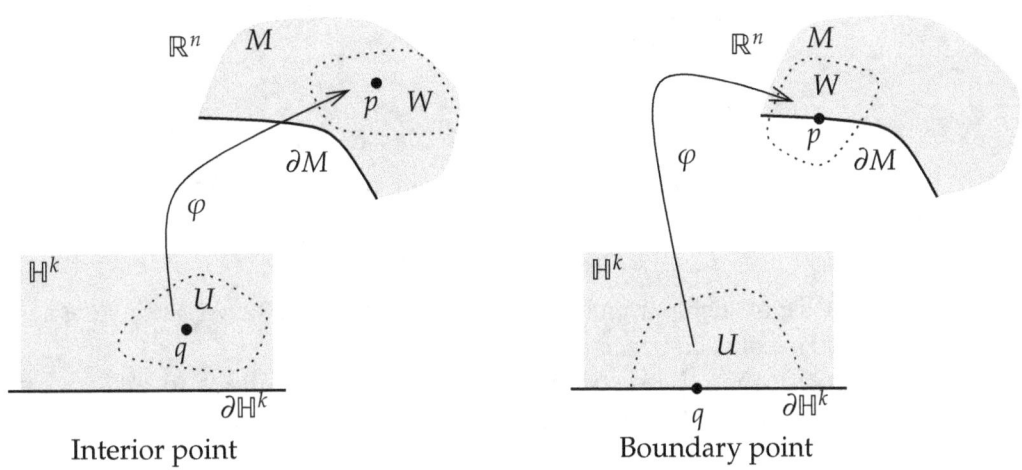

Figure C.1: Parametrization at an interior and a boundary point of a submanifold.

Completely correctly, we should say *submanifold of* $\mathbb{R}^n$. Sometimes people (including me) say *manifold* when they mean *submanifold*. A manifold is a more abstract concept, but all submanifolds are manifolds. The word *embedded* has to do with the topology on M, and this has to do with the condition $\varphi(U) = M \cap W$ and φ being open. The condition means that φ is a homeomorphism onto $M \cap W$. It is important that W is an open set in $\mathbb{R}^n$. For our purposes here, all submanifolds will be embedded. We have also made some economy in the definition. If q is not on the boundary of $\mathbb{H}^k$, then we might as well have used $\mathbb{R}^k$ instead of $\mathbb{H}^k$. A submanifold is something that is locally like $\mathbb{R}^k$, and if it has a boundary, then near the boundary it is locally like $\mathbb{H}^k$ near a point of $\partial \mathbb{H}^k$.

We also remark that submanifolds are often defined in reverse rather than by parametrizations, that is, by starting with the (relatively) open sets $M \cap W$, and the maps φ^{-1}, calling those *charts*, and calling the entire set of charts an *atlas*. The particular version of the definition we have chosen makes it easy to evaluate integrals in the same way that parametrizing curves makes it easy to evaluate integrals.

Examples of such submanifolds are domains with smooth boundaries as in Definition 2.2.1; we can take the inclusion map $x \mapsto x$ as our parametrization. The domain is then the submanifold M and ∂M is the boundary of the domain. Domains are the key application for our purposes. Another example is a smooth curve.

If M is an embedded submanifold-with-boundary of dimension k, then ∂M is also an embedded submanifold of dimension $k - 1$. Simply restrict the parametrizations to the boundary of $\mathbb{H}^k$.

We also need to define an orientation.

Definition C.2. Let $M \subset \mathbb{R}^n$ be an embedded submanifold-with-boundary of dimension $k \geq 2$. Suppose a set of parametrizations can be chosen such that each point of M is in the image of one of the parametrizations, and if $\varphi \colon U \to M$ and $\widetilde{\varphi} \colon \widetilde{U} \to M$ are two parametrizations such that $\varphi(U) \cap \widetilde{\varphi}(\widetilde{U}) \neq \emptyset$, then the *transition map* (automatically smooth) defined by

$$\widetilde{\varphi}^{\,-1} \circ \varphi$$

on $\varphi^{-1}\big(\varphi(U) \cap \widetilde{\varphi}(\widetilde{U})\big)$ (in other words, wherever it makes sense) is orientation preserving, that is,

$$\det D\big(\widetilde{\varphi}^{\,-1} \circ \varphi\big) > 0$$

at all points. The set of such parametrizations is the *orientation* on M, and we usually take the maximal set of such parametrizations.

If M is oriented, then the restrictions of the parametrizations to $\partial \mathbb{H}^k$ give an orientation on ∂M. We say this is the *induced orientation* on ∂M.

For dimensions $k = 0$ (isolated points) and $k = 1$ (curves) we must define orientation differently. For $k = 0$, we simply give each point an orientation of $+$ or $-$. For $k = 1$, we need to allow parametrization by open subsets not only of $\mathbb{H}^1 = [0, \infty)$, but also $-\mathbb{H}^1 = (-\infty, 0]$. The definition is the same otherwise. To define the orientation of the boundary, if the boundary point corresponds to the 0 in $[0, \infty)$ we give this boundary point the orientation $-$, and if it corresponds to the 0 in $(-\infty, 0]$, then we give this point the orientation $+$. The reason for this complication is that unlike in $\mathbb{R}^k$ for $k \geq 2$, the set $\mathbb{H}^1 = [0, \infty)$ cannot be "rotated" (in $\mathbb{R}^1$) or mapped via an orientation preserving map onto $-\mathbb{H}^1 = (-\infty, 0]$, but in $\mathbb{R}^2$ the upper-half-plane $\mathbb{H}^2$ can be rotated to the lower-half-plane $-\mathbb{H}^2 = \{x \in \mathbb{R}^2 : x_2 \leq 0\}$. For computations, it is often useful for compact curves with endpoints (boundary) to just give one parametrization from $[0, 1]$ or perhaps $[a, b]$, then a corresponds to the $-$ and b corresponds to the $+$.

The fact that the transition map is smooth does require a proof, which is a good exercise in basic analysis. It requires a bit of care at boundary points.

An orientation allows us to have a well-defined integral on M, just like a curve needs to be oriented in order to define a line integral. However, unlike for curves, not every submanifold of dimension more than one is *orientable*, that is, admits an orientation. A classical nonorientable example is the Möbius strip.

Now that we know what we integrate "on", let us figure out what "it" is that we integrate. Let us start with 0-forms. We define 0-forms as smooth functions (possibly complex-valued). Sometimes we need a function defined just on a submanifold. A function f defined on a submanifold M is smooth when $f \circ \varphi$ is smooth on U for every parametrization $\varphi \colon U \to M$. Equivalently, one can prove that f is the restriction of some smooth function defined on some neighborhood of M in $\mathbb{R}^n$.

A 0-form ω defined on a 0-dimensional oriented submanifold M is integrated as

$$\int_M \omega \overset{\text{def}}{=} \sum_{p \in M} \epsilon_p \omega(p),$$

where ϵ_p is the orientation of p given as $+1$ or -1. To avoid problems of integrability, one can assume that ω is compactly supported (it is nonzero on at most finitely many points of M) or that M is compact (it is a finite set).

The correct definition of a 1-form is that it is a "smooth section" of the dual of the vector bundle $T\mathbb{R}^n$. It is something that eats a vector field, and spits out a function. We use the pairing notation $\langle \omega, v \rangle$ instead of the functional notation $\omega(v)$ to indicate linearity in v (and in ω). The 1-form dx_k is supposed to be the object that does

$$\left\langle dx_k, \frac{\partial}{\partial x_k} \right\rangle = 1, \qquad \left\langle dx_k, \frac{\partial}{\partial x_\ell} \right\rangle = 0 \quad \text{if } \ell \neq k.$$

For our purposes here, just suppose that a 1-form in $\mathbb{R}^n$ is an object of the form

$$\omega = g_1 dx_1 + g_2 dx_2 + \cdots + g_n dx_n,$$

where $g_1, g_2, \ldots, g_n$ are smooth functions. That is, a 1-form is at each point a linear combination of $dx_1, dx_2, \ldots, dx_n$ that varies smoothly from point to point. Suppose M is a one-dimensional submanifold (possibly with boundary), $\varphi \colon U \to M$ is a parametrization compatible with the orientation of M, and g_k is supported in $\varphi(U)$. Define

$$\int_M \omega \overset{\text{def}}{=} \sum_{k=1}^{n} \int_U g_k\big(\varphi(t)\big) \varphi'_k(t) \, dt,$$

where the integral $\int_U \cdots dt$ is evaluated with the usual positive orientation (left to right) as $U \subset \mathbb{R}$, and φ_k is the kth component of φ.

Generally, a 1-form has support bigger than just $\varphi(U)$. In this case, one needs to use a so-called partition of unity to write ω as a locally finite sum

$$\omega = \sum_{\iota} \omega_{\iota},$$

where each ω_{ι} has support in the image of a single parametrization. By locally finite, we mean that on each compact neighborhood only finitely many ω_{ι} are nonzero. Define

$$\int_M \omega \overset{\text{def}}{=} \sum_{\iota} \int_M \omega_{\iota}.$$

The definition makes sense only if this sum actually exists. For example, if ω is compactly supported, then this sum is only finite, and so it exists.

Higher degree forms are constructed out of 1-forms and 0-forms by the so-called wedge product. Given a k-form ω and an ℓ-form η,

$$\omega \wedge \eta$$

is a $(k + \ell)$-form. We require the wedge product to be bilinear at each point: If f and g are smooth functions, then

$$(f\omega + g\eta) \wedge \xi = f(\omega \wedge \xi) + g(\eta \wedge \xi), \qquad \text{and} \qquad \omega \wedge (f\eta + g\xi) = f(\omega \wedge \eta) + g(\omega \wedge \xi).$$

The wedge product is not commutative; we require it to be anticommutative on 1-forms. If ω and η are 1-forms, then

$$\omega \wedge \eta = -\eta \wedge \omega.$$

The negative keeps track of orientation. When ω is a k-form and η is an m-form,

$$\omega \wedge \eta = (-1)^{km} \eta \wedge \omega.$$

We wedge together the basis 1-forms to get all k-forms. A k-form is then an expression

$$\omega = \sum_{\ell_1=1}^{n} \sum_{\ell_2=1}^{n} \cdots \sum_{\ell_k=1}^{n} g_{\ell_1,\ldots,\ell_k} \, dx_{\ell_1} \wedge dx_{\ell_2} \wedge \cdots \wedge dx_{\ell_k},$$

where $g_{\ell_1,\ldots,\ell_k}$ are smooth functions. We can simplify even more. Since the wedge is anticommutative on 1-forms,

$$dx_\ell \wedge dx_m = -dx_m \wedge dx_\ell, \qquad \text{and} \qquad dx_\ell \wedge dx_\ell = 0.$$

In other words, every form $dx_{\ell_1} \wedge dx_{\ell_2} \wedge \cdots \wedge dx_{\ell_k}$ is either zero, if any two indices from $\ell_1, \ldots, \ell_k$ are equal, or can be put into the form $\pm dx_{\ell_1} \wedge dx_{\ell_2} \wedge \cdots \wedge dx_{\ell_k}$, where $\ell_1 < \ell_2 < \cdots < \ell_k$. Thus, a k-form can always be written as

$$\omega = \sum_{1 \leq \ell_1 < \ell_2 < \cdots < \ell_k \leq n} g_{\ell_1,\ldots,\ell_k} \, dx_{\ell_1} \wedge dx_{\ell_2} \wedge \cdots \wedge dx_{\ell_k}.$$

In general, just like 1-forms are linear functionals of vector fields, k-forms are alternating multilinear functions of k vector fields. To simplify matters, let us note how k vectors are plugged into $dx_{\ell_1} \wedge dx_{\ell_2} \wedge \cdots \wedge dx_{\ell_k}$. Consider vector fields $X_1, \ldots, X_k$ given by $X_j = \sum_{m=1}^{n} c_{mj} \frac{\partial}{\partial x_m}$. As k-forms are alternating and multilinear, instead of plugging in the k-tuple $(X_1, \ldots, X_k)$, we write it as taking the wedge product $X_1 \wedge \cdots \wedge X_k$, where the wedge has the same properties as for forms. Then

$$\left\langle dx_{\ell_1} \wedge dx_{\ell_2} \wedge \cdots \wedge dx_{\ell_k}, X_1 \wedge \cdots \wedge X_k \right\rangle = \det \left(\begin{bmatrix} c_{\ell_1 1} & c_{\ell_1 2} & \cdots & c_{\ell_1 k} \\ c_{\ell_2 1} & c_{\ell_2 2} & \cdots & c_{\ell_2 k} \\ \vdots & \vdots & \ddots & \vdots \\ c_{\ell_k 1} & c_{\ell_k 2} & \cdots & c_{\ell_k k} \end{bmatrix} \right).$$

That is, each dx_{ℓ_j} picks out the ℓ_jth row out of the matrix of all coefficients and we take the determinant. Here is an explicit example for $k = 2$:

$$\left\langle dx_1 \wedge dx_2, \left(a\frac{\partial}{\partial x_1} + b\frac{\partial}{\partial x_2}\right) \wedge \left(c\frac{\partial}{\partial x_1} + d\frac{\partial}{\partial x_2}\right)\right\rangle = ad - bc.$$

Consider an oriented k-dimensional submanifold M (possibly with boundary), a parametrization $\varphi\colon U \to M$ from the orientation, and a k-form ω supported in $\varphi(U)$ (that is each $g_{\ell_1,\dots,\ell_k}$ is supported in $\varphi(U)$). Denote by $t \in U \subset \mathbb{R}^k$ the coordinates on U. Define

$$\int_M \omega \overset{\text{def}}{=} \sum_{1 \le \ell_1 < \ell_2 < \cdots < \ell_k \le n} \int_U g_{\ell_1,\dots,\ell_k}\big(\varphi(t)\big) \det D(\varphi_{\ell_1}, \varphi_{\ell_2}, \dots, \varphi_{\ell_k})\, dt$$

where the integral $\int_U \cdots dt$ is evaluated in the usual orientation on $\mathbb{R}^k$ with dt the standard measure on $\mathbb{R}^k$ (think $dt = dt_1 dt_2 \cdots dt_n$), and $D(\varphi_{\ell_1}, \varphi_{\ell_2}, \dots, \varphi_{\ell_k})$ denotes the derivative of the mapping whose mth component is φ_{ℓ_m}.

Similarly as before, if ω is not supported in the image of a single parametrization, then using a partition of unity, write

$$\omega = \sum_{\ell} \omega_\ell$$

as a locally finite sum, where each ω_ℓ has support in the image of a single parametrization of the orientation. Then

$$\int_M \omega \overset{\text{def}}{=} \sum_{\ell} \int_M \omega_\ell.$$

Again, the sum has to exist, such as when ω is compactly supported and the sum is finite.

The only nontrivial differential forms on $\mathbb{R}^n$ are $0, 1, 2, \dots, n$ forms. The only n-forms are of the form

$$f(x)\, dx_1 \wedge dx_2 \wedge \cdots \wedge dx_n.$$

The form $dx_1 \wedge dx_2 \wedge \cdots \wedge dx_n$ is called the volume form. Integrating it over a domain (an n-dimensional submanifold) gives the standard volume integral.

More generally, one defines integration of k-forms over k-chains, which are just linear combinations of smooth submanifolds, but we do not need that level of generality.

In computations, we can avoid sets of zero measure (k-dimensional), so we can ignore the boundary of the submanifold. Similarly, if we parametrize several subsets we can leave out a measure zero subset. Let us give a few examples of computations.

Example C.3: Consider the circle $S^1 \subset \mathbb{R}^2$. We use a parametrization $\varphi \colon (-\pi, \pi) \to S^1$ where $\varphi(t) = (\cos(t), \sin(t))$, so the circle is oriented counter-clockwise. Let $\omega(x_1, x_2) = P(x_1, x_2)\, dx_1 + Q(x_1, x_2)\, dx_2$, then

$$\int_{S^1} \omega = \int_{-\pi}^{\pi} \Big(P\big(\cos(t), \sin(t)\big)\big(-\sin(t)\big) + Q\big(\cos(t), \sin(t)\big) \cos(t) \Big)\, dt.$$

We can ignore the point $(-1, 0)$ as a single point is of 1-dimensional measure zero.

Example C.4: A domain $U \subset \mathbb{R}^n$ is an oriented submanifold. We use the parametrization $\varphi \colon U \to U$, where $\varphi(x) = x$. Then

$$\int_U f(x)\, dx_1 \wedge dx_2 \wedge \cdots \wedge dx_n = \int_U f(x)\, dx_1\, dx_2 \cdots dx_n = \int_U f(x)\, dV,$$

where dV is the standard volume measure.

Example C.5: Finally, consider M the upper hemisphere of the unit sphere $S^2 \subset \mathbb{R}^3$ as a submanifold with boundary. That is, consider

$$M = \big\{ x \in \mathbb{R}^3 : x_1^2 + x_2^2 + x_3^2 = 1, x_3 \geq 0 \big\}.$$

The boundary is the circle in the (x_1, x_2)-plane:

$$\partial M = \big\{ x \in \mathbb{R}^3 : x_1^2 + x_2^2 = 1, x_3 = 0 \big\}.$$

Consider the parametrization of M using the spherical coordinates

$$\varphi(\theta, \psi) = \big(\cos(\theta)\sin(\psi), \sin(\theta)\sin(\psi), \cos(\psi) \big)$$

for U given by $-\pi < \theta < \pi, 0 < \psi \leq \pi/2$. After a rotation this is a subset of a half-plane with the points corresponding to $\psi = \pi/2$ corresponding to boundary points. We miss the points where $\theta = \pi$, including the point $(0, 0, 1)$, but the set of those points is a 1-dimensional curve, and so a set of 2-dimensional measure zero. For the purposes of integration, we can ignore it. Let

$$\omega(x_1, x_2, x_3) = P(x_1, x_2, x_3)\, dx_1 \wedge dx_2 + Q(x_1, x_2, x_3)\, dx_1 \wedge dx_3 + R(x_1, x_2, x_3)\, dx_2 \wedge dx_3.$$

Then

$$\int_M \omega = \int_{-\pi}^{\pi} \int_0^{\pi/2} \bigg[P\big(\varphi(\theta, \psi)\big) \bigg(\frac{\partial \varphi_1}{\partial \theta} \frac{\partial \varphi_2}{\partial \psi} - \frac{\partial \varphi_2}{\partial \theta} \frac{\partial \varphi_1}{\partial \psi} \bigg)$$

$$+ Q\big(\varphi(\theta, \psi)\big) \bigg(\frac{\partial \varphi_1}{\partial \theta} \frac{\partial \varphi_3}{\partial \psi} - \frac{\partial \varphi_3}{\partial \theta} \frac{\partial \varphi_1}{\partial \psi} \bigg)$$

$$+ R\big(\varphi(\theta, \psi)\big) \bigg(\frac{\partial \varphi_2}{\partial \theta} \frac{\partial \varphi_3}{\partial \psi} - \frac{\partial \varphi_3}{\partial \theta} \frac{\partial \varphi_2}{\partial \psi} \bigg) \bigg]\, d\theta\, d\psi$$

$$= \int_{-\pi}^{\pi} \int_0^{\pi/2} \bigg[P\big(\cos(\theta)\sin(\psi), \sin(\theta)\sin(\psi), \cos(\psi)\big)\big(-\cos(\psi)\sin(\psi)\big)$$

$$+ Q\big(\cos(\theta)\sin(\psi), \sin(\theta)\sin(\psi), \cos(\psi)\big) \sin(\theta)\sin^2(\psi)$$

$$+ R\big(\cos(\theta)\sin(\psi), \sin(\theta)\sin(\psi), \cos(\psi)\big)\big(-\cos(\theta)\sin^2(\psi)\big) \bigg]\, d\theta\, d\psi.$$

The induced orientation on the boundary ∂M is the counter-clockwise orientation used in Example C.3, because that is the parametrization we get when we restrict to the boundary, $\varphi(\theta, \pi/2) = (\cos(\theta), \sin(\theta), 0)$.

The derivative on k-forms is the *exterior derivative*, which is a linear operator that eats k-forms and spits out $(k+1)$-forms. For a k-form

$$\omega = g_{\ell_1,\dots,\ell_k}\, dx_{\ell_1} \wedge dx_{\ell_2} \wedge \cdots \wedge dx_{\ell_k},$$

define the exterior derivative $d\omega$ as

$$d\omega \overset{\text{def}}{=} dg_{\ell_1,\dots,\ell_k} \wedge dx_{\ell_1} \wedge dx_{\ell_2} \wedge \cdots \wedge dx_{\ell_k} = \sum_{m=1}^{n} \frac{\partial g_{\ell_1,\dots,\ell_k}}{\partial x_m}\, dx_m \wedge dx_{\ell_1} \wedge dx_{\ell_2} \wedge \cdots \wedge dx_{\ell_k}.$$

Then define d on every k-form by extending it linearly.

For example,

$$d\left(P\, dx_2 \wedge dx_3 + Q\, dx_3 \wedge dx_1 + R\, dx_1 \wedge dx_2\right)$$
$$= \frac{\partial P}{\partial x_1}\, dx_1 \wedge dx_2 \wedge dx_3 + \frac{\partial Q}{\partial x_2}\, dx_2 \wedge dx_3 \wedge dx_1 + \frac{\partial R}{\partial x_3}\, dx_3 \wedge dx_1 \wedge dx_2$$
$$= \left(\frac{\partial P}{\partial x_1} + \frac{\partial Q}{\partial x_2} + \frac{\partial R}{\partial x_3}\right) dx_1 \wedge dx_2 \wedge dx_3.$$

You should recognize the divergence of the vector field (P, Q, R) from vector calculus. All the various derivative operations in $\mathbb{R}^3$ from vector calculus make an appearance. If ω is a 0-form in $\mathbb{R}^3$, then $d\omega$ is like the gradient. If ω is a 1-form in $\mathbb{R}^3$, then $d\omega$ is like the curl. If ω is a 2-form in $\mathbb{R}^3$, then $d\omega$ is like the divergence.

A quick computation gives the *Leibniz rule*: If ω is a p-form and η is a q-form, then

$$d(\omega \wedge \eta) = d\omega \wedge \eta + (-1)^p \omega \wedge d\eta.$$

Because partial derivatives commute, we find that for every ω,

$$d(d\omega) = 0.$$

The equation is sometimes written as $d^2 = 0$. If $\Lambda^k(M)$ denotes the k-forms on an n-dimensional submanifold M, then we get a so-called complex

$$0 \xrightarrow{d} \Lambda^0(M) \xrightarrow{d} \Lambda^1(M) \xrightarrow{d} \Lambda^2(M) \xrightarrow{d} \cdots \xrightarrow{d} \Lambda^n(M) \xrightarrow{d} 0.$$

The initial d from the trivial space to $\Lambda^0(M)$ is just the zero map. One can study the topology of M by computing from this complex the so-called *de Rham cohomology* groups $H_{dR}^0(M), \dots, H_{dR}^n(M)$, that is, $\frac{\ker(d \colon \Lambda^k \to \Lambda^{k+1})}{\mathrm{im}(d \colon \Lambda^{k-1} \to \Lambda^k)}$, which measure global solvability of the differential equation $d\omega = \eta$ for an unknown ω. These groups are isomorphic (the de Rham theorem) to the singular cohomology groups $H^0(M, \mathbb{R}), \ldots, H^n(M, \mathbb{R})$. Variations on this idea abound and one appears in chapter 4, but we digress.

We now state *Stokes' theorem*, sometimes called the *generalized Stokes' theorem* to distinguish it from the classical Stokes' theorem you know from vector calculus, which is a special case.*

Theorem C.6 (Stokes). *Suppose $M \subset \mathbb{R}^n$ is an embedded compact smooth oriented $(k + 1)$-dimensional submanifold-with-boundary, ∂M has the induced orientation, and ω is a smooth k-form defined on M. Then*

$$\int_{\partial M} \omega = \int_M d\omega.$$

One can get away with less regularity, both on ω and M (and ∂M) including "corners." In $\mathbb{R}^2$, it is easy to state in more generality, see Green's theorem (Theorem B.2). The classical Stokes' theorem is just the generalized Stokes' theorem with $n = 3$, $k = 2$. Classically instead of using differential forms, the line integral is an integral of a vector field instead of a 1-form ω, and its derivative $d\omega$ is the curl operator.

To at least get a flavor of the theorem, we prove it in a simpler setting, which however is often almost good enough, and it is the key idea in the proof. Suppose $U \subset \mathbb{R}^n$ is a domain such that for each $k = 1, \ldots, n$, there exist two smooth functions α_k and β_k and U as a set is given by

$$(x_1, \ldots, x_{k-1}, x_{k+1}, \ldots, x_n) \in \pi_k(U),$$
$$\alpha_k(x_1, \ldots, x_{k-1}, x_{k+1}, \ldots, x_n) \le x_k \le \beta_k(x_1, \ldots, x_{k-1}, x_{k+1}, \ldots, x_n),$$

where $\pi_k(U)$ is the projection of U onto the $(x_1, \ldots, x_{k-1}, x_{k+1}, \ldots, x_n)$ components. Orient ∂U as usual. Write $x' = (x_1, \ldots, x_{k-1}, x_{k+1}, \ldots, x_n)$, and let dV_{n-1} be the volume form for $\mathbb{R}^{n-1}$. Consider the $(n-1)$-form

$$\omega = f \, dx_1 \wedge \cdots \wedge dx_{k-1} \wedge dx_{k+1} \wedge \cdots \wedge dx_n.$$

Then $d\omega = (-1)^{k-1} \frac{\partial f}{\partial x_k} dx_1 \wedge \cdots \wedge dx_n$. By the fundamental theorem of calculus,

$$\int_U d\omega = \int_U (-1)^{k-1} \frac{\partial f}{\partial x_k} dV_n$$

$$= (-1)^{k-1} \int_{\pi_k(U)} \int_{\alpha_k(x')}^{\beta_k(x')} \frac{\partial f}{\partial x_k} dx_k \, dV_{n-1}$$

$$= (-1)^{k-1} \left(\int_{\pi_k(U)} f(x_1, \ldots, x_{k-1}, \beta_k(x'), x_{k+1}, \ldots, x_n) \, dV_{n-1} \right.$$

$$\left. - \int_{\pi_k(U)} f(x_1, \ldots, x_{k-1}, \alpha_k(x'), x_{k+1}, \ldots, x_n) \, dV_{n-1} \right) = \int_{\partial U} \omega.$$

The $(-1)^{k-1}$ disappears due to orientation on ∂U. Any $(n - 1)$-form can be written as a sum of forms like ω for various k. Integrating each one of them in the correct direction provides the result.

---

*Interestingly, Stokes had nothing to do with proving either version aside from liking the theorem.

D Basic Terminology and Results from Algebra

We quickly review a few basic definitions and a result or two from commutative algebra needed in chapter 6. See a book such as Zariski–Samuel [ZS] for a complete reference.

Definition D.1. A set G is called a *group* if it has a binary operation $x * y$ defined on it and it satisfies the following axioms:

(G1) If $x \in G$ and $y \in G$, then $x * y \in G$.

(G2) *(associativity)* $(x * y) * z = x * (y * z)$ for all $x, y, z \in G$.

(G3) *(identity)* There exists an element $1 \in G$ such that $1 * x = x * 1 = x$ for all $x \in G$.

(G4) *(inverse)* For every element $x \in G$ there exists an element $x^{-1} \in G$ such that $x * x^{-1} = x^{-1} * x = 1$.

A group G is called *abelian* if it also satisfies:

(G5) *(commutativity)* $x * y = y * x$ for all $x, y \in G$.

A subset $K \subset G$ is a *subgroup* if K is a group with the same operation as the group G. For groups G and H, a function $f : G \to H$ is a *group homomorphism* if it respects the group law: $f(a * b) = f(a) * f(b)$. If f is bijective, then it is a *group isomorphism*.

An example of a group is a group of automorphisms. Let $U \subset \mathbb{C}$ be open and suppose G is the set of biholomorphisms $f : U \to U$. Then G is a group under composition, but G is not necessarily abelian: If $U = \mathbb{C}$, then $f(z) = z + 1$ and $g(z) = -z$ are members of G, but $(f \circ g)(z) = -z + 1$ and $(g \circ f)(z) = -z - 1$.

Definition D.2. A set R is called a *commutative ring* if it has two operations defined on it, addition $x + y$ and multiplication xy, and if it satisfies the following axioms:

(A1) If $x \in R$ and $y \in R$, then $x + y \in R$.

(A2) *(commutativity of addition)* $x + y = y + x$ for all $x, y \in R$.

(A3) *(associativity of addition)* $(x + y) + z = x + (y + z)$ for all $x, y, z \in R$.

(A4) There exists an element $0 \in R$ such that $0 + x = x$ for all $x \in R$.

(A5) For every element $x \in R$ there exists an element $-x \in R$ such that $x + (-x) = 0$.

(M1) If $x \in R$ and $y \in R$, then $xy \in R$.

(M2) *(commutativity of multiplication)* $xy = yx$ for all $x, y \in R$.

(M3) *(associativity of multiplication)* $(xy)z = x(yz)$ for all $x, y, z \in R$.

(M4) There exists an element $1 \in R$ (and $1 \neq 0$) such that $1x = x$ for all $x \in R$.

(D) *(distributive law)* $x(y + z) = xy + xz$ for all $x, y, z \in R$.

The ring R is called an *integral domain* if in addition to being a commutative ring:

(ID) $xy = 0$ implies that $x = 0$ or $y = 0$.

The ring R is called a *field* if in addition to being a commutative ring:

(F) For every $x \in R$ such that $x \neq 0$ there exists an element $1/x \in R$ such that $x(1/x) = 1$.

In a commutative ring R, the elements $u \in R$ for which there exists an inverse $1/u$ as above are called *units*.

If R and S are rings, a function $f : R \to S$ is a *ring homomorphism* if it respects the ring operations, that is, $f(a + b) = f(a) + f(b)$ and $f(ab) = f(a)f(b)$, and such that $f(1) = 1$. If f is bijective, then it is called a *ring isomorphism*.

Namely, a commutative ring is an abelian additive group (by additive group we just mean we use + for the operation and 0 for the respective identity), with multiplication thrown in. If the multiplication also defines a group on the set of nonzero elements, then the ring is a field. A ring that is not commutative is one that does not satisfy commutativity of multiplication. Some authors define ring without asking for the existence of 1.

A ring that often comes up in this book is the ring of holomorphic functions. Let $\mathcal{O}(U)$ be the set of holomorphic functions defined on an open set U. Pointwise addition and multiplication give a ring structure on $\mathcal{O}(U)$. The set of units is the set of functions that never vanish in U. The set of units is a multiplicative group.

Given a commutative ring R, let $R[x]$ be the set of polynomials

$$P(x) = c_k x^k + c_{k-1} x^{k-1} + \cdots + c_1 x + c_0,$$

where $c_0, \ldots, c_k \in R$. If $c_k \neq 0$, the integer k is the *degree* of the polynomial and c_k is the *leading coefficient* of $P(x)$. If the leading coefficient is 1, then P is *monic*. If R

is a commutative ring, then so is $R[x]$. Similarly, we define the commutative ring $R[x_1, \ldots, x_n]$ of polynomials in n indeterminates.

The most basic result about polynomials, Theorem B.29 the fundamental theorem of algebra, which states that every nonconstant polynomial over $R = \mathbb{C}$ has a root, is really a theorem in one complex variable.

Definition D.3. Let R be a commutative ring. A subset $I \subset R$ is an *ideal* if $f \in R$ and $g, h \in I$ implies that $fg \in I$ and $g + h \in I$. In short, $I \subset R$ is an additive subgroup such that $RI = I$. Given a set of elements $S \subset R$, the *ideal generated by* S is the intersection I of all ideals containing S. If $S = \{f_1, \ldots, f_k\}$ is a finite set, we say I is *finitely generated*, and we write $I = (f_1, \ldots, f_k)$. A *principal ideal* is an ideal generated by a single element. An integral domain where every ideal is a principal ideal is called a *principal ideal domain* or a PID. A commutative ring R is *Noetherian* if every ideal in R is finitely generated.

It is not difficult to prove that "an ideal generated by S" really is an ideal, that is, the intersection of ideals is an ideal. If an ideal I is generated by $f_1, \ldots, f_k$, then every $g \in I$ can be written as

$$g = c_1 f_1 + \cdots + c_k f_k,$$

for some $c_1, \ldots, c_k \in R$. Clearly the set of such elements is the smallest ideal containing $f_1, \ldots, f_k$.

Theorem D.4 (Hilbert basis theorem). *If R is a Noetherian commutative ring, then $R[x]$ is Noetherian.*

As the proof is rather short, we include it here.

Proof. Suppose R is Noetherian, and $I \subset R[x]$ is an ideal. Starting with the polynomial f_1 of minimal degree in I, construct a (possibly finite) sequence of polynomials $f_1, f_2, \ldots$ such that f_k is the polynomial of minimal degree from the set $I \setminus (f_1, \ldots, f_{k-1})$. The sequence of degrees $\deg(f_1), \deg(f_2), \ldots$ is by construction nondecreasing. Let c_k be the leading coefficient of f_k.

As R is Noetherian, there exists a finite k such that $(c_1, c_2, \ldots, c_m) \subset (c_1, c_2, \ldots, c_k)$ for all m. Suppose for contradiction there exists a f_{k+1}, that is, the sequence of polynomials did not end at k. In particular, $(c_1, \ldots, c_{k+1}) \subset (c_1, \ldots, c_k)$ or

$$c_{k+1} = a_1 c_1 + \cdots a_k c_k.$$

As the degree of f_{k+1} is at least the degree of f_1 through f_k, we can define the polynomial

$$g = a_1 x^{\deg(f_{k+1}) - \deg(f_1)} f_1 + a_2 x^{\deg(f_{k+1}) - \deg(f_2)} f_2 + \cdots + a_k x^{\deg(f_{k+1}) - \deg(f_k)} f_k.$$

The polynomial g has the same degree as f_{k+1}, and in fact it also has the same leading term, c_{k+1}. On the other hand, $g \in (f_1, \ldots, f_k)$ while $f_{k+1} \notin (f_1, \ldots, f_k)$ by construction. The polynomial $g - f_{k+1}$ is also not in $(f_1, \ldots, f_k)$, but as the leading terms canceled, $\deg(g - f_{k+1}) < \deg(f_{k+1})$, but that is a contradiction, so f_{k+1} does not exist and $I = (f_1, \ldots, f_k)$. $\square$

Definition D.5. An element $f \in R$ is *irreducible* if f is not a unit and whenever $f = gh$ for two elements $g, h \in R$, then either g or h is a unit. An integral domain R is a *unique factorization domain* (UFD) if up to multiplication by a unit, every nonzero nonunit has a unique factorization into irreducible elements of R.

One version of a result called the Gauss lemma says that just like the property of being Noetherian, the property of being a UFD is retained when we take polynomials.

Theorem D.6 (Gauss lemma). *If R is a commutative ring that is a UFD, then $R[x]$ is a UFD.*

The proof is not difficult, but it is perhaps beyond the scope of this book.

E Results From Real Analysis

E.1 Measure theory review

The beginning of this course does not require the Lebesgue integral, however, knowing it may make some of the earlier results easier to understand and the exercises easier to work. In some of the later chapters, the Lebesgue integral does become necessary in several places. To make the first reading of the entire book easier for a student who has not had a course on measure theory yet, we present the basic ideas of the Lebesgue integral and list the results that make it so useful. We avoid getting into the details of the definition and simply state the useful results without proof. A reader who is interested can consult, for example, [R1] or [R3].

Given a set X, we designate a collection $\mathcal{M}$ of subsets of X, called the *measurable sets*. The collection $\mathcal{M}$ should be a *σ-algebra*, meaning that it is closed under taking complements, countable unions, and countable intersections. On these measurable sets we define a measure, that is, a function $\mu \colon \mathcal{M} \to [0, \infty]$, such that $\mu(\emptyset) = 0$ and μ is *σ-additive*, that is, the measure of a union of countably many disjoint sets is the sum of the measures. If X is the euclidean space $\mathbb{R}^n$, there always exists a measure called the *Lebesgue measure* that will agree with the standard n-dimensional volume on simple sets such as rectangles. A complication is that not all subsets of $\mathbb{R}^n$ can then be measurable. We say that $(X, \mathcal{M}, \mu)$ is a *measure space*.

A function $f \colon X \to \mathbb{R}$ is *measurable* if its sublevel sets are measurable. Since one generally has to work hard to produce a nonmeasurable function in the measure spaces we consider, the reader may be forgiven for assuming every function in this book is measurable. A function is *simple* if its support is of finite measure and it only has finitely many values, in which case the integral is defined as

$$\int_X f \, d\mu \overset{\text{def}}{=} \sum_{y \in f(X)} y \, \mu\bigl(f^{-1}(y)\bigr).$$

That is, on the set where $f(x) = y$ we define the integral as the value of the function times the measure of the set and then we add these up. If the function is actually a step function and the measure was the Lebesgue measure, this is the same as would be done for the Riemann integral. The integral of a nonnegative measurable f is defined by

$$\int_X f \, d\mu \stackrel{\text{def}}{=} \sup_{\substack{\varphi \leq f \\ \varphi \text{ is simple}}} \int_X \varphi \, d\mu.$$

The integral of any real-valued measurable function is then defined by writing $f = f_+ - f_-$ for nonnegative functions f_+ and f_-, as long as the integrals of at least one of these is not infinite, and writing

$$\int_X f \, d\mu \stackrel{\text{def}}{=} \int_X f_+ \, d\mu - \int_X f_- \, d\mu.$$

Similarly the integral of complex-valued measurable functions is defined by writing $f = u + iv$, that is,

$$\int_X f \, d\mu \stackrel{\text{def}}{=} \int_X u \, d\mu - i \int_X v \, d\mu.$$

The most common class of functions we deal with is the class of L^1-*integrable* (or simply L^1) functions, which are the functions such that

$$\int_X |f| \, d\mu < \infty.$$

For any function where the integral is defined we obtain the most basic estimate

$$\left| \int_X f \, d\mu \right| \leq \int_X |f| \, d\mu.$$

For the purposes of integration, we often allow nonnegative functions to take on the value ∞ at some points. In general, we allow our functions to be undefined on a set of measure zero if we are integrating them since changing a function on a measure zero set does not change the integral.

There are a couple of things to notice about the definition. First, because step functions are simple functions with respect to the Lebesgue measure, the integration is a generalization of the Riemann integral on the real line and on $\mathbb{R}^n$ in general in the sense that the two integrals agree when they are both defined.

Second, many more functions (all measurable functions in fact) can be limits of simple functions, and the integral is defined as a limit of such integrals, one would, rightly, expect that limits can easily pass under the integral and we no longer need to worry about integrability of the limit.

Besides integration, one often forgotten feature of this setup is that it applies to series. For a countable set X such as $\mathbb{N}$ we can define the so-called *counting measure*, where every set $S \subset X$ is measurable and $\mu(S)$ is simply the number of elements in S. If $z_n = f(n)$ is a function defined on X, then we write

$$\sum_{n \in X} z_n = \int_X f \, d\mu.$$

So the following theorems also apply to series, where being L^1 simply means that the series is absolutely summable.

We say that something happens *almost everywhere* if the set where it does not happen is of measure zero. Similarly we may say that this something happens for *almost every* $x \in X$. Note that if a sequence of measurable functions converges to a function f almost everywhere, then this function can be assumed to be measurable (it is equal almost everywhere to a measurable function). We have the following three theorems, which despite appearances are actually just equivalent to each other, but each statement is useful in different situations.

Theorem E.1.1 (Fatou's lemma). *Let* $(X, \mathcal{M}, \mu)$ *be a measure space and* $\{f_n\}$ *a sequence of nonnegative measurable functions that converges almost everywhere to a function* f. *Then*

$$\int_X f \, d\mu \le \liminf_{n \to \infty} \int_X f_n \, d\mu.$$

Theorem E.1.2 (Monotone convergence theorem). *Let* $(X, \mathcal{M}, \mu)$ *be a measure space,* $\{f_n\}$ *a sequence of nonnegative measurable functions where* $f_n \le f_{n+1}$ *almost everywhere for all* n, *and suppose* $\{f_n\}$ *converges to* f *almost everywhere. Then*

$$\int_X f \, d\mu = \lim_{n \to \infty} \int_X f_n \, d\mu.$$

Theorem E.1.3 (Dominated convergence theorem). *Let* $(X, \mathcal{M}, \mu)$ *be a measure space,* $\{f_n\}$ *a sequence of measurable complex-valued functions converging almost everywhere to* f, *and* $g \colon X \to [0, \infty]$ *an* L^1 *function such that for all* n, $|f_n(x)| \le g(x)$ *for almost every* x. *Then*

$$\int_X f \, d\mu = \lim_{n \to \infty} \int_X f_n \, d\mu.$$

A common application of this theorem is differentiating under the integral.

Theorem E.1.4 (Differentiation under the integral). *Let* $U \subset \mathbb{R}$ *be open and* $(X, \mathcal{M}, \mu)$ *be a measure space. Suppose* $f \colon U \times X \to \mathbb{C}$ *is such that for each* $t \in U$, $x \mapsto f(t, x)$ *is* L^1, *for almost every* x, $\frac{\partial f}{\partial t}$ *exists on all of* U, *and* $g \colon X \to [0, \infty]$ *is an* L^1 *function such that* $\left|\frac{\partial f}{\partial t}(t, x)\right| \le g(x)$ *for all* $t \in U$ *and almost every* $x \in X$. *Then for all* $t \in U$,

$$\frac{d}{dt} \int_X f(t, x) \, d\mu(x) = \int_X \frac{\partial f}{\partial t}(t, x) \, d\mu(x).$$

A measure space is *σ-finite* if it is a countable union of sets of finite measure. For example, the euclidean space with Lebesgue measure is σ-finite because it is a union of balls, which are of finite measure. We often want to write an integral over a product space as an iterated integral, such as writing an integral over a subset of $\mathbb{R}^n$ using n one-dimensional integrals. If $(X, \mathcal{M}, \mu)$ and $(Y, \mathcal{N}, \nu)$ are product spaces, we can define a product measure space by requiring that $\mu \times \nu(A \times B) = \mu(A)\nu(B)$ (we again skip details). First, for nonnegative functions we obtain the following simple theorem where no integrability needs to be checked, and we are allowing things to be infinite if needed.

Theorem E.1.5 (Tonelli). *Suppose $(X, \mathcal{M}, \mu)$ and $(Y, \mathcal{N}, \nu)$ are σ-finite measure spaces and $f \colon X \times Y \to \mathbb{R}$ is a nonnegative measurable function. Then:*

(i) *For almost every $x \in X$, $y \mapsto f(x, y)$ is measurable, and for almost every $y \in Y$, $x \mapsto f(x, y)$ is measurable.*

(ii) *The functions $y \mapsto \int_X f(x, y)\, d\mu(x)$ and $x \mapsto \int_Y f(x, y)\, d\nu(y)$ are measurable.*

(iii)

$$\int_{X \times Y} f(x, y)\, d(\mu \times \nu) = \int_Y \left(\int_X f(x, y)\, d\mu(x) \right) d\nu(y)$$

$$= \int_X \left(\int_Y f(x, y)\, d\nu(y) \right) d\mu(x).$$

In general there is the Fubini theorem. A measure is *complete,* if every subset of a measure zero set is also measurable. A measure can be completed by simply throwing those sets in, but it is a minor technicality that the product of two measure spaces is not in general complete and must be completed. This is an issue for measurability of the functions involved, but the functions that one usually considers in applications are easily shown measurable in all of these measure spaces and their completions.

Theorem E.1.6 (Fubini). *Suppose $(X, \mathcal{M}, \mu)$ and $(Y, \mathcal{N}, \nu)$ are complete measure spaces and $f \colon X \times Y \to \mathbb{R}$ is L^1-integrable. Then:*

(i) *For almost every $x \in X$, $y \mapsto f(x, y)$ is L^1-integrable, and for almost every $y \in Y$, $x \mapsto f(x, y)$ is L^1-integrable.*

(ii) *The functions $y \mapsto \int_X f(x, y)\, d\mu(x)$ and $x \mapsto \int_Y f(x, y)\, d\nu(y)$ are L^1-integrable.*

(iii)

$$\int_{X \times Y} f(x, y)\, d(\mu \times \nu) = \int_Y \left(\int_X f(x, y)\, d\mu(x) \right) d\nu(y)$$

$$= \int_X \left(\int_Y f(x, y)\, d\nu(y) \right) d\mu(x).$$

Tonelli theorem is often applied in tandem with the Fubini theorem. Tonelli establishes integrability and Fubini is used to write the integral we need as iterated integral, or swap the order of integrations.

The Tonelli and Fubini theorems are useful in simplifying the development of the power series by using the counting measure as we mentioned above. They are also useful for swapping a series summation and integration such as

$$\int_0^1 \sum_{n=1}^{\infty} a_n(x)\, dx = \sum_{n=1}^{\infty} \int_0^1 a_n(x)\, dx.$$

Here are also a couple of useful estimates for the Lebesgue integral. First, we have an infinite-dimensional version of Cauchy–Schwarz.

Theorem E.1.7 (Cauchy–Schwarz). *Suppose $(X, \mathcal{M}, \mu)$ is a measure space and $f: X \to \mathbb{C}$ and $g: X \to \mathbb{C}$ are so-called L^2 functions on X, that is, they are measurable and $\int_X |f|^2 d\mu < \infty$ and $\int_X |g|^2 d\mu < \infty$. Then*

$$\left| \int_X f \bar{g} d\mu \right|^2 \leq \int_X |f|^2 d\mu \int_X |g|^2 d\mu.$$

Next, we have the integral version of the inequality resulting from a convex combination of the values of a convex function.

Theorem E.1.8 (Jensen's inequality). *Suppose $(X, \mathcal{M}, \mu)$ is a probability measure space, that is, $\mu(X) = 1$, $\varphi: \mathbb{R} \to \mathbb{R}$ is convex, and f is measurable. Then*

$$\varphi \left(\int_X f d\mu \right) \leq \int_X \varphi \circ f d\mu$$

E.2   Classical convexity

A set $S \subset \mathbb{R}^n$ is *convex* (or as we will say in the main text *geometrically convex*) if for every $x, y \in S$ and every $\lambda \in [0,1]$ the point $(1 - \lambda)x + \lambda y$ is in S. Interior and closure of convex sets are convex. Arbitrary intersection of convex sets is convex, and increasing unions are convex. Given a set S the *convex hull* of S is the intersection of all convex sets containing S. The closed convex hull is the closure of that.

A *hyperplane* $H \subset \mathbb{R}^n$ is the set of solutions x of the equation $x \cdot a = b$ for some $a \in \mathbb{R}^n$ and $b \in \mathbb{R}$. A *closed half-space* is the set of points defined by $x \cdot a \geq b$. A function of the form $x \mapsto x \cdot a + b$ is called a *real affine function*.

Theorem E.2.1 (Supporting hyperplane theorem). *If $S \subset \mathbb{R}^n$ is convex and $x_0 \in \partial S$, then there exists a supporting hyperplane through x_0. That is, there exists an $a \in \mathbb{R}^n$ and $b \in \mathbb{R}$ such that $x_0 \cdot a = b$ (x_0 is on the hyperplane), and $x \cdot a \geq b$ for all $x \in S$ (S is in the closed half-space defined by that hyperplane).*

Note that the supporting hyperplanes need not be unique.

Theorem E.2.2 (Minkowski). *If $S, T \subset \mathbb{R}^n$ are two nonempty disjoint convex sets. Then there is a separating hyperplane, that is, there exists an $a \in \mathbb{R}^n$ and a $b \in \mathbb{R}$ such that $x \cdot a \geq b$ for all $x \in S$ and $x \cdot a \leq b$ for all $x \in T$.*

We can also put these together:

Corollary E.2.3. *A closed convex set $S \subset \mathbb{R}^n$ is the intersection of all closed half-spaces containing S. More generally, for any set S, the closed convex hull of S is the intersection of all closed half-spaces that contain S.*

A point $x_0 \in S$ is called an *extreme point* if for every $x, y \in S$ and $\lambda \in [0,1]$ such that $(1 - \lambda)x + \lambda y = x_0$ we have $x_0 = x$ or $x_0 = y$. A point $x_0 \in S$ is called an *exposed point* if there is an affine linear function whose restriction to S achieves a strict maximum at x_0, in other words, if there is a supporting hyperplane which intersects S at exactly one point.

Theorem E.2.4 (Straszewicz). *Let $S \subset \mathbb{R}^n$ be closed and convex. Then the set of exposed points of S is dense in the set of extreme points of S.*

Theorem E.2.5 (Krein–Milman). *Let $K \subset \mathbb{R}^n$ be compact and convex, then it is the convex hull of its extreme points.*

What is useful a couple of times for us in this book is that a compact convex set has exposed points. Or more generally, the convex hull of a compact set has exposed points.

Given a convex set S, a function $f \colon S \to \mathbb{R}$ is *convex* if for every $x, y \in S$ and $\lambda \in [0,1]$, we have

$$f\big((1 - \lambda)x + \lambda y\big) \le (1 - \lambda)f(x) + \lambda f(y).$$

Alternatively, f is convex if its *epigraph* is a convex set, where

$$\text{epigraph } f \overset{\text{def}}{=} \big\{(x, y) \in S \times \mathbb{R} : f(x) \le y\big\}.$$

So arguments about convex sets translate to convex functions. For example, the supporting hyperplane theorem shows a couple of rather interesting facts. First, convex functions have "tangent" hyperplanes although not unique by considering a supporting hyperplane of the epigraph:

Proposition E.2.6. *Suppose $S \subset \mathbb{R}^n$ is a convex set, $f \colon S \to \mathbb{R}$ a convex function, and $x_0 \in S$. Then there exists an affine function g such that $g(x) \le f(x)$ for all $x \in S$ and $g(x_0) = f(x_0)$.*

This proposition has the following consequence:

$$f(x_0) = \sup\big\{g(x_0) : g \text{ is an affine function such that } g(x) \le f(x) \text{ for all } x \in S\big\}.$$

E.3   Smooth bump functions and partitions of unity

The function $f(x) = 0$ for $x \le 0$ and $f(x) = e^{-1/x}$ for $x > 0$ is a smooth (C^∞) function that is zero for all $x \le 0$ and positive (and less than 1) for $x > 0$. The function $g(x) = \frac{f(x)}{f(x) + f(1-x)}$ is a smooth function that is zero for all $x \le 0$ and 1 for all $x \ge 1$. By modifying such examples we obtain the following *bump function*:

Theorem E.3.1 (Bump function). *Suppose $U \subset \mathbb{R}^n$ is open and $K \subset U$ is compact. Then there exists a smooth function $\varphi \colon \mathbb{R}^n \to [0,1]$ such that φ is compactly supported in U and $\varphi \equiv 1$ on a neighborhood of K.*

By support, supp φ, we mean the closure of the set $\{x : \varphi(x) \neq 0\}$, and by compactly supported in U we mean that supp φ is a compact subset of U. Another variant of a bump function is the Urysohn lemma:

Theorem E.3.2 (Smooth Urysohn lemma). *Suppose $U \subset \mathbb{R}^n$ is open and $A, B \subset U$ are disjoint closed (in subspace topology) subsets. Then there exists a smooth function $\varphi: U \to [0,1]$ such that $\varphi = 0$ on A and $\varphi = 1$ on B.*

These functions are used usually for localizing some problem, or extending a smooth function to all of U or all of $\mathbb{R}^n$. One can also ask for such bump functions to glue together nicely. Suppose $U \subset \mathbb{R}^n$ is open and $\{U_\iota\}_{\iota \in I}$ is an *open cover* of U, that is, $U = \bigcup_\iota U_\iota$.

Theorem E.3.3 (Smooth partition of unity). *Suppose $U \subset \mathbb{R}^n$ is open and $\{U_\iota\}_{\iota \in I}$ is an open cover of U, then there exists a partition of unity subordinate to the cover. That is, there exist a family $\{\varphi_\kappa\}_{\kappa \in K}$ of smooth compactly supported functions $\varphi_\kappa: \mathbb{R}^n \to [0,1]$ such that:*

(i) *For each $\kappa \in K$, there is some $\iota \in I$ such that supp $\varphi_\kappa \subset U_\iota$.*

(ii) *For each point $x \in U$, there is a neighborhood on which all but finitely many φ_κ vanish.*

(iii) *For every $x \in U$, we have $\sum_{\kappa \in K} \varphi_\kappa(x) = 1$.*

Further Reading

[BER] M. Salah Baouendi, Peter Ebenfelt, and Linda Preiss Rothschild, *Real submanifolds in complex space and their mappings*, Princeton Mathematical Series, vol. 47, Princeton University Press, Princeton, NJ, 1999. MR1668103

[B] Albert Boggess, *CR manifolds and the tangential Cauchy-Riemann complex*, Studies in Advanced Mathematics, CRC Press, Boca Raton, FL, 1991. MR1211412

[C] E. M. Chirka, *Complex analytic sets*, Mathematics and its Applications (Soviet Series), vol. 46, Kluwer Academic Publishers Group, Dordrecht, 1989. MR1111477

[D] John P. D'Angelo, *Several complex variables and the geometry of real hypersurfaces*, Studies in Advanced Mathematics, CRC Press, Boca Raton, FL, 1993. MR1224231

[GR] Robert C. Gunning and Hugo Rossi, *Analytic functions of several complex variables*, Prentice-Hall Inc., Englewood Cliffs, N.J., 1965. MR0180696

[H] Lars Hörmander, *An introduction to complex analysis in several variables*, 3rd ed., North-Holland Mathematical Library, vol. 7, North-Holland Publishing Co., Amsterdam, 1990. MR1045639

[K] Steven G. Krantz, *Function theory of several complex variables*, 2nd ed., The Wadsworth & Brooks/Cole Mathematics Series, Wadsworth & Brooks/Cole Advanced Books & Software, Pacific Grove, CA, 1992. MR1162310

[L] Jiří Lebl, *Guide to Cultivating Complex Analysis, Working the Complex Field*. https://www.jirka.org/ca/ or https://jirilebl.github.io/ca/.

[R1] Walter Rudin, *Principles of mathematical analysis*, 3rd ed., McGraw-Hill Book Co., New York-Auckland-Düsseldorf, 1976. International Series in Pure and Applied Mathematics. MR0385023

[R2] _____, *Function theory in the unit ball of $\mathbf{C}^n$*, Grundlehren der Mathematischen Wissenschaften [Fundamental Principles of Mathematical Science], vol. 241, Springer-Verlag, New York, 1980. MR601594

[R3] _____, *Real and complex analysis*, 3rd ed., McGraw-Hill Book Co., New York, 1987. MR924157

[W] Hassler Whitney, *Complex analytic varieties*, Addison-Wesley Publishing Co., Reading, Mass.-London-Don Mills, Ont., 1972. MR0387634

[ZS] Oscar Zariski and Pierre Samuel, *Commutative algebra, Volume I*, The University Series in Higher Mathematics, D. Van Nostrand Company, Inc., Princeton, New Jersey, 1958. With the cooperation of I. S. Cohen. MR0090581

Index

List of Notation

Notation	Description	Page
$\langle \omega, v \rangle$	pairing of a form and a vector (evaluating ω on v)	16, 219
$d\bar{z}, d\bar{z}_j$	$d\bar{z} = dx - i\,dy$, $d\bar{z}_j = dx_j - i\,dy_j$	16
δ_j^k	Kronecker delta, $\delta_j^j = 1$, $\delta_j^k = 0$ if $j \neq k$	16
$\mathbb{N}_0$	nonnegative integers $\{0, 1, 2, \ldots\}$	18
z^α	$z_1^{\alpha_1} z_2^{\alpha_2} \cdots z_n^{\alpha_n}$	18
dz	$dz_1 \wedge dz_2 \wedge \cdots \wedge dz_n$	18
$\|\alpha\|$	$\alpha_1 + \alpha_2 + \cdots + \alpha_n$	18
$\alpha!$	$\alpha_1! \alpha_2! \cdots \alpha_n!$	18
$\dfrac{\partial^{\|\alpha\|}}{\partial z^\alpha}$	$\dfrac{\partial^{\alpha_1}}{\partial z_1^{\alpha_1}} \dfrac{\partial^{\alpha_2}}{\partial z_2^{\alpha_2}} \cdots \dfrac{\partial^{\alpha_n}}{\partial z_n^{\alpha_n}}$	18
$\displaystyle\sum_\alpha c_\alpha z^\alpha$	multiindex power series	19
$\mathcal{O}(U)$	the set/ring of holomorphic functions on U	27
$f \circ g$	composition, $p \mapsto f\big(g(p)\big)$	29, 203
$Df, Df(a)$	Jacobian matrix / derivative (at a)	30, 57, 63
$D_\mathbb{R}f, D_\mathbb{R}f(a)$	real Jacobian matrix / real derivative (at a)	31, 57
f^{-1}	function inverse	32
$\subset\subset$	compact or relatively compact subset	33
$f^{-1}(S)$	pullback of a set, $\{q : f(q) \in S\}$	33
S^{2n-1}	unit sphere in $\mathbb{C}^n$, $S^{2n-1} = \partial \mathbb{B}_n$	34
$T_p\mathbb{R}^n, T_pM$	tangent space	55
$\dfrac{\partial}{\partial x_j}\Big|_p$	tangent vector	55
$T\mathbb{R}^n, TM$	tangent bundle	57
∇r	gradient of r	59
$O(\ell)$	big-oh notation	60
$\mathbb{C}T_p\mathbb{R}^n, \mathbb{C}T_pM$	complexified tangent space	62
$T_p^{(1,0)}\mathbb{R}^n, T_p^{(1,0)}M$	holomorphic tangent vectors	63
$T_p^{(0,1)}\mathbb{R}^n, T_p^{(0,1)}M$	antiholomorphic tangent vectors	63
$D_\mathbb{C}f, D_\mathbb{C}f(a)$	complexified real derivative	63

Notation	Description	Page
$\mathscr{L}(X_p, X_p)$	Levi form	66
v^*, A^*	conjugate transpose	67
$P_r(\theta)$	Poisson kernel for the unit disc	79
$dV, dV(w)$	volume measure, euclidean volume form	86, 157
$f * g$	convolution	86
$\widehat{K}, \widehat{K}_{\mathscr{F}}$	hull of K with respect to $\mathscr{F}$	90
$\text{dist}(x, y)$	distance of two points, sets, or a point and a set	95
$\widehat{K}_U$	holomorphic hull of K	98
C^ω	real-analytic	104
$[v]^2$	$v_1^2 + \cdots + v_n^2$	118
$dA, dA(w)$	area form, $dA = dx \wedge dy$	130
$d\psi$	exterior derivative	132, 223
$\partial\psi$	holomorphic part of exterior derivative	132, 138
$\bar{\partial}\psi$	antiholomorphic part of exterior derivative	132, 138
$H^{(p,q)}(U)$	Dolbeault cohomology groups	138
$\widehat{dx_j}$	removed one-form, $dx_1 \wedge \widehat{dx_2} \wedge dx_3 = dx_1 \wedge dx_2$	157
$A^2(U)$	Bergman space, $\mathbb{O}(U) \cap L^2(U)$	160
$L^2(U)$	square integrable functions	160
$\|f\|_{A^2(U)}, \|f\|_{L^2(U)}$	L^2 norm	160
$\langle f, g \rangle$	L^2 inner product	160
$K_U(z, \bar{\zeta})$	Bergman kernel of U	162
U^*	complex conjugate of a domain	162, 198
$H^2(\partial U)$	Hardy space	165
$S_U(z, \bar{\zeta})$	Szegö kernel of U	166
(f, p)	germ of f at p	167
$_n\mathbb{O}_p, \mathbb{O}_p$	ring of germs of holomorphic functions at p	168
(A, p)	germ of a set A at p	168
Z_f	zero set of f, that is, $f^{-1}(0)$	168, 182
$\text{ord}_a f$	order of vanishing of f at a	169

Notation	Description	Page
$\mathcal{O}(U)[z_n]$	polynomial in z_n with coefficients in $\mathcal{O}(U)$	169
$\mathcal{O}_0[z_n]$	polynomial in z_n with coefficients in $\mathcal{O}_0$	169
$(f), (f_1, \ldots, f_n)$	ideal generated by f or by $f_1, \ldots, f_n$	180, 227
$I_p(X)$	ideal of germs at p vanishing on X	183
$V_p(I)$	the common zero set of germs in I	183
Γ_f	graph of f	184
X_{reg}	regular points of X	185
X_{sing}	singular points of X	185
$\dim_p X$	dimension of X at p	186
$\dim X$	dimension of X	186
$\Sigma_q(U, r), \Sigma_q$	Segre variety	198
$A \setminus B$	set subraction	203
$\overline{S}$	topological closure	203
∂X	boundary of X (topological, or manifold)	203, 217
$f : X \to Y$	a function from X to Y	203
$x \mapsto F(x)$	a function of x	203
$f\|_S$	restriction of f to S	203
$X \overset{\text{def}}{=} Y$	define X to be Y	203
$\int_\gamma f(z)\, dz$	path integral	204
$\omega \wedge \eta$	wedge product of differential forms	220